CATASTROPHIC DIPLOMACY

CATASTROPHIC DIPLOMACY

US Foreign Disaster Assistance in the American Century

★ ★ ★

JULIA F. IRWIN

THE UNIVERSITY OF NORTH CAROLINA PRESS

Chapel Hill

This book was published with the assistance of the Luther H. Hodges Jr. and Luther H. Hodges Sr. Fund of the University of North Carolina Press.

Designed by Jamison Cockerham
Set in Arno, Sentinel, and Irby
by codeMantra

Cover photograph courtesy of Sueddeutsche Zeitung Photo via Alamy Stock Photo.

Manufactured in the United States of America

LIBRARY OF CONGRESS CATALOGING-IN-PUBLICATION DATA
Names: Irwin, Julia, author.
Title: Catastrophic diplomacy : US foreign disaster
assistance in the American century / Julia F. Irwin.
Description: Chapel Hill : The University of North Carolina Press, [2024]. |
Includes bibliographical references and index.
Identifiers: LCCN 2023023194 | ISBN 9781469676234 (cloth ; alk. paper) |
ISBN 9781469677231 (paperback ; alk. paper) | ISBN 9781469676241 (ebook)
Subjects: LCSH: Disaster relief—United States—History—20th century. |
Disaster relief—Government policy—United States. | Humanitarian
assistance—United States—History—20th century. | Humanitarian
assistance, American—Foreign countries—History—20th century. |
United States—Foreign relations—History—20th century. | BISAC:
HISTORY / World | SOCIAL SCIENCE / Disasters & Disaster Relief
Classification: LCC HV555.U6 1795 2024 | DDC 363.34/80973—dc23/eng/20230624
LC record available at https://lccn.loc.gov/2023023194

for

STEVE

Contents

Illustrations

American soldiers standing in the ruins of a house in
Agadir, Morocco, following the 1960 earthquake · *226*

A US Air Force transport plane delivering a
truck and other assistance to Skopje, Yugoslavia,
following the 1963 earthquake · *250*

Maps

Acknowledgments

When I embarked on this project, I imagined that writing a second book would be easier and faster than the first. That was . . . not the case. This book took me an awfully long time to research, write, and complete. But fortunately, I had a lot of help. Over the years, numerous people provided me with assistance, inspiration, and camaraderie. I'm beyond grateful to everyone who supported this project—and me—along the way.

My research for this book was made possible by generous grants from the University of South Florida, including an Office of Research and Innovation New Research Grant, a Humanities Institute Summer Grant, and a Global Citizenship Program Faculty Fellowship. These funds enabled me to travel to multiple repositories throughout the United States and Europe, where I benefited from the aid and expertise of countless archivists and librarians. I received personal assistance from Grant Mitchell and Mélanie Blondin at the Archives of the International Federation of Red Cross and Red Crescent Societies; David Langbart at the National Archives and Records Administration; Matthew Schaefer of the Herbert Hoover Presidential Library; Annette Amerman at the Marine Corps Archives; Fabio Ciccarello at the Food and Agricultural Organization of the United Nations Archives; and Fabrizio Bensi and Daniel Palmieri at the Archives of the International Committee of the Red Cross. To these individuals, and to dozens more working behind the scenes, I offer my sincere thanks.

As I was conceiving and drafting this book, I presented portions of my research to many different audiences, all of whom helped me to sharpen my ideas and refine my arguments. I feel especially fortunate for the intellectual community I have in the Society for Historians of American Foreign Relations. In multiple conference panels and roundtables, and in countless informal chats and conversations, SHAFR colleagues and friends offered me constructive critiques and invaluable feedback over the years. This book would have looked very different without their collective engagement and encouragement.

I also benefited enormously from opportunities to present my work at specialized conferences and workshops on humanitarian history. These include *L'Humanitaire S'Exhibe, 1867–2016*, organized by the Université de Fribourg, the Université de Genève, the Graduate Institute of Geneva, and the University of Manchester; *Histories of the Red Cross/Red Crescent Movement since 1919*, organized by members of the Resilient Humanitarianism project and the International Federation of Red Cross/Red Crescent Societies; *New Approaches to Medical Care: Humanitarianism and Violence during the "Long" Second World War*, organized by the University of Manchester; *Toward a Global History of Development*, held at the Radcliffe Institute for Advanced Study; *Histories of Humanitarianism: Religious, Philanthropic, and Political Practices in the Modernizing World*, organized by the German Historical Institute and the University of Maryland; and *Humanitarianisms in Context*, sponsored by the Zentrum für Zeithistorische Forschung Potsdam. Toward the end of drafting this book, in 2019, I also had the immense fortune to participate in the final Global Humanitarianism Research Academy. During the two weeks I spent in Mainz and Geneva, I learned more than I could have possibly imagined from brilliant graduate students, early career scholars, and colleagues. Thanks to everyone at the University of Cologne, the University of Exeter, and the Leibniz-Institut für Europäische Geschichte for inviting me and supporting my visit.

At a variety of other conferences and workshops, audiences asked me questions and offered comments on my papers and draft chapters, shaping this project in fruitful ways. These include the annual meetings of the American Historical Association, Organization of American Historians, and the American Association for the History of Medicine; a workshop on *Rethinking American Grand Strategy*, held at Oregon State University; the North American History Seminar at the Institute of Historical Research, University of London; the Centre for Geopolitics, University of Cambridge; the Washington History Seminar, National History Center; the Center for the Study of Force and Diplomacy, Temple University; the American Political History Institute Seminar Series, Boston University; the Harvard International & Global History Seminar, Harvard University; the Centre for

Imperial & Global History, Exeter University; the Foreign Policy Seminar Series, Department of History, University of Connecticut; the Vassar College History Department; the New College of Florida History Department; the Oklahoma State University History Department; and the University of South Florida Geosciences Colloquium. My thanks to everyone who attended these talks and gave me space to test out my ideas.

While researching and writing this book, I published related articles in several journals, including *Diplomatic History, Moving the Social, First World War Studies, Isis, Journal of Advanced Military Studies,* and *American Historian.* I also contributed book chapters to the volumes *Rethinking American Grand Strategy, The Development Century: A Global History, L'humanitaire s'exhibe,* and the *Cambridge History of America and the World.* I am very appreciative to the editors and peer reviewers for each of these publications, who helped me hone my arguments and allowed me to share pieces of my research in its earlier stages. Additionally, a very, very hearty thanks to the individuals who served as peer reviewers for this book. These people engaged deeply with my manuscript, offering insightful comments, meaningful critiques, and terrific suggestions for improving it. Each of you has my enduring appreciation.

Several students assisted me in researching this book, and I am very thankful for their help. I owe a particular debt of gratitude to Rebecca Siwiec, who worked with me for three years during her time as a USF undergraduate, finding, reading, analyzing, and summarizing scores of newspaper articles. At the University of Michigan, Erin McGlashen tracked down papers I needed from the Ford Presidential Library. My ability to write this book was also made possible by the tireless work of our incredible History Department staff. And of course, I would be remiss if I did not thank my colleagues in the USF History Department for their collective support over the years.

I have such a long list of fellow historians to thank—it's truly difficult to where to begin. Each and every one of the following people has left an imprint on this book in some way, shape, or form. I am especially grateful to Daniel Immerwahr for chatting with me about this project ad nauseum during conferences and over games of *Twilight Struggle* and for all his astute insights and probing questions. He, Ryan Irwin, Megan Black, Gretchen Heefner, David Milne, and Emily Conroy-Krutz have been not only brilliant writing group members but also very good friends. Over the years, I've also valued the collegiality and camaraderie of many other scholars of US foreign relations and international history, among them Brooke Blower, Jeff Byrne, Chris Capozzola, Matthew Connelly, Amanda Demmer, Thomas Field, Anne Foster, Petra Goedde, Andrew Johnstone, Paul Kramer, Adriene

Lentz-Smith, Jana Lipman, Michele Louro, Shaul Mittelpunkt, Kaeten Mistry, Harvey Neptune, Lien-Hang Nguyen, Chris Nichols, Aaron O'Connell, Marc-William Palen, Jason Parker, Elisabeth Piller, Rob Rakove, Kyle Romero, Ilaria Scaglia, Brad Simpson, Sarah Snyder, and Rebecca Herman Weber. I am grateful to Kristin Hoganson, Andrew Preston, Amy Sayward, Erez Manela, Richard Immerman, Mary Dudziak, Ara Keys, and David Engerman for their mentorship and unwavering support.

I've benefited from opportunities to discuss disasters with Alvita Akiboh, Lisa Covert, Pierre Fuller, Andy Horowitz, Scott Knowles, Sönke Kunkel, Alexander Poster, Jacob A.C. Remes, Eleonora Rohland, J. Charles Schencking, Lukas Schemper, Ian Seavey, Spencer Segalla, Jenny Leigh Smith, and all of my *Journal of Disaster Studies* coeditors. Likewise, I've learned so much about humanitarian history from conversations with Emily Baughan, Cedric Cotter, Eleanor Davey, Sébastien Farré, Romain Fathi, Jean-François Fayet, Kimberly Lowe Frank, Jaclyn Granick, Laure Humbert, Charlie Laderman, Joshua Mather, Daniel Maul, Marian Moser-Jones, Melanie Oppenheimer, Kevin O'Sullivan, Johannes Paulmann, Davide Rodogno, Silvia Salvatici, Pierre-Yves Saunier, Bertrand Taithe, Andrew Thompson, Boyd van Dijk, Neville Wylie, and Olivier Zunz. A special thanks to Fabian Klose, whose insights about humanitarian intervention and related subjects were particularly valuable to me.

Debbie Gershenowitz is an editor extraordinaire. She began talking with me about this project when it was barely off the ground. Over the years, she listened to my ideas, offered perceptive insights and editorial suggestions, and worked tirelessly with me to navigate the road from manuscript to book. I feel very fortunate to count her as not only my editor but also my good friend. At the University of North Carolina Press, Madge Duffey, JessieAnne D'Amico, Erin Granville, Laura Dooley, and others answered many questions and offered unparalleled support as I prepared the book for publication. It has been a pleasure to work with them.

Here in Tampa, I'm lucky to be part of a wonderful community of friends. Amy Rust and Scott Ferguson are dining companions par excellence, and absolutely terrific people. I've thoroughly enjoyed getting to know Katrin Pesch and Tim Ridlen in recent years, and I look forward to many more dinner parties ahead. Darcie Fontaine and Brian Connolly are fantastic colleagues and even better friends. Jamie McElman, you always bring such a big smile to my face! You, too, JoEllen Irizarry. Still more love goes out to Sari Altschuler, Kellan Anfinson, Jan Awai, Jennifer Bosson, Devon Brady, Gena Camoosa, Dave Davisson, Matt King, Karl Petersen, Tom Pluckhahn, Bernd

Reiter, Jennifer Rogers, Brenton Wiernick, Becky Zarger, and the entire Friday happy hour crew, past and present. Beyond Tampa, Kirsten Weld has been my sounding board and close confidante over more Scrabble games than I can possibly count.

I can always depend on my dad, Lee Irwin, for great conversation and to share my love of food and travel. I feel immensely fortunate for the time we spent together in Switzerland and France as I worked on this book. Mona Rozovich is an absolute treasure, and I'm so glad she came into my dad's life (and mine). My mom, Ann Irwin, died while I was writing this book. She was an incredibly kind and loving person, and I will always remember her affection. From Brooklyn to Taipei, I've made so many fond memories with Allison Prince and Alan Jou, and I can't wait to get to know their daughter, Nora. Ken and Jane Prince, thank you for your gracious hospitality and for being a second family. My three furry housemates—Stella, Sacco, and the dearly departed Zetti—have brought me immense joy over the years. I can always count on them to remind me of the important things in life, like relaxing and snacks.

For more reasons that I can possibly enumerate, this book is dedicated to Steve Prince. Steve graciously offered his services as this book's developmental editor. He read every word of my manuscript, multiple times (he also ruthlessly slashed many of its original words so that no one else would be subjected to them). His incisive comments and sharp recommendations strengthened this book immeasurably, allowing me to figure out what I actually wanted to say. For that alone, I am grateful—but I'm also appreciative for so much more. For half our lives, Steve has been my best friend and absolute favorite person. Whether we're traveling the world or playing video games on the couch, hiking in the Alps or kayaking in Florida springs, enjoying multicourse gourmet dinners or drinking beers at the Independent, life is immensely more fun with him by my side. Thank you, Steve, for always loving me and for making me so happy. I'm already looking forward to our next adventure.

Abbreviations

ACVFA Advisory Committee on Voluntary Foreign Aid

ARC American National Red Cross

CIA Central Intelligence Agency

FDRC Foreign Disaster Relief Coordinator

FOA Foreign Operations Administration

ICA International Cooperation Administration

IMF International Monetary Fund

MATS Military Air Transport Service

OFDA Office of Foreign Disaster Assistance

US United States

USSR Union of Soviet Socialist Republics

USAID United States Agency for International Development

CATASTROPHIC DIPLOMACY

INTRODUCTION

The Politics of Disaster, the Politics of Aid

In November 1966, Sam Krakow, the longtime director of international services for the American Red Cross, made a disparaging observation to the organization's president, General James F. Collins. "The political implications of international disaster relief," he lamented to his boss, a retired army commander, "have just been discovered." Krakow criticized US diplomats for politicizing the aid the United States provided to their host countries following major catastrophes. "The American Government," he groused, "is most anxious to get credit for its action in foreign disaster relief."[1] Amid the geopolitical tumults of the global Cold War and decolonization, as Krakow saw it, humanitarian assistance had become little more than a crude tool of US foreign relations. Policymakers had begun placing American diplomatic and strategic interests over the needs of disaster victims around the world.

Although Krakow identified the politicization of US international disaster relief as a recent phenomenon, the use of foreign disaster assistance as an instrument of US foreign relations had a much longer history than he knew or cared to admit. By the dawn of the twentieth century, US policymakers had already come to embrace the "political implications of international disaster relief." In the early 1900s, US diplomatic and military officials began devoting increased attention to catastrophes in other nations and empires. At the same time, they started to contribute rising material and financial resources to the survivors of these disasters, grasping the strategic, diplomatic, economic, and moral potential of American humanitarian assistance.

From this point forward, responding to catastrophes caused by earthquakes, tropical cyclones, floods, and other natural hazards became a fixture of US foreign relations. Over the next half-century, working in close partnership with key American voluntary organizations, the US federal government and armed forces steadily expanded their involvement in this humanitarian field. In the process, US foreign disaster assistance gradually became a more formal instrument of national foreign policy, institutionalized within the federal government's bureaucracy and legal architecture. As US international disaster relief operations increased in number and in scale across the early to mid-twentieth century, they also established precedents for broader, more ambitious foreign aid programs in years ahead.

By the time Krakow recorded his observations, in short, the US government had amassed a long history of responding to calamities abroad. Rarely had its actions *not* been political in some way, shape, or form.

Tracing the history of US foreign disaster assistance from the early twentieth century through the mid-1970s, *Catastrophic Diplomacy* recovers the origins of this humanitarian practice and the complex motivations that lay behind it. Throughout the twentieth century, disaster aid served as a consistent and flexible tool of US foreign policy. It functioned as a means of projecting American power and influence globally and as a vehicle for preserving order and control abroad. By assisting survivors of international calamities, US officials sought not only to ameliorate distant suffering but also to promote the diplomatic and strategic interests of the United States. Disaster relief, recovery, and reconstruction operations abroad repeatedly punctuated the history of twentieth-century US foreign relations. Attention to these humanitarian efforts has much to reveal about how Americans interacted with other nations and empires, particularly during moments of extreme and unexpected global upheaval.

To tell this story, *Catastrophic Diplomacy* analyzes the official humanitarian operations, and evolving roles, of three key pillars of the US humanitarian aid system. They include, first, the State Department and the staff of US diplomatic, consular, and development missions; second, the Departments of War, Navy, and Defense and the service personnel of the US Armed Forces; and third, the US government's preferred partners in the American voluntary sector, among them the American Red Cross and various missionary societies, philanthropies, and aid organizations.

Examining their collaborative responses to hundreds of disasters in dozens of nations and empires across these years, and evaluating the successes, the shortcomings, and the complex politics of these aid operations, *Catastrophic Diplomacy* tracks the steady rise of US foreign disaster aid as an instrument of US foreign policy. In the process, it demonstrates the importance of international disaster assistance—and humanitarian aid more broadly—to US relations with the twentieth-century world.

✪ ✪ ✪

This book centers on catastrophes triggered by sudden geological, climatological, hydrological, or meteorological phenomena—events that today are often categorized as rapid-onset natural disasters. Its focus is on humanitarian emergencies that arose abruptly in the wake of hurricanes, earthquakes, tsunamis, flash floods, and other natural hazards, events that rapidly claimed numerous lives and wrought enormous physical destruction. This book does not consider, at least not as its primary emphasis, the effects of war, civil strife, nuclear and industrial accidents, or other anthropogenic or "man-made" calamities. It also largely excludes US responses to famine, drought, epidemics, and other catastrophes whose effects were felt gradually over months and years, events that are today classified as creeping or slow-onset disasters. Although it touches on the food shortages, disease outbreaks, and other longer-term consequences of more abrupt catastrophes, it is sudden natural hazards—and the humanitarian emergencies they precipitate—that lie at the heart of this book.

This distinction may at first seem arbitrary. Famine and drought, after all, have as much potential to disrupt and devastate the societies they afflict as more rapid-onset disasters. So do conflicts, refugee crises, and other catastrophes caused by deliberate human action. The humanitarian crises that follow events such as tsunamis, typhoons, volcanic eruptions, or floods, moreover, are never solely the product of natural forces; they are *always* rooted in a combination of environmental and human factors. The lines separating one type of disaster from another, in other words, are decidedly blurry. These categories, moreover, are not set in stone. They are social and cultural constructions, whose meanings differ across time and place.[2]

Even as this book concentrates primarily on sudden, so-called natural disasters, then, it acknowledges from the outset the subjective aspects of this categorization. Indeed, I join other disaster studies scholars in insisting that there is really no such thing as a natural disaster. These sorts of catastrophes occur only when an environmental hazard affects, and then harms, an already vulnerable population, overwhelming its capacity to cope effectively. To put it another way, it is humans, not nature, who are ultimately responsible for the severity of any catastrophic event. The choices people make before, during, and after a natural hazard occurs determine its eventual impacts upon their societies.[3]

Given these caveats and qualifications, why confine the focus to these types of humanitarian crises? Because there are also some compelling reasons for doing so.

The first is conceptual. Catastrophes triggered suddenly by earthquakes, tropical cyclones, and other natural hazards have historically been understood within American culture—and by many other societies throughout the world—as unique or exceptional events. This was certainly true throughout the early to mid-twentieth-century United States, the period covered in this book.[4] The individuals who populate its pages employed the term "natural disaster" regularly and without hesitation, understanding it as a meaningful category and descriptor. They described these events as "acts of God" or "acts of nature," rarely as "acts of humankind." For many of the subjects of this book, in short, there *was* such thing as a natural disaster.

A second reason, closely related to the first, is grounded in ideology. The belief that some disasters are natural or inevitable is, at its heart, an ideological one. It conveniently ignores the underlying political, socioeconomic, and environmental factors that leave some populations at greater risk from natural hazards than others—a concept disaster studies scholars call "vulnerability."[5] Such a worldview renders "natural" catastrophes as apolitical crises, unpredictable emergencies divorced from the conditions that produce them and the complex contexts in which they arise.

Deeply ingrained in twentieth-century American culture, these ideological assumptions profoundly influenced how US officials understood and responded to "natural" disasters abroad. Focused principally on the spectacular devastation and sudden destruction associated with these events, they concentrated their energies on delivering short-term relief to those who were immediately affected. In choosing to prioritize the emergency human-itarian response over prevention and preparedness activities, however, US officials did little to mitigate the myriad factors that created vulnerability to natural hazards in the first place. They failed to address, or even acknowledge, the root causes of catastrophe. Attention to these types of disasters thus has much to reveal about the limitations and shortcomings of US foreign assistance.

Material factors provide the third justification for this book's focus. The ways Americans physically responded to emergencies triggered by natural hazards were necessarily distinct from their responses to other humanitarian crises. The act of delivering life-saving rations, emergency shelter, or medical treatment immediately after an earthquake or hurricane struck, for instance, differed in fundamental ways from establishing a stable, predictable food supply during longer periods of drought or famine. Likewise, contributing humanitarian relief to disaster survivors during times of peace differed, le-gally and politically, from aiding victims of conflict and postwar upheaval.

Together, these conceptual, ideological, and material distinctions had very real historical consequences. They led twentieth-century Americans to channel foreign disaster aid through dedicated agencies, institutions, and bureaucratic channels. Their humanitarian responses to these catastrophes were guided by an identifiable set of methods and philosophies. They were governed by particular laws and policies. Although it certainly overlapped with other types of international aid, then, US foreign disaster assistance evolved along its own, specific trajectory. It has a history all its own.

★   ★   ★

As much as it is a book about natural hazards and the disasters they precip-itated, *Catastrophic Diplomacy* is above all a study of how the United States responded to those crises. It is a history of US foreign assistance, grounded in broader scholarship on both disasters and international humanitarianism.

Disaster studies is a thriving field, spanning anthropology, sociology, po-litical science, and other disciplines—including, not least, history. Within the past two decades, historians of the United States have offered political, cultural, legal, and environmental perspectives of catastrophes occurring within that country. Scholars of other nations and empires have been equally attuned to the historic effects of disaster in the places they study. Global historians, meanwhile, have begun charting the development of an international system of disaster management, which took form over the twentieth century.[6]

Scholarship on humanitarianism, likewise, has grown tremendously in recent years. Historians have studied US foreign assistance, other nations' humanitarian traditions, and the international humanitarian system, analyzing both governmental and nongovernmental forms of aid. They have examined wartime efforts to provide relief to soldiers, civilians, and prisoners of war, postwar operations to aid to refugees and displaced persons, and programs to assist victims of famine, epidemics, and other peacetime crises. Together, this work sheds critical light on the politics, economics, culture, and ethics of humanitarian aid.[7]

Drawing valuable insights from both these fields, *Catastrophic Diplomacy* presents fresh perspectives on the entwined histories of disasters, peacetime humanitarianism, and US foreign assistance. Most fundamentally, it demonstrates the salience of these topics to the history of American foreign relations. Analyzing disaster relief operations as both a reflection and a manifestation of US global power, it argues for the centrality of humanitarian concerns and activities to twentieth-century American foreign policy. At the same time, it underlines the importance of environmental agents and forces to US international history.

Catastrophic Diplomacy also locates the US government's entry into the field of foreign assistance far earlier than is commonly perceived. Commencing in the early 1900s, decades before the Marshall Plan, US diplomats, military personnel, and other government officials played an active role in planning, administering, and financing American humanitarian aid operations. These early twentieth-century disaster relief efforts served as a precedent, and sometimes a model, for future US foreign assistance programs. Spanning the first three quarters of the twentieth century, this book also traces continuities across both world wars, conflicts that too often serve as bookends in both humanitarian history and the history of US foreign affairs. It considers how the United States' political, military, and humanitarian involvement in these momentous conflicts shaped the evolving system of US foreign disaster assistance, highlighting the confluences between wartime and peacetime aid.

Attuned to the close ties among US diplomats, the US military, and American voluntary organizations, *Catastrophic Diplomacy* underscores the centrality of state-private partnerships to the histories of American humanitarianism and US foreign relations. In so doing, it analyzes a form of governance that scholars have termed "the associational state."[8] Throughout the twentieth century, as this book's chapters reveal, US government officials tapped quasi-governmental and nongovernmental entities to act as the state's humanitarian auxiliaries. Government officials depended on these organizations to carry out the nation's foreign disaster relief efforts and to project American influence abroad. In exchange, American voluntary organizations gained invaluable material benefits, including tax breaks and financial subsidies, access to surplus commodities and government property, privileged information about other nations, and official lines of communication and transportation. As these partnerships reveal, the distinction often made between state and nonstate actors is overdrawn. To understand the workings of twentieth-century US foreign policy, this book argues, requires attention to the associational state.

Although *Catastrophic Diplomacy* is principally a history of disasters and humanitarian relief, it also explores the relation between those subjects and another major category of foreign aid: development assistance. Over the past two decades, historians have written widely on modernization and international development, focusing particularly on long-term technical, agricultural, economic, and military assistance projects. They have devoted considerably less attention, however, to short-term, emergency aid efforts for the victims of disasters and other humanitarian crises. Demarcating humanitarian relief from development assistance (and its earlier analogs, civilizing and technical missions), existing scholarship tends to treat these activities as discrete categories of foreign aid. Relief, this literature suggests, focuses on the *restorative*—saving lives, reducing suffering, and returning societies to a precrisis state—while development emphasizes the *transformative*, the construction of a better future.[9]

Catastrophic Diplomacy aims to challenge, or at least muddy, these distinctions. Building on the rich historiography of development and modernization, it explores the material and conceptual intersections between disaster relief and development assistance.[10] Its chapters analyze several long-term, far-reaching US disaster recovery and reconstruction operations in other nations. These sorts of comprehensive aid projects, I argue, should be understood as both analogs and precursors to other twentieth-century international development initiatives. With an eye toward these parallels, the book calls attention to places where different types of foreign aid activities coexisted, coevolved, and converged, emphasizing the *development* of relief.

✪   ✪   ✪

Catastrophic Diplomacy tells the history of US foreign disaster assistance through an overarching analysis of US responses to hundreds of global catastrophes, interspersed with in-depth and illustrative case studies. Its chapters trace the system of US foreign disaster assistance as it evolved across the first three quarters of the twentieth century while exploring how particular relief operations intersected with contemporary US foreign policy concerns. The events and episodes that populate this book's pages reflect a key organizing principle: they were selected not for the magnitude of any given disaster but instead for the character of the US response to that catastrophe. Rather than focusing only on the world's most destructive disasters, in other words, *Catastrophic Diplomacy* examines the US humanitarian aid operations— major and minor—that followed a wide spectrum of global calamities.

The disasters that prompted official US foreign assistance efforts in these decades ranged widely in their severity. Several were immense, truly horrific catastrophes, which claimed or uprooted hundreds of thousands of lives. Most were comparatively less destructive. Nonetheless, those catastrophes that occasioned US relief efforts all tended to be fairly sizable emergencies— substantial enough, at least, to capture the attention of US observers and then compel them to act. Whether or not Americans became aware of any given disaster, however, was also a function of preexisting diplomatic, economic, and cultural ties between the United States and the place that crisis occurred. Such personal and political connections ensured that certain catastrophes garnered greater American concern than others.

Just as the enormity of these catastrophes varied, so did the scale of US foreign disaster assistance efforts that followed them. Sometimes, US officials made only a token contribution to disaster victims; on other occasions, they provided more substantive levels of aid. Whatever amount they contributed, US officials most often restricted their aid to temporary forms of relief, intended to meet survivors' most pressing, essential needs. This humanitarian relief normally took the form of cash grants, food, clean water, clothing, temporary shelter, and life-saving medical assistance, contributed in the days and weeks after a catastrophe's onset. American personnel also regularly took part in search-and-rescue missions or aid distributions, usually concluding these efforts once the emergency period was declared complete.

As a rule, US officials believed their assistance should be restricted to this short-term relief phase. They typically expected foreign governments and societies to shoulder the primary burdens for recovery and reconstruction. Such assumptions reflected prevailing beliefs about charity and self-help, commonplace in the twentieth-century United States, which held that too much aid would be detrimental to disaster survivors, fostering their dependency and idleness. These fears not only applied to individuals, moreover, but also extended to nations and governments. Emphasizing the values of self-reliance and self-sufficiency, US officials usually preferred to keep their humanitarian involvement brief.

There were, however, noteworthy exceptions to this pattern. Several of the US disaster assistance efforts this book recounts went far beyond emergency relief, evolving into long-term programs of recovery and reconstruction. These humanitarian operations lasted for months—and sometimes years—beyond the initial onset of the disasters that precipitated them, cost hundreds of thousands or even millions of dollars, and involved scores of US diplomatic, military, and voluntary sector personnel. The decision to take on these sorts of projects was initially somewhat arbitrary: when US officials found themselves with extra cash or supplies after relief efforts concluded, they occasionally repurposed that aid for rehabilitation or rebuilding activities. In the decades following the Second World War, by contrast, contributing recovery and reconstruction aid to other nations became an increasingly conscious and even proactive choice, steered by US foreign policy strategies and international development agendas. Whether planned or impromptu, each of these more comprehensive disaster aid operations constituted a significant encroachment, on the part of the United States, into the political, economic, and humanitarian affairs of other nations.

Whatever form their assistance took, US officials consistently viewed their aid not only as a means to address the suffering of disaster survivors but also—and perhaps more importantly—as a tool for promoting American diplomatic and strategic interests. For this reason, US foreign disaster assistance flowed far more often through bilateral channels than multilateral ones. Government officials and their partners in the American voluntary sector tended to maintain tight control over the disaster aid they contributed to other nations. They took concerted steps to ensure that this assistance was identified with the United States and used in the manner they prescribed, strongly preferring to keep American humanitarian aid in American hands. Foreign disaster assistance was never intended solely to ameliorate distress. It was also a conscious display of public diplomacy.

American foreign disaster assistance spread far and wide across the first three quarters of the twentieth century. Yet it did not flow equally or evenly to all parts of the globe. Nor did it reach victims of each and every catastrophic earthquake, tropical storm, flood, or other hazard that happened to strike.

In the wake of many catastrophes, including some of the most destructive disasters to occur during these decades, the United States contributed little or no humanitarian aid at all. Although there were certainly correlations between the human and material toll of a disaster and the scale of any subsequent US aid effort, the severity of a particular calamity was never the only consideration influencing government reactions to it. A host of additional factors together determined the magnitude and priorities of any humanitarian response.

When US officials learned a catastrophe had occurred in some other nation or empire, one of their first considerations was how that disaster affected American interests. Diplomats, military strategists, and other government officials had to decide whether the afflicted site was important to US foreign policy—and if so, what political or strategic benefits the United States stood to gain by providing humanitarian assistance. Domestic pressures further complicated these calculations. When disaster struck abroad, US government officials fielded letters from constituents, appeals from immigrant communities, and lobbying from the agricultural sector, religious organizations, US multinational corporations, and other interest groups, all calling on them to act in some way. While some of these groups strongly supported foreign disaster assistance, others invariably opposed it, urging their elected leaders to focus on problems closer to home. Behind any US foreign disaster assistance effort was an attempt to balance these multifaceted, and often contradictory, international and domestic concerns.

Once US officials made the decision to respond, a second factor influencing the distribution of aid was what we might call the United States' "humanitarian geography"—the spatial elements that determined whether American aid could reach a particular place in a timely manner. In part, this was a function of the physical location of the United States and its overseas territories, but it was also shaped by other dynamics: the United States' global diplomatic and military footprint, world-shrinking transportation and communication technologies, and the shifting winds of geopolitics. Together, these elements determined the United States' humanitarian reach and influenced officials' decisions about whether and how to respond to specific foreign catastrophes. Far from fixed, these spatial elements transformed significantly across the first three quarters of the twentieth century, remapping American humanitarian geography in the process.

A third consideration affecting the allocation of US foreign disaster assistance was rooted in international norms regarding the obligations that states owe their citizens. Ordinarily, in accordance with contemporary laws and customs, US officials expected the governments and civil societies of disaster-stricken countries to assume responsibility for aiding their own populations. Only following particularly momentous catastrophes, where the needs of sufferers overwhelmed a nation's capacity to minister to them, did governments tend to seek and accept offers of foreign assistance. Accordingly, only in these more extreme cases did US officials typically extend offers of humanitarian aid.

Not limited to nation-states, these expectations about governmental responsibility also extended to colonial empires. American officials generally assumed, however naively, that imperial powers had the primary duty to provide disaster aid within the territories they administered. They therefore sent relatively little disaster aid to the colonies of other empires, particularly outside the Western Hemisphere. Even as decolonization accelerated in the decades after the Second World War, postcolonial networks and dynamics remained entrenched, informing US decisions about whether to provide disaster aid to newly independent nations. The US government and its partners in the American voluntary sector, similarly, treated Guam, Hawai'i (until its statehood), the Philippines (until its independence), Puerto Rico, and other overseas territories as domestic dependencies in times of disaster. American responses to catastrophes in these locations were determined by the laws and policies that governed disaster assistance in US insular possessions—procedures that differed greatly from those guiding US foreign disaster aid.

A fourth factor influencing the global flow of US disaster aid hinged on the legal matter of national sovereignty. Respecting the sovereign authority of other governments, US officials delivered disaster assistance to foreign countries almost exclusively by invitation or with express permission, not by force. Although US officials debated whether to flout this norm from time to time, it was usually only with a foreign government's consent that US officials delivered whatever disaster assistance they pledged.

This raises a critical point about terminology. Although US disaster assistance operations arguably constituted interventions, for humanitarian purposes, into other nations' affairs, it would be problematic to classify these episodes as "humanitarian interventions." This label has a specific political genealogy and denotes a distinct type of action. Although scholars debate the precise definition, most agree that a "humanitarian intervention" represents an incursion across another state's borders *without that state's permission* to prevent or mitigate human suffering or human rights abuses. A humanitarian intervention, in other words, necessarily entails a violation of sovereignty and a lack of consent on the part of the intervenee, in a way the episodes in this book simply did not.[11]

And yet, this is not to say that the sorts of power dynamics inherent to humanitarian interventions were absent from US foreign disaster relief operations. Quite the contrary. A fair number of these humanitarian responses involved the deployment of US military troops in other nations. Although US military personnel often provided life-saving material and logistical support to disaster survivors, many of the tasks they performed were less obviously humanitarian in nature: restoring and enforcing order, preventing looting, and regulating the actions and movements of foreign populations. Even when US disaster relief efforts did not involve the military, they tended to be grounded in profoundly uneven power relationships between American officials and the foreign recipients of their aid. As they operated within other nations and empires, US diplomats and American aid workers often exercised considerable authority and control over local disaster victims.

American efforts to aid disaster survivors, in short, routinely coexisted with attempts to police, coerce, and govern them. These dynamics invite a consideration of how the history of US foreign disaster assistance intersects with the history of humanitarian interventions and "humanitarian invasions," to borrow historian Timothy Nunan's provocative term.[12] Moreover, they compel us to think more critically about both the militarization of humanitarian aid and the entangled relationship between humanitarianism and human rights.[13]

Returning to the issue of where US disaster aid flowed, a fifth determining factor was tied to the socioeconomic disparities between the United States and the places it assisted. Throughout the twentieth century, the United States contributed much of its foreign disaster assistance to what have been variously called developing countries, the Third World, and the Global South. That it did so is a potent reminder about the entrenched links among disasters, vulnerability, and global wealth inequality. Although natural hazards posed risks to all countries, they were far more likely to trigger catastrophes in poorer nations, where such issues as substandard housing, overcrowding, and systemic racism all heightened the vulnerability of local populations. Wealthier nations, by contrast, funneled far more capital into preparedness and mitigation activities, actions that prevented many disasters from ever occurring. When they did experience catastrophes, moreover, richer countries were better able to assist their citizens through domestic channels, without relying on external aid. Once again, there were clear exceptions to this overarching trend. Still, because of these disparities, poorer nations were more likely than their wealthier counterparts to require and request US disaster assistance.

In a conspicuous corollary to these socioeconomic and demographic patterns, racial, cultural, and class prejudices permeated many US foreign disaster relief efforts. Throughout the early to mid-twentieth century, the diplomats, troops, and aid workers responsible for carrying out US humanitarian operations were predominantly white, male US citizens from comfortable financial backgrounds. Whether consciously or subliminally, these officials' interactions with foreign disaster survivors were indelibly shaped by their own positions in society. Many US officials harbored deeply racist views about the nonwhite populations they assisted. Others displayed elitist attitudes toward poor and working-class disaster survivors. Still others exhibited strident cultural and national chauvinism, confident in their superiority as Americans. To be sure, these sentiments were by no means universal; some US officials held far more cosmopolitan and egalitarian worldviews and behaved with genuine solidarity toward the individuals they assisted. Even so, many of the US officials profiled in this book acted with suspicion, condescension, and even disdain toward the very people they were supposed to help. These deep-seated attitudes and prejudices further informed the types of aid US officials provided to other nations, the expectations they placed on relief recipients, and the duration and scope of US foreign assistance efforts.

The sixth and final factor determining the distribution of US foreign disaster aid is perhaps best described as pure happenstance. There is, and always has been, a randomness to the incidence of catastrophes. Although US officials understood that an earthquake, tropical storm, or flood might occur in a certain place at some point, it was difficult—if not impossible—for them to know precisely when or where a natural hazard would strike. It was equally challenging to anticipate the severity of the catastrophe it might trigger or to estimate just how disruptive it would be. As a result, government officials were regularly caught by surprise when a catastrophe suddenly overwhelmed another country. Whether they chose to contribute aid to survivors of any given disaster was therefore in many ways a reactive decision, made largely after the fact.

The contingent nature of disasters also meant that American humanitarian responses to them were rather capricious. On some occasions, US officials ended up with far more money and supplies than they anticipated, prompting them to experiment with novel aid projects in an effort to dispense with the surplus. In other cases, they found themselves with much less funding than they hoped, leading them to test alternative, workaround solutions for assisting survivors. For all the planning and thought that went into them, US foreign disaster aid operations were often highly improvised affairs, shaped by the whims of donors, by chance timing, and by the environment itself.

Guided by a complex mix of foreign policy interests, domestic pressures, and ideological assumptions, and steered by the limitations and the possibilities of geography, technology, economics, and geopolitics, US officials in the twentieth century regularly judged foreign catastrophes momentous enough to warrant their nation's assistance. Between the early 1900s and the mid-1970s, the US government, the US military, and their partners in the American voluntary sector provided humanitarian aid to survivors of hundreds of sudden disasters in all corners of the world. As they did so, they made catastrophic diplomacy a fixture of twentieth-century US international relations and laid the foundations for the contemporary system of US foreign disaster assistance.

Before moving further, a few caveats about this book's scope are in order. For one, it bears noting that the United States was hardly alone in responding to international disasters during these years. American officials acted alongside the governments and citizens of other countries, and the personnel of many international organizations, to assist survivors of global catastrophes. The United States was just one among many players in a crowded humanitarian field. Although this book is principally a history of US bilateral disaster assistance, it remains attuned to the broader international context in which the US government and its auxiliaries operated, situating American catastrophic diplomacy against the aid efforts of other nations, empires, and the international humanitarian system.

Second, this book centers primarily on the experiences and concerns of US government and military officials and their partners in the American voluntary sector, the three-pillared system of US foreign disaster assistance. Wherever possible, its chapters give voice to individual relief recipients, foreign government officials, and ordinary American citizens. Yet admittedly, the book concentrates less on how US disaster aid was perceived and received by either foreign populations or the US public and more on what this humanitarian assistance meant to the architects of twentieth-century US foreign policy.

Finally, *Catastrophic Diplomacy* is a history of US disaster aid to other nations and empires, not to US states or territories. Within both the federal government and the American voluntary sector, responses to domestic and international catastrophes were governed by separate policies and legislation and administered by different personnel, bureaus, and agencies. Although they overlapped in some respects, the United States' systems of domestic and foreign disaster assistance were fundamentally distinct—and this book consciously foregrounds the latter. Where pertinent, its chapters reference landmark domestic catastrophes and federal disaster legislation, highlighting points of interplay between domestic and global humanitarian action. Still, the focus is on US foreign disaster assistance and its place in US international relations: the history of catastrophic diplomacy in the American Century.

❋   ❋   ❋

Divided into three parts, *Catastrophic Diplomacy* traces the history of US foreign disaster assistance across the early twentieth century, the decades spanning the First and Second World Wars, and the first three decades following the Second World War. While calling attention to some of the ruptures and key turning points across three-quarters of a century, the story this book tells is principally one of continuity and gradual evolution. Its chapters chart the steady transformation of US foreign disaster assistance, from a common, if ad hoc, activity during the early part the twentieth century into an increasingly routine and formal instrument of US foreign relations, a process solidified by the mid-1970s.

Serving as the book's prologue, chapter 1 examines the US government's very limited role in international disaster relief throughout most of the nineteenth century, then explains how and why this trend began to change as the century drew to a close. Commencing at the dawn of the twentieth century, the three remaining chapters in part I trace the history of American catastrophic diplomacy through late 1916, on the eve of US entry into the First World War. As the United States came to play a more active role in global affairs during these years, the US government began contributing disaster aid regularly to other nations and empires. Grasping the diplomatic potential of these humanitarian actions, members of the Roosevelt, Taft, and Wilson administrations embraced foreign disaster assistance as an expedient instrument of US foreign relations.

Demonstrating these developments and their material effects, the chapters in part I analyze the contours and complicated politics of US relief efforts in Martinique, Chile, Jamaica, Italy, China, the Ottoman Empire, and multiple other sites. They also chart the consolidation of a distinct, three-pillared system of US foreign disaster assistance during these years, comprising the

State Department and its agents, the US Armed Forces, and the government's principal partner in the voluntary sector, the American Red Cross. Working in close collaboration during the early twentieth century, these groups were to oversee the nation's official responses to global catastrophes for decades to come.

Spanning three decades, part II begins with US entry into the First World War in April 1917 and concludes in the aftermath of the Second World War, in 1947. Its five chapters reveal many continuities between early twentieth-century US foreign disaster relief efforts and their interwar counterparts, showing how methods and practices introduced in the previous generation became more routinized and systematic in these years. Through two world wars and the trauma of the Great Depression, and amid ongoing debates over the United States' proper role in international affairs, US officials provided disaster aid to dozens of other nations and empires. Building on foundations laid in the early 1900s, catastrophic diplomacy now occupied a central place in US foreign policy planning.

While tracing these threads and through-lines, the chapters in part II also highlight some important changes across these decades. They analyze the geopolitical, technological, economic, and cultural shifts that gradually transformed the conduct of US international disaster aid between the two world wars. They also recount US officials' experiments with novel aid techniques and tactics, approaches that were to become increasingly commonplace during the second half of the twentieth century. Finally, examining some of the most extensive US disaster assistance operations during these years—in Guatemala, China, Japan, the Dominican Republic, Nicaragua, and Chile—these chapters emphasize the entwined histories of humanitarian relief, reconstruction aid, and development assistance.

The four chapters in part III move the narrative into the third quarter of the twentieth century, from 1948 to 1976. As they confronted the geopolitical tumults of the global Cold War and decolonization, US policymakers seized the political potential of catastrophic diplomacy as never before. While dramatically expanding the frequency, scale, and reach of US foreign disaster assistance, government officials also asserted greater control over its planning and execution. Building on developments set in motion during the Second World War era, they made US foreign disaster assistance a more formal, official instrument of US foreign policy, centralized under the aegis of the federal government and institutionalized within its bureaucratic and legal architecture. In the process, the postwar American state and military assumed a far more central role in the financing, coordination, and administration of foreign disaster aid. Yet even amid these sweeping changes, the American voluntary sector remained a key pillar of the US humanitarian system. Both the American Red Cross and other voluntary aid organizations, housed under the State Department's new Advisory Committee on Voluntary Foreign Aid, collaborated closely with the US government and military. Together, they aided survivors of catastrophes all over the world.

As they track these developments, the chapters in part III analyze the United States' official responses to dozens of international disasters. In these humanitarian operations, methods and practices first tested during the interwar years—gifts and sales of surplus commodities, military airlifts and advisory missions, comprehensive reconstruction projects—grew ever more commonplace, becoming part of the US foreign disaster aid routine. At the same time, disaster assistance became increasingly enmeshed with the problems of international development, further blurring the conceptual lines between these categories of foreign aid. Tracing these patterns from the late 1940s to the mid-1970s, part III recovers the origins of the United States' contemporary system of foreign disaster assistance.

Functioning as an epilogue, the book's final chapter concludes with a reflection on American catastrophic diplomacy since the mid-1970s, charting both the continuities and the changes that have defined the past half-century. With an eye toward lessons and consequences, it suggests how the history of catastrophic diplomacy in the American Century can inform the US government's responses to global disasters today, as we confront the current and future humanitarian crises of the Climate Century.

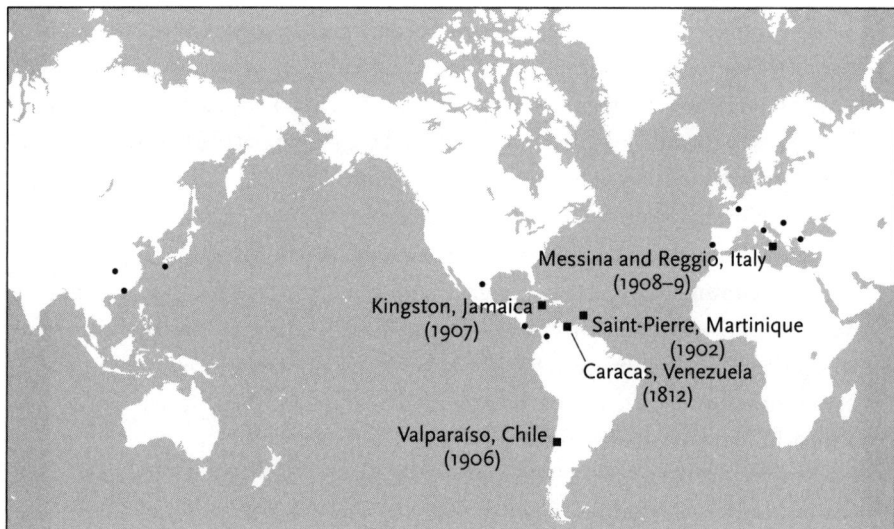

Messina and Reggio, Italy
(1908–9)

Kingston, Jamaica
(1907)

Saint-Pierre, Martinique
(1902)

Caracas, Venezuela
(1812)

Valparaíso, Chile
(1906)

**Recipients of official US foreign disaster assistance
referenced in part I, 1812 and 1902–16**

PART ONE

The Three Pillars of US Foreign Disaster Assistance

1

From Venezuela to Martinique

1812–1902

George Kennan (the elder) is typically remembered for his expertise on the Russian Empire, not the Caribbean. Yet on May 14, 1902, the prominent American writer and explorer found himself aboard the USS *Dixie*, a US Navy cruiser, bound for the French colony of Martinique. Kennan's trip to this tropical isle was no pleasure cruise. He was traveling to Martinique as a correspondent for *The Outlook*, a popular weekly magazine, to report on a tremendous catastrophe that had recently befallen the island.[1]

One week earlier, on May 8, the volcano Mount Pelée erupted violently. The blast destroyed the city of Saint-Pierre and killed 30,000 people, including the US consul and his family. Fifty thousand inhabitants fled the city, becoming refugees whose urgent need for food and shelter overwhelmed the island's available resources.[2] Kennan was heading south to inspect and document the results of this overseas disaster—and, notably, the United States' official response to it.

The Mount Pelée volcanic eruption was remarkable not only for its intensity but also for the reaction to it on the part of the US government. In its wake, the Roosevelt administration, State Department, and War and Navy Departments, together with chosen partners in the American voluntary sector, carried out a sizable relief operation. Financed by congressional funds and private donations, the US government sent a naval vessel, army and navy personnel, and a large cargo of relief supplies to the French West Indies, endeavoring to aid survivors of the immense disaster while burnishing America's image on the world stage.

The US government's response to the calamity in Martinique was not entirely without precedent; instances of official foreign disaster assistance can be found stretching back to the era of the early Republic. For various reasons, however, policymakers provided little direct assistance to other disaster-stricken nations and empires before the twentieth century. Although the situation had begun to change somewhat by the late nineteenth century, the federal government's actions in the Windward Islands in 1902 marked a decided turning point, signaling a growing embrace of foreign disaster assistance within the US foreign policy establishment. Looking back to the nineteenth century, this chapter examines the reasons—geopolitical, technological, and cultural—for this historic shift.

✪ ✪ ✪

On March 26, 1812, nine decades before the eruption in Martinique, a major earthquake occurred in Caracas, Venezuela, and its environs. The resulting disaster proved devastating. Estimates at the time put the death toll at 30,000. Thousands more suffered from injuries and uprootedness.[3] But if this disaster constituted a momentous event for Venezuela, it also marked a critical moment in the history of US foreign affairs. In the aftermath of this catastrophe, some 2,000 miles from Washington, the US government carried out its first official foreign disaster relief operation.

In the early nineteenth century, news traveled slowly from Venezuela to the United States. Only in late April, one month after the earthquake, did reports of the disaster begin to reach North America. Through newspapers and broadsides, Americans learned of the "melancholy matter" and the "fear and despair, grief and ruin" in Venezuela.[4] In Washington, the news prompted US Representative Nathaniel Macon to introduce a resolution to Congress to provide supplies for the "relief of Caraccas." Although some debate ensued over the specifics of the appropriation, the central question of whether to provide federal aid to Venezuela garnered considerable support.[5] Ultimately, both houses of Congress voted to allow President James Madison to purchase and send provisions "for the relief of citizens who have suffered by the late earthquake" and authorized a monetary contribution of $50,000 from the Treasury. On May 8, 1812—ninety years, to the day, before the volcanic eruption in Martinique—the US government passed its first relief measure for a foreign "natural" disaster.[6]

While this allocation was undoubtedly motivated in part by a desire, in Representative John C. Calhoun's words, "to aid the cause of humanity," a strategic rationale also lay behind this unprecedented governmental largesse.[7] Less than a year before the earthquake, Venezuelan revolutionaries had launched a war for independence from Spain. Just a generation removed from their own revolution, many US officials proved eager to encourage this anticolonial rebellion. Citizens of the young United States, however, understood that any overt support for Venezuelan independence risked inviting Spanish hostility—and, quite possibly, a reprisal. While this alone would have been a risky move, a looming threat of war with Britain made provoking Spain seem all the more foolhardy.[8]

In this charged political context, US officials widely agreed, sending disaster aid to Caracas offered an ostensibly neutral way to support Venezuelan revolutionaries while nurturing fledgling Pan-American sentiments.[9] As Representative John Rhea put it, the relief appropriation would serve "the interests of the United States" by allowing the nation "to cultivate amity with and conciliate the South American provinces."[10]

As Rhea and many of his fellow congressmen recognized, providing aid to survivors of disasters caused by earthquakes and other natural hazards held considerable potential as an instrument of foreign affairs. By contributing relief, government officials could demonstrate their compassion for foreign nations and populations—and, assuming all went well, accrue diplomatic goodwill as a result. Through the channels of humanitarian assistance, moreover, government officials could exert their influence on other nations' affairs. In 1812, US policymakers had discovered the potential value of catastrophic diplomacy.

In spite of the political benefits to be reaped from foreign disaster relief, the US government's assistance to Venezuela represented an anomaly. Ninety years passed before Congress made a comparable financial appropriation for disaster aid to another foreign country. Until the late nineteenth century, moreover, federal outlays for *any* overseas humanitarian crisis proved few and far between. Before the 1890s, Congress occasionally debated measures for the relief of foreign suffering while passing just three: for Ireland and Scotland in 1847, France and Germany in 1871, and Ireland (again) in 1880. Not one of these successful aid measures, significantly, followed a sudden cataclysm; all three represented responses to enduring humanitarian crises caused by famine or war. Furthermore, Congress did not empower the federal government to provide cash or purchase supplies for these international humanitarian ventures as it had done for Venezuelans in 1812. The relief legislation merely authorized US Navy vessels to transport privately donated contributions overseas, a much more limited commitment.[11]

The US government did occasionally provide foreign disaster aid through other channels during the nineteenth century. If diplomatic or military personnel happened to be stationed nearby when a devastating catastrophe occurred, they sometimes offered their assistance or contributed what supplies they had on hand. When an earthquake and tsunami struck Peru in 1868, for example, the US minister in Lima provided food, clothing, and medical assistance to disaster survivors, while the commander of the US Pacific Fleet loaned several ships to the Peruvian government to transport additional aid.[12] Comparable scenarios occurred in such places as China, Colombia, El Salvador, Guatemala, Haiti, Italy, and Japan.[13] Sporadic exceptions aside, however, the government and its agents provided little disaster aid to other countries throughout most of the nineteenth century.

Given that policymakers recognized the potential for foreign disaster relief to serve "the interests of the United States" as early as 1812, what accounted for the government's reluctance to provide such aid during the subsequent three quarters of the century? Two broad factors—the first logistical, the second cultural—account for this relative dearth of federal relief.

First, before the late nineteenth century, the federal government was not well positioned or equipped to respond quickly to disasters in most parts of the world. The State Department and the War and Navy Departments were not sizable institutions. US diplomats, consular officials, and troops, therefore, were unlikely to be in close proximity and ready to respond when a disaster suddenly overwhelmed another nation. Even when officials were close enough to assist in a timely fashion, they typically lacked sufficient stockpiles of material supplies for disaster relief.

Although such aid could ostensibly be shipped from the United States, existing transportation and communication technologies hindered a swift response. As the example of the Caracas earthquake attests, it often took weeks before policymakers learned about a catastrophic event in another part of the world. It would have then taken additional time to ship aid from the United States to the disaster-stricken country. Whatever material aid the government sent would arrive long after the immediate emergency had subsided, minimizing both its humanitarian and its diplomatic effectiveness. Geography and technology, coupled with the government's limited resources and small overseas footprint, thus conspired to constrain policymakers from responding to disasters abroad.

A second and equally important factor was the conviction, widespread in US culture throughout much of the nineteenth century, that aiding survivors of foreign disasters was not an appropriate activity for the government to undertake. Relief for victims of catastrophes in other nations, most Americans concurred, was a task better left to churches and charitable societies, not the state.[14] Federal officials, in turn, ceded considerable authority to these private entities, entrusting them to deliver aid on the United States' behalf. In time, the government would forge more formal partnerships with the American voluntary sector, a hallmark of the associational state. For most of the nineteenth century, however, policymakers were content to let private citizens take the reins when foreign disasters struck.

Domestically, the situation was somewhat different. During the early to mid-nineteenth century, Congress passed several dozen relief appropriations following catastrophes within the United States and its territories, such as the New Madrid earthquakes of 1811–12 and a great fire that consumed Alexandria, then part of the District of Columbia, in 1827.[15] Yet "while the federal government did on occasion provide relief to victims of catastrophe before the Civil War," as historian Gareth Davies has observed, "much more commonly it did nothing."[16] If this was the case domestically, it was truer still when it came to disasters overseas.

Beginning in the last quarter of the nineteenth century, several broad historical trends, many of them on the surface unrelated to the US government's humanitarian capabilities, contributed to a gradual erosion of this state of affairs. By the dawn of the twentieth century, the cumulative effect of these developments was to fundamentally transform the United States and its role in the world, while also positioning government officials to deliver disaster relief more readily than their predecessors.

To begin with, during the 1880s and 1890s, the government increased its involvement in international affairs and extended its global reach considerably. Founded as a cluster of settler colonies, the United States swelled into a vast continental empire in the century after its independence, extending its borders through invasion, conquest, and unbridled territorial expansion. During the last two decades of the nineteenth century, the US empire continued to spread, now moving overseas. Across these years, the number of US diplomatic missions and consular posts grew steadily, as did the number of personnel employed by both services. At the same time, policymakers took tentative steps to modernize and expand the US Navy, a process that accelerated in the late 1890s.

All the while, US policymakers were becoming increasingly assertive, and at times aggressive, in their relations with other nations. After a series of diplomatic disputes with Germany, Chile, and Great Britain, US officials went on to intervene in Cuba's War for Independence, declaring war on the Spanish Empire in 1898. Following the United States' victory in that conflict, the American empire enlarged substantially, as the federal government annexed Puerto Rico, the Philippines, Guam, the Hawai'ian Islands, Eastern Samoa, and the Wake Islands. As the twentieth century dawned, the United States had emerged, by many measures, as a great power.

As US government officials became more active in world affairs, American citizens were simultaneously busy forging new transnational connections. During the last two decades of the nineteenth century, the number of American missionaries in the world significantly increased, particularly in Asia, but also in Latin America, the Middle East, and Africa. American corporations, refineries, mining operations, and financial houses, meanwhile, invested extensively overseas and established new foreign subsidiaries, their global spread fueling, and fueled by, a period of tremendous US economic productivity and industrial growth.

Facilitating this increased global connectedness were world-shrinking technological advances, chief among them transoceanic cables and swifter, steam-powered oceangoing ships. With these innovations in communication and transportation, information spread more quickly than ever before, while money and material goods moved across the world at a more rapid pace. In addition, the formation of international news agencies and the mass circulation of such outward-looking periodicals as *Atlantic Monthly*, *Harper's Weekly*, the *Nation*, and the *Christian Herald* brought greater coverage of global events to a more cosmopolitan American reading public.

Finally, in the late nineteenth century there evolved a growing willingness, on the part of both the US government and American citizens, to provide relief for catastrophes *within* the United States and its territories. In the decades after the Civil War, Congress began passing domestic disaster relief appropriations with increasing frequency. The War Department, too, more regularly committed the material aid and manpower of the US Armed Forces to assist US cities, states, and territories stricken by catastrophes.[17] In the 1880s and 1890s, citizens also organized major, nationwide disaster relief efforts in a way they rarely had before. Their aid reached survivors of many domestic and insular disasters, including floods on the Mississippi and Ohio Rivers, an earthquake in Charleston, tornados in Illinois, and hurricanes in the Sea Islands, Galveston, and Puerto Rico. In responding to these and other catastrophes, Americans displayed both a newfound confidence in their ability to mitigate suffering and a budding sense of their obligation to do so. Such convictions reflected the emerging ideals of the so-called Progressive Era, which privileged the power of expertise and placed faith in human interventions to improve both society and the natural world.[18]

Although churches, fraternal orders, and spontaneously formed relief committees raised and delivered much of the nation's nongovernmental disaster assistance, privately donated relief also started reaching disaster survivors via a new quasi-voluntary, quasi-state aid organization: the American National Red Cross (ARC). Established in 1881 by Clara Barton, the ARC quickly gained a national reputation for its responses to dozens of disasters within the United States and its territories. During the 1890s, Barton also led the ARC in providing aid during several humanitarian crises overseas.[19]

In 1900, acknowledging the role the ARC had come to play in both domestic and international humanitarianism, Congress granted the organization its first federal charter. The charter recognized the American National Red Cross as the government's chosen instrumentality for humanitarian assistance in times of war, established to "carry out the purposes" of the Geneva Convention in the United States. Significantly, it also charged the ARC to "carry on a system of national and international relief in time of peace" and with "mitigating the sufferings caused by pestilence, famine, fire, floods, and other great national calamities."[20] Policymakers had long looked to private citizens and organizations to provide both domestic and foreign disaster aid on the nation's behalf. Now, in 1900, they took decisive steps to formalize the federal government's reliance on the American voluntary sector.

Together, these late nineteenth-century developments had important implications for how the US government responded to disasters abroad. For one, they made US officials more aware of international catastrophes than ever before. Thanks to novel transportation and communications technologies, news about foreign disasters was transmitted to the United States more swiftly than in the past, reaching policymakers within days or even hours, rather than weeks or months.

These developments not only made policymakers more cognizant of foreign disasters but also enabled them to respond more quickly. Due to the nation's expanding global footprint, there was now a greater likelihood that US diplomats, troops, and private citizens would be in the vicinity and available to deliver humanitarian assistance when catastrophes occurred. This was especially true after 1898, when the annexations of Puerto Rico, the Philippines, and other territories gave the government new staging grounds for launching relief operations in the Caribbean and Southeast Pacific, a direct (if unintended) consequence of US imperial expansion. Even if no Americans were nearby, swifter ships and larger fleets, both military and commercial, lessened the time required to transport relief supplies from the United States and its territories to other countries.

While technological and geographical developments enhanced the government's ability to respond to foreign catastrophes during the last two decades of the nineteenth century, policymakers were also becoming more philosophically predisposed to aiding other nations. In part, this was an effect of the United States' changing role in world affairs. In an era that found policymakers increasingly attuned to the importance of cultivating and managing the nation's global image, foreign disaster relief offered a simple way to burnish the nation's reputation abroad. By providing generous aid to survivors of catastrophes and delivering it rapidly, officials stood to demonstrate the government's compassion, its economic power, and its military capabilities—not only to the recipients of aid, significantly, but also to any other nation that witnessed the United States' humanitarian actions.

At the same time, many of the cultural forces then compelling Americans to respond more actively to disasters within the United States and its territories were simultaneously leading them to reconsider what role they ought to play when catastrophes occurred abroad. Though the feeling was by no means universal, a growing number of US government officials and citizens came to believe they possessed the power, the resources, and the moral responsibility to aid innocent sufferers in distant lands. The shared sense of civilizing mission that compelled many Americans to embark on imperial ventures and missionary work also inspired in them a conviction that they should aid survivors of foreign catastrophes. They came to feel they could—and should—ameliorate the effects of disasters wherever they struck, regardless of national borders.[21]

During the last two decades of the nineteenth century, confluent political, technological, and cultural developments thus laid the groundwork for foreign disaster relief to become a more regular part of US foreign affairs. As the century drew to a close, the effects of these shifts became increasingly apparent, as both the US government and American citizens began providing greater levels of humanitarian assistance to other nations than in previous generations.

In the late 1890s, attesting to this trend, Congress passed a slate of international relief measures. The first, in 1897, authorized US Navy ships to transport privately donated food to famine victims in India, while the others, in 1898 and 1899, provided direct federal relief to Cubans during and after the Spanish-American War.[22] In late 1899, the US government also undertook a major disaster relief effort in Puerto Rico, a recently annexed US territory, after a catastrophic hurricane struck that island. Although this relief operation occurred within a new outpost of the American empire, not in a foreign country, the actions that the US military government took in Puerto Rico—and the messy politics of that episode—closely mirrored the US government's future responses to disasters in many sovereign nations.[23]

During the 1890s, the humanitarian activities of private citizens were also multiplying. Responding to a growing number of foreign disasters in these years, American church groups and charitable societies sent monetary and material contributions abroad. Missionaries and other expatriates, meanwhile, organized relief operations in the places they resided. Following a tsunami in the region of Sanriku, Japan, in June 1896, for instance, three American Baptist missionaries in the area spent nearly a month distributing cash, food, and clothing to disaster survivors. Similarly, in early 1899, US missionaries and businessmen in Chefoo, China, organized a relief appeal for victims of the Yellow River floods.[24] In addition to responding to disasters caused by natural hazards, American citizens aided victims of many other international humanitarian crises during the 1890s, including victims of famine in Russia and India, religious and political upheaval in the Ottoman Empire, and conflict in Cuba.[25]

By the time Theodore Roosevelt became president of the United States in September 1901, the US government, the US military, and their partners in the voluntary sector were primed to provide far more foreign disaster aid than their nineteenth-century predecessors, and to make foreign disaster aid a more regular part of US international affairs. As the twentieth century dawned, a sudden and momentous catastrophe in the Windward Islands presented them with just such an opportunity for humanitarian action.

✪　✪　✪

When Mount Pelée blew its top on May 8, 1902, triggering a major disaster in Saint-Pierre, few US citizens knew it at the time. The volcanic blast severed telegraph wires and other normal lines of communication, complicating the transmission of up-to-date or reliable information. As a result, news of the cataclysm in Martinique did not reach the continental United States until the following day, May 9, via a telegraph from Louis Aymé, the US consul in Guadeloupe.[26]

As reports of the disaster surfaced, they captivated the public. For weeks, the "St. Pierre Horror" remained headline news.[27] Early accounts detailed the total destruction, providing lurid descriptions of this "disaster of Pompeiian magnitude." They told of a city "buried under a storm of fire from heaven," leaving nothing behind but "a mass of corpse-strewn ruins."[28] While the disaster eventually faded from the front pages, articles in *McClure's, Frank Leslie's Popular Monthly, Scribner's,* and other periodicals brought stories of the devastation to millions of readers for months to come.[29]

Stirred into action, Americans took concerted steps to assist the survivors of this grave calamity. In towns and cities across the United States, civic and religious leaders organized relief committees and campaigned for funds. Citizens responded enthusiastically to their appeals, giving more than $100,000 in less than two weeks.[30] Many Americans also offered their services to the US government. William R. Corwine, a leading figure in the New York Merchants' Association, informed the War Department that his association could purchase supplies on behalf of the US government, promising to do so "with greater expedition than they could be purchased and shipped by the Government itself."[31] Louis Klopsch, editor of the widely circulated newspaper the *Christian Herald,* volunteered as well. He pledged 1,000 barrels of flour to disaster survivors and issued, through the pages of his periodical, "an urgent appeal to the Christian people of the country enlisting their help and sympathy."[32]

But it was not just private citizens who reacted; the catastrophe also gripped the federal government's attention. Coming on the heels of the Spanish-American War and subsequent US occupations of Puerto Rico and Cuba, the disaster presented the United States with another opportunity to flex its imperial muscle in the Caribbean Basin. By delivering swift and generous aid to Martinique, officials understood, they could exhibit the United States' financial and military prowess to rival empires in Europe while also cultivating the friendship of France, a potential ally.

Motivated by these diplomatic goals, government officials responded promptly to the catastrophe. On receiving Aymé's initial communiqué,

Secretary of State John Hay instructed the consul to proceed to Martinique to inspect the destruction. Meanwhile, Secretary of War Elihu Root and Navy Secretary William Moody ordered two US Navy vessels to head to the scene, an action facilitated by the nation's newly expanded humanitarian geography. One, a tugboat, departed from San Juan, Puerto Rico, by now a US territory under civilian government; the other, a cruiser, left from Santo Domingo in the Dominican Republic, where it was stationed to protect US citizens and American property during a period of political instability.[33] President Roosevelt, meanwhile, sent telegrams to his French counterparts, expressing the "profound sympathy of the American people in the appalling calamity."[34]

The president then did something rather out of the ordinary: he urged Congress to allocate hundreds of thousands of dollars for the relief of disaster survivors.[35] Not since Venezuela's 1812 earthquake had the US government made a monetary relief appropriation of this sort, intended for victims of a sudden catastrophe in another nation or empire. Although precedents for such legislation certainly existed, particularly in recent years, Roosevelt's suggestion nonetheless represented a departure from the typical foreign policy actions of the previous century. In pressing Congress to make a large financial commitment for this French colony, the president embraced an expanded role for the federal government in foreign disaster relief.

Despite its relative novelty, Roosevelt's proposal quickly gained traction. On Saturday May 10, the Senate passed a bill appropriating $100,000 "For the Relief of Citizens of the West Indies" and authorizing the secretary of war to employ US Navy ships to carry out aid operations.[36] Over the next two days, as the bill awaited consideration by the House, additional requests for assistance reached the State Department and White House. On Sunday, Aymé arrived in Martinique's capital, Fort de France. After seeing the destruction, he implored the US government to ship "codfish, flour, beans, rice, salt meats, [and] biscuits" to feed survivors.[37] On the same day, France's foreign minister requested the Roosevelt administration's help in transporting refugees to safer areas.[38] By Monday, such appeals convinced Roosevelt to formally request a congressional relief appropriation. Declaring, "One of the greatest calamities in history has fallen upon our neighboring island of Martinique," Roosevelt urged Congress to act.[39]

In the House of Representatives, Roosevelt's proposal sparked a lively debate over whether the government possessed the responsibility—or the right—to assist the citizens and colonial subjects of another nation. The initial response to his motion was very positive, with the House Committee on Appropriations doubling the Senate's proposed allotment to $200,000.[40]

The committee also added language to the bill authorizing the president to "procure and distribute" assistance, "to employ any vessels of the US Navy," and to "take such other steps as he shall deem advisable" to carry out its provisions.[41]

In both the amount it authorized and the power it afforded the executive branch, the terms of this legislation constituted a deviation from most nineteenth-century precedents. Even so, most House members responded to it favorably. There were a few naysayers, though. Alabama congressman Oscar Underwood fiercely opposed the bill, declaring, "I do not think it wise or expedient or proper to pass this class of legislation." For Underwood, the issue was not whether the United States, as a people, ought to provide foreign disaster relief. "Our own citizens to-day are raising funds from their own pockets to contribute aid to our suffering neighbors," he noted, adding, "It is right, it is generous, it is just that they should do so." His objections stemmed, instead, from his belief that the US government did not possess the authority to commit taxpayers' dollars in this way. "What right have we to be generous with the money of our constituencies to help a foreign people?" Underwood asked. "I know of none."[42]

Although Underwood succeeded in rallying a handful of allies, the majority of his fellow congressmen rejected his logic. A representative from Texas asked how it would be "possible for the American people to act in this emergency except through their Government and their chosen representatives here?" while his colleague from New Jersey wondered what to do with those Americans who were "too mean" to give aid. "Is there any other way of making them contribute," he inquired, "except by taking it out of the National Treasury?" A representative from Indiana pondered, somewhat more philosophically, "how far the brotherhood of man extends" and whether "it is limited by territorial or international lines."[43]

Ultimately, these proponents of government-funded foreign disaster relief won the day. The House voted to pass the relief measure by overwhelming margins, while the Senate passed the amended version of the legislation with little debate. On May 13, five days after the volcanic eruption in Martinique, "An Act for the Relief of Citizens of the French West Indies" became law.[44]

With the passage of this legislation, Congress made a considerable financial commitment to survivors of a catastrophe in another empire. By authorizing the president to take whatever steps he deemed necessary to carry out the humanitarian operation, Congress granted broad new powers to the executive branch over foreign disaster relief. As the House debate made clear, doubts about the government's proper place in foreign disaster relief certainly remained. The passage of the relief measure, however, suggested that most of the nation's elected representatives now viewed foreign disaster relief as a legitimate activity for the federal government to undertake.

Crucially, the government's decision to assume this humanitarian role did not mean that officials rejected the tradition of collaborating with private citizens. Desiring a partnership with the American voluntary sector, not a rivalry, government officials took active steps to ensure this outcome. Specifically, Roosevelt appointed roughly a dozen prominent businessmen and merchants to spearhead a national campaign to collect monetary and material donations from private citizens. The Associated Committee for Relief of the West Indies, as the group was called, eventually raised more than $150,000 for the cause.[45]

On the government's side, responsibility for carrying out the congressional mandate fell to the War Department. Once Congress passed its relief legislation, Root and the adjutant general of the army, H. C. Corbin, set to work. They commanded US troops in Puerto Rico to procure "tents, flour, bacon, and other ration articles" from local US Army stores and assigned Captain J. T. Crabbs the task of delivering them from San Juan to Martinique.[46] On Wednesday, May 15, Crabbs and a small crew of army officers arrived in Fort de France and began distributing this aid.

Root and Corbin also requisitioned the USS *Dixie*, a navy cruiser, to deliver aid from the continental United States. On Tuesday, May 14, the *Dixie* set sail from New York with a dozen military officers and army physicians and 900 tons of supplies, a quantity sufficient to feed and clothe an estimated 50,000 people for thirty-six days.[47] Also on board were three scientists from the National Geographic Society and reporters from more than twenty newspapers and magazines—George Kennan among them.[48] New York businessman William Corwine soon departed for the French West Indies as well, traveling there on the steamship *Fontabelle*. Going as a representative of both the New York Merchants' Association and the Associated Committee for Relief, Corwine headed to Martinique distribute the food and supplies the committee had collected.[49]

On the ground in Martinique, the relief operation proved a relatively swift and uncomplicated affair. Crabbs and his fellow officers, arriving quickly from nearby Puerto Rico, participated in relief and rescue efforts during the early days after the eruptions. Otherwise, Americans were not very involved in administering relief. By the time the *Dixie* reached Martinique on May 21,

French colonial officials were well in command of assistance efforts. Ceding authority to these individuals, US personnel turned over the *Dixie*'s stores to French naval officers. Three days later, French officials told Aymé "that there were now sufficient supplies in the colony" and that "nothing further need be sent."[50] Landing in Martinique on May 26, Corwine found the situation well under control, concurring with other observers that the "immediate necessities had been provided for."[51]

Shortly thereafter, US emergency relief operations drew to a close. On May 31, the crew of the *Dixie* set sail for the United States, along with most of the American journalists and scientists who accompanied them.[52] Kennan, Corwine, and other remaining US citizens followed on the *Fontabelle* a week later.[53]

The US relief operation for Martinique may have been short-lived, but policymakers sensed that in that brief time, they had achieved remarkable diplomatic success. Messages of thanks poured into Washington. Throughout Martinique, one US military officer reported, he "heard the people murmur their blessings on the American Government for sending assistance to them in this emergency."[54] Such messages were echoed by the governor of Martinique, the French ambassador to the United States, and French president Émile Loubet.[55] By assisting Martinique, it seemed, US officials had succeeded in their goals of cultivating Franco-American amity and demonstrating the beneficence of US power in the Caribbean Basin.

For many US officials, these assorted accolades confirmed something their predecessors realized ninety years prior: aiding disaster survivors in foreign countries was decidedly in the United States' diplomatic interest. Nine decades after the Caracas earthquake, the federal government had come full circle.

2

Casting the
Three Pillars

1902–1908

In hindsight, the US government's response to the 1902 catastrophe in Martinique constituted a landmark event whose patterns were to recur across many future US foreign disaster assistance operations. The episode reflected a maturing consensus that the federal government possessed the authority to administer disaster relief abroad. American actions vis-à-vis Martinique were also predicated on a set of norms that were to guide (though not dictate) subsequent assistance efforts. They included the beliefs that US officials should provide disaster aid only with an affected country's permission; that US aid should normally be limited to short-term, emergency forms of relief; and that American assistance should flow mainly through bilateral channels, with Americans maintaining considerable latitude over how that aid was distributed. Perhaps most significantly, the US response to the Mount Pelée catastrophe depended on the involvement and collaboration of three key pillars of the US humanitarian system: the State Department and its agents, the War and Navy Departments and their personnel, and the government's chosen partners in the American voluntary sector.

Across the remaining six years of Theodore Roosevelt's presidency, building on precedents set in Martinique and the trends of the previous two decades, government officials took several actions that further facilitated the conduct of foreign disaster assistance. These included the acquisition of new overseas bases, the continued modernization and expansion of the US Navy, and reforms of the US Army and the diplomatic and consular corps. Embracing a form of governance that historians term "the associational state," government officials also forged a closer and more formal partnership with the American Red Cross, a move that fundamentally transformed the state's relationship with the voluntary sector and the administration of US international disaster relief operations.[1]

Concurrent with these developments, government officials participated in a rising number of disaster responses during these years in such places as Chile, Hong Kong, Jamaica, and Italy. Together, these relief efforts demonstrated the growing tendency, on the part of the government and its auxiliaries, to contribute disaster aid to other nations and empires. At the same time, these episodes highlighted difficulties that officials confronted in conducting foreign disaster aid operations, including financial and logistical challenges and the diplomatic risks of intervening in other nations' affairs. They also cast light on the ugly specter of racial, class, and cultural prejudices, revealing the condescending attitudes that many US officials held toward aid recipients. Not confined to the early twentieth century, these dynamics resurfaced time and again in future US foreign disaster assistance efforts.

✪　✪　✪

Theodore Roosevelt's presidency brought dramatic changes to the conduct of US foreign affairs. Although never explicitly on the president's agenda, many of the Roosevelt administration's foreign policy initiatives indirectly enhanced the government's capabilities for responding to international catastrophes. Together with developments set in motion during the last quarter of the nineteenth century, these actions further increased the federal government's readiness and willingness to act when disasters struck abroad.

In the most general sense, the Roosevelt administration's active and interventionist approach to world affairs carried over into the humanitarian sphere, where it bolstered the idea that the federal government had a responsibility to respond to foreign disasters. As president, Roosevelt acted with the conviction that the United States possessed "international duties no less than international rights."[2] During his presidency, guided by these assumptions, the government intervened repeatedly in other nations' political, economic, and social affairs. In many instances, these incursions entailed the use (or threatened use) of military force, as in the ongoing Philippine-American War, Roosevelt's Corollary to the Monroe Doctrine, and the landing of US Marines in multiple Caribbean nations. In other instances, US intervention took the form of diplomatic engagement and negotiations, including Roosevelt's efforts to broker peace between Russia and Japan in 1905 and Secretary of State Elihu Root's work in negotiating the Hague Convention of 1907.

The notion that the US government had an obligation to participate in the humanitarian affairs of other nations was a logical outgrowth of this interventionist political culture. In the early twentieth century, foreign disaster aid formed part of a growing constellation of international duties the government now claimed as its own.

At a more specific level, several of the Roosevelt administration's hallmark foreign policy initiatives carried important, if unintentional, consequences for US foreign disaster aid. The imperial policies the United States pursued during the first few years of the twentieth century, for one, further extended the government's humanitarian reach. Following the wars of 1898, US officials had annexed Puerto Rico, the Philippines, and several other territories, a land grab whose indirect effect was to establish new sites for launching relief missions. During Roosevelt's first term in office, the government again expanded its humanitarian geography when it assumed jurisdiction and control over two additional overseas territories: Guantánamo Bay and the Panama Canal Zone. Both these sites—but especially the Canal Zone—soon became key staging grounds for US foreign disaster relief missions in the Caribbean Basin and, in time, throughout the entire Western Hemisphere.

The Roosevelt administration's reforms of the US Armed Forces and US diplomatic and consular corps also had a direct bearing on foreign disaster aid, for these services came to play a critical role in carrying out the nation's overseas humanitarian missions. Arguably the most significant of these projects was the expansion and modernization of the US Navy and Marine Corps. Between 1899 and 1908, spurred by liberal congressional spending, the size of the navy more than doubled while the Marine Corps expanded threefold. During these same years, an ambitious capital shipbuilding program transformed the US Navy into the world's second most powerful, behind only

Great Britain's Royal Navy.[3] Roosevelt's first secretary of war, Elihu Root, also spearheaded a series of reforms to modernize the US Army. Most notable of these "Root reforms" were the creation of the Army General Staff and the reorganization of the National Guard, the Army Medical Department, and other branches.[4] During Roosevelt's second term, now serving as secretary of state, Root similarly set out to professionalize the diplomatic and consular services, implementing competitive entrance examinations, merit-based promotions, and a major reorganization of the State Department.[5]

Although the principal aim of these collective reforms was to improve the core institutions of US international relations, they also brought fundamental changes to the federal agencies tasked with conducting responses to foreign catastrophes. Together, they strengthened the government's ability to deliver aid and administer disaster assistance operations abroad while also increasing the state's control over these humanitarian activities.

Of all these developments, the action with the most direct and significant consequences for US foreign disaster assistance was the consolidation, in 1905, of a unique partnership between the federal government and the American Red Cross. Through this special relationship, the government secured the services of a humanitarian auxiliary responsible for financing, planning, and superintending its foreign disaster assistance operations. By drawing the ARC more fully into the government's foreign policy apparatus, US officials also tethered the American voluntary sector more closely to the state, thereby aligning privately funded humanitarian initiatives with the government's foreign policy agendas.

Founded in 1881, the ARC had received its first federal charter in 1900, legally establishing it as the federal government's instrument for domestic and international humanitarian aid, including foreign disaster relief.[6] In practice, however, it took several more years and a major reorganization before the organization fully stepped into this latter role. In the years after receiving its first charter, the ARC experienced a period of bitter infighting. This power struggle pitted the organization's aging president, Clara Barton, and her allies against a faction of critics, who argued that the ARC must become more professional, better managed, and better organized.

Ultimately, these upstarts won the day, precipitating a wholesale restructuring of the ARC. In May 1904, Barton succumbed to her critics' pressure and resigned the ARC presidency. With her departure, a new cadre of leaders set out to remake the organization. They quickly submitted a revised charter to Congress, which Roosevelt signed into law in January 1905.[7]

While keeping much of the ARC's existing structure intact, the 1905 congressional charter substantially strengthened the ARC's special relationship with the government.[8] Under its terms, the ARC retained its status as a voluntary organization, delegated by the federal government to carry out domestic and foreign relief efforts for victims of wars and "great national calamities." Officially, the ARC preserved much of its legal and financial autonomy under this arrangement. The organization continued to rely solely on private donations, not federal funds, to finance its relief efforts, while the majority of its leadership still hailed from the private sector.

At the same time, however, the new charter granted the US government a far more direct role in the ARC's operations. It did so in two principal ways. First, it empowered the president of the United States to appoint one-third of the ARC's governing central committee and to fill those positions with federal officials from the Departments of State, War, Navy, Treasury, and Justice. Second, the new charter required the War Department to audit the ARC's finances and activities each year and to submit the results of this annual report to Congress. With these new arrangements, the ARC's already close relationship with the federal government grew even tighter. The government, in turn, vested considerable authority for foreign assistance in a unique, quasi-governmental organization. It also established a powerful voice for the state in the ARC's humanitarian activities.

The 1905 ARC charter marked another critical milestone in the evolving US foreign disaster aid system. From the Roosevelt presidency until well into the post–Second World War era, the ARC served as the US government's official auxiliary for foreign disaster aid. Across these decades, the ARC's leaders collaborated closely with US diplomatic, military, and other government officials to plan and carry out the nation's official foreign disaster relief operations. Throughout much of this time, it also served as the principal source of funding for these efforts.[9]

Bolstering the ARC's identity as the government's preferred partner in foreign disaster aid, the Roosevelt administration and its successors vigorously promoted the ARC and its relief campaigns to the public. They also greatly facilitated the ARC's humanitarian work, helping the organization send money and aid supplies overseas and to maintain communications with other disaster-stricken countries. Perhaps most notably, from 1905 on, government officials played a direct role in the ARC's governance and decision-making, influencing the organization's actions in response to international catastrophes. Although the ARC's civilian leaders and staff continued to aver the organization's neutrality and independence, the complex array of formal and informal ties that bound their organization to the US government complicated any such claims to impartiality, blurring the lines between state and nonstate humanitarian assistance.

Certainly, the ARC did not displace the myriad other American charitable organizations, churches and missionary associations, and philanthropies involved in international disaster relief initiatives. In future catastrophes, many of these groups continued to carry out their own aid efforts as well. Yet the fact remains that US officials designated the American Red Cross—and the American Red Cross alone—as the state's preferred instrument for foreign disaster relief. The consolidation of this alliance in 1905 had profound consequences, determining how the federal government financed, organized, and orchestrated the nation's official responses to international catastrophes for decades to come. It also increased the federal government's control over humanitarian decision-making, as officials in the State Department, War and Navy Departments, and the White House assumed an influential voice in determining which nations should be prioritized for foreign disaster aid.

By Roosevelt's second term in office, with the ARC's new charter freshly ratified, the outlines of a coherent system of US foreign disaster assistance were increasingly visible. With the three pillars of foreign disaster assistance reformed, expanded, and in place, officials were poised to respond to international calamities more consistently and to test the possibilities and limits of their catastrophic diplomacy throughout the world.

The world's next major catastrophe, however, occurred at home. In April 1906, the federal government, the US military, and the newly reorganized ARC faced the first serious test of their *domestic* disaster assistance capabilities after a momentous earthquake and fire destroyed San Francisco. The ARC's response to this catastrophe did much to cement its reputation as the nation's official disaster relief agency. After the quake, Roosevelt urged the public to donate to the ARC, identifying it as "the national organization best fitted to undertake such relief work."[10] In San Francisco, flush with more than $8 million in donations, ARC personnel played a central role in providing relief and recovery assistance to tens of thousands of survivors. They also experimented with novel techniques including ration card systems and work programs for disaster refugees, policies intended—in their minds, at least—to promote disaster survivors' self-sufficiency and to make humanitarian relief a more scientific, professional endeavor.[11]

The US Army's response to the catastrophe was equally noteworthy, previewing the critical role the military stood to play in American disaster relief and recovery efforts overseas. Cooperating closely with the ARC, army officers and troops engaged in many humanitarian tasks, including extinguishing fires, overseeing food and clothing distributions, and tackling sanitary, health, and medical problems. The army also installed and superintended twenty-one refugee camps, sheltering thousands of San Franciscans under direct military authority.[12] While delegating much of the responsibility for relief work to the army and ARC, federal government officials in Washington steadfastly promoted those efforts. At Roosevelt's urging, Congress made a considerable financial commitment to the disaster, appropriating $2.5 million to procure and ship emergency relief supplies to California.[13]

The San Francisco earthquake and fire, as historian Ted Steinberg has argued, has long stood in collective memory as an "archetype" of disaster, as "the event that defines calamity in the popular imagination."[14] By the same token, the unified response to that crisis by the US military, the ARC, and the federal government became a touchstone for disaster *relief* in the early twentieth century, a model emulated in many future crises. To be sure, the response was not without controversy or criticism.[15] Nonetheless, the methods of disaster assistance the federal government, military, and ARC together employed in San Francisco became key reference points for subsequent disaster relief efforts, not only within the United States but far beyond the nation's borders.

✪   ✪   ✪

Four months after the San Francisco earthquake and fire, the Roosevelt administration turned its attention to another seismic cataclysm—this one 6,000 miles from northern California. On August 16, 1906, a strong earthquake occurred in Valparaíso, Chile, one of South America's largest seaports, killing several thousand people and leaving an estimated 100,000 homeless.[16] Reports of the catastrophe quickly reached the United States. Many Americans regarded the situation with an understandable empathy, born of their own nation's recent calamity, and reasoned that their nation had a special obligation to assist Chile. Calling the situation in Valparaíso "sickeningly like that of San Francisco," editorialists insisted the United States must extend "its aid and sympathy to Chile."[17] Others saw in the disaster a diplomatic opportunity, a fortuitous occasion to cultivate better hemispheric relations through humanitarian aid. As a group of Massachusetts citizens advised Roosevelt, it appeared to be "a propitious moment [for the United States] to prove sympathy with sister republics."[18] A New York man, likewise, informed the president, "Now is the time for this country to show its good will to South America."[19]

Roosevelt and his advisers shared these convictions. While they felt the United States had a moral responsibility to assist Chile, they also viewed the crisis as a chance to demonstrate the US commitment to Pan-American solidarity, one of the administration's chief foreign policy concerns. A generous response to the catastrophe, Roosevelt and his advisers imagined, might even serve to counteract Latin American resentment over the United States' recent incursions in Panama, Cuba, the Dominican Republic, and other sites. Guided by these diplomatic and humanitarian motivations, Roosevelt quickly wrote to Chilean officials expressing his "intense sympathy" for the suffering.[20]

His administration ran into difficulties, however, in endeavoring to provide tangible, material aid to Valparaíso. The Senate and House of Representatives were not in session when the quake occurred, so Roosevelt could not request a congressional relief appropriation, as he had after the eruption of Mount Pelée in 1902. The distance between Chile and the United States or any of its overseas territories, moreover, impeded the government from shipping relief supplies rapidly to Valparaíso, as it had to Martinique four years prior.

Faced with these hurdles, the Roosevelt administration responded to the disaster in Valparaíso through other channels at its disposal. In a fortuitous turn of events, Secretary of State Elihu Root happened to be in Rio de Janeiro at the time of the earthquake, attending the Third Pan-American Conference. Recognizing that an official visit to the disaster zone would represent a meaningful diplomatic gesture, Root traveled to Chile after his work in Brazil concluded, arriving in Santiago two weeks after the quake.[21] Meeting with Chile's president and other dignitaries in the capital, Root expressed his country's compassion for Chilean sufferers, presenting American empathy as a harbinger of closer US-Chilean relations. "It brings us nearer to you to be able to mourn with you in your misfortune," he proclaimed, "to sympathise with you in your distress."[22] Parlaying these declarations into a more concrete demonstration of US concern, Root and his entourage also traveled to Valparaíso to survey the ruins personally. Recounting the trip to the State Department, US Envoy John Hicks reported that he could "not speak in too high terms to express the good results of Mr. Root's visit to Chile." This action, Hicks believed, would "greatly assist in bringing about a good and kindly feeling between the two countries."[23]

Although they appreciated Root's mission to Chile for its symbolic value, the Roosevelt administration hoped to make a more tangible contribution to the relief effort. In the absence of congressional funding or authorization, State Department officials looked to the American Red Cross to spearhead the official US relief effort for Chile. Eager to demonstrate their utility to the US government, the ARC's leadership soon launched a national appeal for Chilean relief. Roosevelt himself assumed the role of booster for this campaign, prevailing on US citizens, "in this time of woe of our sister republic," to make "a generous response" to the organization.[24]

Government officials and ARC leaders assumed the organization would bring in large sums for Valparaíso, just as it had for San Francisco four months prior. Yet much to their chagrin, contributions remained decidedly sluggish. A week after the earthquake, the ARC had managed to collect only a few hundred dollars.[25] Some potential donors surely felt overextended after contributing so much to California so recently; others simply cared less about the situation in Chile, feeling few personal ties with that distant nation. Whatever the reasons, the ARC ultimately raised just over $12,000, less than a tenth of private American contributions for Martinique four years earlier. Although additional American donations reached Valparaíso through other channels—including a $15,000 contribution from the citizens of San Francisco—the tepid results of the ARC campaign sorely disappointed the organization's leadership and the Roosevelt administration.[26]

Despite these financial setbacks, the Roosevelt administration continued to collaborate with the ARC to carry out the official US relief effort in Chile. The question now was how best to administer that operation on the ground. Like many contemporary US citizens, ARC leaders held deeply condescending attitudes toward South Americans, and they distrusted Chileans to distribute US funds in a manner they deemed professional. American Red Cross leaders privately judged their sister society, the Chilean Red Cross, as "not sufficiently organized . . . for the American Red Cross simply to turn over this money to them."[27] More broadly, they expressed concern that "the science of relief [in Chile] has not been developed to such an extent as it has in this country."[28]

Skeptical of Chilean approaches to humanitarian aid—and confident of their own expertise in this area—ARC leaders desired to retain control over the funds they contributed. They realized, however, that it would be impractical to send one of the ARC's own representatives to Valparaíso, given the considerable distance. Instead, taking advantage of the organization's tight alliance with the State Department, they named the US envoy in Santiago, John Hicks, as the ARC's special representative. The State Department readily endorsed his appointment.[29]

American Red Cross leaders charged Hicks with receiving and distributing the organization's contribution at his discretion, leaving it "to his judgment how it would be expended."[30] Though ostensibly impartial, Hicks's determinations were steered by his own sense of cultural, class, and racial superiority. In his communications with Washington, Hicks voiced considerable disdain toward the Chileans he assisted. He described disaster victims as "ragged and filthy, their Indian blood showing in its most unpleasant form." Echoing ARC leaders, he also disparaged Chilean authorities, complaining that they did not know how to distribute aid effectively. Hicks admitted that it would have been easiest to "turn the whole amount over to the government and thus get rid of the labor and responsibility." Yet, distrustful of both Chilean officials and Chilean relief applicants, he explained that he "could not reconcile" this action "with my sense of duty under the circumstances."[31]

Instead, retaining tight control over ARC funds, Hicks established a system to evaluate relief applicants and determine who, in his estimation, truly deserved American aid. As he explained to the State Department, this approach enabled him to dole "out the money at my discretion ... according to my best judgment."[32] From his offices at the American legation, Hicks and an American consul and vice-consul met, interviewed, and inspected each and every relief applicant. They also demanded written "testimonials as to the character of the sufferers." The oversight did not end there. Assuming that Chileans "might waste the money if they had so much at once," Hicks and his associates distributed their assistance only in small sums, requiring relief recipients to make follow-up visits if they desired additional aid.[33] Although Hicks did coordinate his efforts with trusted Chilean government officials and charitable agencies, he and his American colleagues maintained the ultimate authority over the US relief operation.

Hicks served as arbiter for relief claims for several months after the disaster struck, until May 1907, when his funds dried up and brought the US assistance effort to a close.[34] This would not be the last time the ARC relied on US diplomats and consuls to act in this capacity. Such arrangements became increasingly commonplace in coming years, reflecting the tight partnership between the US government and its humanitarian auxiliary. Likewise, it was not the only time that racial, classist, and nativist prejudices shaped the direction of US foreign disaster relief efforts. These forces, too, became a familiar specter, rearing their heads across many future US aid operations.

In late 1906 and early 1907, as Hicks was busy overseeing the distribution of ARC funds in Chile, two abrupt calamities presented the Roosevelt administration with additional opportunities to test the government's foreign disaster assistance apparatus. Unlike in Valparaíso, American officials relied primarily on the US Navy to respond to both catastrophes, a choice made possible by the United States' newly expanded humanitarian geography.

The first and more limited of these relief operations occurred in September 1906, when a typhoon struck the British colony of Hong Kong, killing 5,000 people. In the storm's wake, the commander of the US Asiatic Fleet—which was stationed in the region to protect US interests in the Philippines, China, and Japan—offered his aid. The governor of Hong Kong quickly accepted. American crewmen spent the next several days rescuing survivors and salvaging wreckage before departing for other duties.[35] The second and more substantial of these relief efforts unfolded several months later in another British colony, this one much closer to the continental United States. By the time this operation concluded, it had sparked a charged diplomatic incident, exposing the political risks associated with US foreign disaster aid.[36]

On January 14, 1907, a powerful earthquake occurred in Kingston, the capital of the British colony of Jamaica. Across the city, buildings crumbled and collapsed. Raging fires contributed further to the destruction. Hundreds, and perhaps thousands, of people died, and many more suffered serious injuries.[37] As news reached the United States, editorialists once again called for a sizable US response, proclaiming that Jamaica "has the right to look to America, by reason of its nearness, for its 'first aid.'"[38] Whether or not Jamaicans—or any other foreign nationals—had a "right" to US disaster relief remained a matter for debate. Yet no matter one's stance on that philosophical question, the US government was undoubtedly better positioned to deliver material aid rapidly to Kingston than it had been to Valparaíso, given Jamaica's proximity to the United States and its Caribbean territories. A swift and generous response to the calamity, moreover, held the potential to advance two of the Roosevelt administration's chief foreign policy goals: showcasing US power in the Caribbean and demonstrating the nation's commitment to Anglo-American amity.

In light of these considerations, reports of the devastation in Kingston triggered a prompt reaction in Washington. On January 16, Roosevelt and Root sent messages of sympathy and offers of assistance to their counterparts in Great Britain and Jamaica.[39] The following day, without waiting for contributions to roll in (or not), ARC leaders voted to use $5,000 from their reserve funds to purchase food for Jamaica and chartered a private steamship to transport the aid.[40] On January 18, the ship left New York for Kingston, where it arrived a week later.[41]

Unlike the previous year, Congress was now in session, and US representatives and senators quickly took up the question of relief legislation. By January 18, both houses of Congress approved a bill authorizing the president to distribute clothing, medicine, food, and other supplies from US military stores to the disaster's survivors.[42] Although at least one representative wondered whether there should be "some limitation on the power to use the resources of the Government for this purpose," the relief act provoked relatively little opposition.[43] As it had in 1902, Congress voted overwhelmingly in favor of allocating federal resources for foreign disaster relief.

Yet this appropriation was of little immediate consequence, for the US Navy had already landed in Kingston with assistance. On January 16, the navy secretary ordered the commander of the new naval base in nearby Guantánamo Bay, Cuba, to send relief supplies to the British colony. He delegated his second in command, Rear Admiral Charles H. Davis, to proceed to Jamaica to oversee the aid distribution.[44] The following day, Davis reached Kingston with a fleet of three ships, carrying food, medical and surgical supplies, and several dozen navy sailors and surgeons.[45] A few days later, the US Army became involved, when the commanding general of the Army of Cuban Pacification—which Roosevelt had deployed the previous September to protect American interests on the island—sent an additional load of tents to Jamaica. The tents sailed from Santiago de Cuba, transported to Kingston by the US Eleventh Infantry.[46]

The official mobilization for Jamaican disaster relief thus proceeded smoothly within the United States and the Caribbean territories it occupied, but in Kingston, the US response sparked a major diplomatic imbroglio. The trouble originated with a series of events occurring shortly after Rear Admiral Davis reached Kingston. On arrival, Davis reported to the offices of British governor of Jamaica, Alexander Swettenham, only to learn that the governor had retired to his residence outside of town. Davis met instead with the governor's secretary, C. H. Bourne, and Kingston's deputy inspector general of police. From these men, he learned that 600 prisoners were presently mutinying at Kingston's penitentiary and that general disorder prevailed throughout the city. In response, Davis offered to land two armed units, one to help put down the prison rebellion and the other to protect the American consulate. He also proposed sending his remaining men ashore to help clear the wreckage. Bourne, assuming the authority to speak on Swettenham's behalf, agreed to these suggestions. With this approval, Davis ordered his men to disembark and get to work.[47]

Some 200 US Navy officers and sailors were soon on the ground in Kingston, unloading relief supplies, recovering bodies, clearing rubble, and demolishing walls and buildings they deemed unsafe. More than fifty armed US troops, meanwhile, joined local police in quelling the prison mutiny. They then established patrols at the US consulate and in other areas of the city.[48] Despite having Bourne's consent, Davis determined it best to secure the governor's formal authorization as well. He therefore made his way to Swettenham's residence, where he and the governor enjoyed a cordial meeting. Swettenham thanked Davis for his help and granted retroactive approval for his actions. However, he also informed Davis that the situation was now under control and that further US assistance was unnecessary.[49]

It was at this point that frictions between Davis and Swettenham began materializing. These tensions stemmed from fundamental disagreements over questions of sovereign authority and humanitarian responsibility. As governor of Jamaica, Swettenham was privately "astonished and disgusted" to learn Davis had landed dozens of US troops—many of them armed—without his permission. Although he conceded Davis had "acted with the best and most generous intentions," Swettenham believed that he and other British colonial officials were perfectly capable of managing the crisis themselves.[50] Davis, for his part, doubted the governor's claim to have the situation under control. Swettenham greatly underestimated the severity of the crisis, the rear admiral reported to his superiors, leaving disaster survivors in dire straits. In Davis's mind, this precarious situation meant that he and his men had an obligation to continue helping Kingston. As he told his commander, "I consider it my duty to remain for the present, at least."[51]

After their meeting, Davis wrote Swettenham a letter communicating his position, stating his intention to "continue my operations unless requested explicitly to withdraw."[52] When he failed to receive a reply by the next morning, Davis chose to land scores of additional US sailors in Kingston, this time without any explicit authorization. The stage was now set for a full-scale confrontation between the US rear admiral and the British governor of Jamaica.

On their second day in Kingston, Davis's men again went to work patrolling the city, demolishing ruined buildings, and clearing away wreckage. Davis also assigned thirty surgeons and sailors to erect and run an emergency hospital, on land belonging to American Jesuits.[53] The troops had been engaged in these activities for the better part of the day when Davis received an acerbic note from Swettenham, demanding that he "reembark . . . all parties which your kindness has prompted you to land." Swettenham rebuked Davis for overstepping his authority in all areas, but particularly for sending US troops into a British colony to maintain order. "I may remind your excellency that not long ago thieves had lodged in and pillaged the town houses of a New York millionaire," Swettenham remarked caustically, "but this fact would not have justified a British admiral in landing an armed party to assist the New York police."[54]

Swettenham's reaction shocked Davis, who believed that he and his men were engaged in vital humanitarian work in Kingston. He saw little choice but to obey the governor's commands, however, reasoning, "As a foreign Naval Officer, I am bound to respect the wishes and requirements of the supreme authority of this island."[55] Davis promptly withdrew his men and departed Jamaica, returning to Guantánamo Bay the following day.[56]

Davis may have left Kingston, but the controversy was far from over. As reports of the dispute surfaced, the episode became a highly public affair. What was supposed to have been a testament to Anglo-American ties instead devolved into a trial of Anglo-American relations. It took many weeks and a flurry of diplomatic notes between Washington and London before Elihu Root, Theodore Roosevelt, and their British counterparts reached an amicable resolution, which involved a formal apology from Swettenham and his resignation as governor.[57]

As this trans-Atlantic political crisis unfolded, the actual emergency in Kingston began to subside. Meanwhile other, less contentious parts of the US relief operation quietly drew to a close. Following Davis's departure, additional US assistance arrived in Kingston, including the ARC's shipment of food and the consignment of tents from the US Army in Cuba.[58] Swettenham, now under considerable pressure from his embattled British superiors, "very graciously accepted" this American aid.[59] By late January, the official US relief effort for Kingston had largely concluded, leaving Jamaican and British officials to oversee the tasks of recovery and rebuilding.

The controversial US response to the Kingston earthquake offered valuable lessons for American officials—if they cared to listen. As Davis's experiences in Jamaica made clear, *any* US intervention in a foreign nation or empire, even for ostensibly benevolent purposes, could easily be perceived by those on the ground as a threat to their autonomy. When military forces were involved, the potential for fomenting resistance and resentment proved greater still.

Yet if this was the moral of the Kingston saga, not everyone wished to learn it. Nearly three months after the earthquake, Theodore Roosevelt steadfastly refused to concede that Davis could have possibly been "in the wrong and Swettenham in the right, or that the American Government is in any way to blame."[60] The episode did, however, convince at least some government officials of the importance of respecting the sovereignty of disaster-stricken nations and ensuring their consent before becoming involved. "Events down there," as Senator Eugene Hale of Maine observed, "are showing that we should think twice before taking possession of a foreign town and undertak[ing] to establish order. What would have happened if Great Britain had sent a fleet to San Francisco," he wondered, "and taken charge there?"[61] Not confined to Jamaica, questions surrounding intervention, authority, and control resurfaced regularly in future US relief operations, generating tensions between US officials and the recipients of their humanitarian aid.

Despite these assorted challenges and complications, the foundations of a new and more consistent approach to US foreign disaster assistance were, by 1907, firmly in place. In the year after the great San Francisco earthquake, the State, War, and Navy Departments, in partnership with the newly restructured American Red Cross, had closed the books on multiple foreign disaster relief operations, gaining valuable experience in the process. Although official US relief operations had been fairly modest up to that point, an immense catastrophe lay just over the horizon, putting the three pillars of US foreign disaster aid to their greatest test to date.

3

Relief and Rebuilding in Italy

1908–1909

On December 28, 1908, a violent earthquake occurred in the Strait of Messina, the narrow channel of water separating the southern toe of Italy from the island of Sicily. The quake, together with an immense tsunami it generated, demolished the cities of Messina, on Sicily, to the west and Reggio Calabria, Italy, to the east, along with dozens of surrounding towns and villages. Tens of thousands of people—more than 100,000 by some estimates—lost their lives. Between 500,000 and 1 million survivors were rendered homeless and bereft of their worldly possessions. By all measures, this was a catastrophe of incredible magnitude, among the most destructive seismic disasters the world had ever witnessed.[1]

It soon sparked the most extensive foreign disaster assistance operation the US government and its partners had yet undertaken.[2] Unfolding over nearly six months, from late December 1908 to mid-June 1909, the American

50

response to the Messina earthquake and tsunami represented the first major trial of the three-pillared system of US foreign disaster aid that coalesced during Theodore Roosevelt's presidency. It also marked a critical chapter in Italian-American relations, its contours shaped by US foreign policy designs in Europe and by contemporary Italian immigration to the United States. Financed by enormous contributions from the American Red Cross and Congress, and coordinated in Washington by the State Department and ARC, the US response to the catastrophe developed into an extensive program of humanitarian assistance in southern Italy, where it was administered by US diplomats and consuls, navy servicemen, and their trusted associates.

While they initially focused their energies on emergency forms of relief, these US officials ultimately broadened the scope of their humanitarian efforts considerably. In Sicily, Calabria, and Rome, they experimented with novel and far-reaching approaches to recovery and reconstruction assistance, including large-scale rebuilding projects and permanent memorials to American assistance. Blurring the lines between humanitarian relief and long-term development, these activities also created new models for future US foreign disaster aid operations, far beyond southern Italy.

By the time the official humanitarian operation concluded, US officials had provided food, clothing, shelter, money, employment, and logistical assistance to tens of thousands of Italian disaster survivors. The US government and ARC spent nearly $2 million on this extensive range of relief and recovery projects, their costliest foreign disaster relief operation to date. In the process, they generated considerable diplomatic goodwill and helped Italians to meet very real humanitarian needs. Underlying their actions and exhibited in their private correspondence, however, were paternalistic and at times deeply antagonistic attitudes toward Italian citizens, Italian government officials, and Italian culture. Like the recent relief efforts in Chile and Jamaica—yet now on a far larger scale—aid efforts in southern Italy laid bare the complex dynamics of American catastrophic diplomacy.

❂   ❂   ❂

As disaster embroiled southern Italy during late December 1908, the US government was undergoing a political transition. In the November 1908 elections, Roosevelt opted not to run for a third presidential term, instead throwing his support behind a handpicked successor, Secretary of War William Howard Taft. By the time the earthquake and tsunami struck the Strait of Messina region, Roosevelt had entered the twilight of his presidency, with Taft's inauguration in early March 1909 fast approaching.

Roosevelt may have been a lame duck, but he and his administration responded actively to the reports of Italy's devastation. On December 29, the day after the earthquake, the president cabled King Vittorio Emanuele III to offer his "sincerest sympathies," and promised that assistance would soon be forthcoming.[3] The Italian government promptly accepted his offer of aid. Over the next few days, Roosevelt and his navy secretary, Truman Newberry, ordered several US Navy ships to proceed to the waters off southern Italy— where, notably, the Italian, British, French, German, and Russian navies had already begun engaging in relief work.[4] The first two vessels they dispatched were the USS *Scorpion*, the station ship of the US embassy in Constantinople, and the USS *Celtic*, a supply ship at port in New York.[5] The *Celtic* carried a cargo of 550,000 rations, originally intended for US Navy personnel but, as Newberry explained to Roosevelt, "perfectly suitable for the relief of the suffering caused by the earthquake disaster."[6]

These were not the only ships Roosevelt and Newberry deployed. In a highly symbolic act, made possible by fortuitous timing, they also ordered the Atlantic Fleet—known more famously as the Great White Fleet—to southern Italy. One year earlier, Roosevelt had sent the fleet's sixteen battleships, multiple auxiliary vessels, and 14,000 sailors on an around-the-world tour, intended to foster international goodwill and to showcase the strength and prowess of an expanded, modernized US Navy.[7] At the time of the earthquake, the Atlantic Fleet had entered the Gulf of Aden, heading toward the Mediterranean. Given the fleet's proximity to Italy, Roosevelt and Newberry instructed Rear Admiral Charles Sperry to send several ships to the disaster zone. Sperry loaded two auxiliary vessels, the *Culgoa* and the *Yankton*, with foodstuffs, medical supplies, and other stores and directed their crews to sail to Italy. Two battleships, the *Illinois* and the fleet's flagship, the *Connecticut*, followed them.[8] Although it took more than a week before the first of these ships reached the Strait of Messina region, six US Navy vessels were on their way.

Within a few days of the Messina earthquake, Roosevelt had committed ships, sailors, and military supplies to Italy. He had done all this, however, without congressional authorization: US senators and representatives had been on holiday recess since mid-December. On January 4, 1909, the day Congress reconvened, Roosevelt sent a message to legislators requesting retroactive approval for his actions and for the $300,000 he had already expended. He then asked for something more. Declaring that "an exceptional emergency exists which demands that the obligations of humanity shall regard no limit of national lines," Roosevelt urged Congress to appropriate an additional $500,000 for Italian relief.[9]

The total amount of $800,000 represented an extraordinary figure for foreign disaster assistance, one that was four times greater than Congress allocated for Martinique in 1902. Yet in spite of the price tag, few members of Congress balked. Both the Senate and House responded enthusiastically to Roosevelt's request. With little debate, they passed a joint resolution, appropriating $800,000 for the "suffering and destitute people of Italy" and granting Roosevelt wide discretion to spend the funds as he saw fit. Congress also authorized the president to employ any "suitable steamships or vessels" for the relief effort.[10] Although congressional proponents of the bill conceded that they had never before passed relief legislation "of such magnitude," they pointed to the recent appropriations for Martinique and Jamaica as direct precedents for their actions.[11] Many US citizens applauded the legislation as well, praising the government for such a "noble act on this terrible calamity."[12]

If support for the relief appropriation was overwhelming, however, it was not universal. Several congressmen opposed the legislation as an over-extension of government authority. Voicing the concerns of this bloc, a senator from Texas stated that he did "not believe that the Federal Government possesses the power to apply the people's money in this way," adding that he was concerned with creating precedents that would "justify all future appropriations of this kind."[13] Some private citizens complained, too. Telling the State Department that it was a "great mistake for Congress to appropriate money for Italy," one letter writer warned that it would "stop private aid" and teach "people to rely on Government instead of themselves."[14] Such critics, however, remained in the minority. By 1909, most Americans had come to accept the idea that their government, at least in some instances, should aid victims of foreign disasters.

While the White House, Congress, and Navy Department worked out the federal government's response to the catastrophe, people throughout the United States organized their own forms of assistance. Within a week of the earthquake, American churches, charitable organizations, benevolent societies, and municipal relief committees raised hundreds of thousands of dollars for Italian relief.[15] Hoping to better coordinate these spontaneous acts of giving, the Roosevelt administration once again turned to the American Red Cross, the United States' official voluntary aid organization.

In the weeks before the disaster in Italy, in fact, the bonds between the federal government and the ARC had grown tighter still. In early December 1908, guided by their recent experiences in Valparaíso and Kingston, ARC leaders established an International Relief Board within their organization. Charged with overseeing foreign disaster assistance and other peacetime aid operations, this specialized division involved key US diplomatic and military officials even more fully in the planning and execution of these efforts. With the first assistant secretary of state as its chairman and several high-ranking officers from the War and Navy Departments filling out its ranks, this administrative unit further strengthened the ties between the federal government, the military, and their humanitarian auxiliary.[16]

Working through the auspices of the new International Relief Board, ARC leaders and US government officials together organized the nation's official voluntary response to the cataclysm. The day after the earthquake, William Howard Taft—who was not only president-elect of the United States but also, notably, the president of the ARC—sent a message to the Italian Red Cross, tendering the ARC's "profound sympathy because of the terrible earthquake."[17] The next day, ARC leaders wired an initial $50,000 to the US ambassador to Italy and launched a nationwide fundraising appeal, which Roosevelt publicized enthusiastically.[18] Eager to streamline voluntary giving, ARC leaders and State Department officials implored US citizens to transmit all donations solely "through the channel of the American Red Cross," emphasizing its special status among aid organizations.[19] As the ARC's leaders informed the governor of Massachusetts, it would be "a pity not to have all funds that are raised in this country sent through the American Red Cross ... the official organization to carry on international relief."[20]

In stark contrast to the appeal for Chile several years earlier, many US citizens heeded the call, a reflection of both the greater severity of the Italian disaster and the more extensive ties between Italy and the United States. By January 5, the ARC had received and sent more than $400,000 to Italy.[21] By the time its relief drive concluded in early February, the ARC had raised well over $1 million for Italian relief, vastly more than any other US voluntary organization.[22]

Within a week of the Messina earthquake and tsunami, the White House, Congress, and ARC thus directed considerable material and financial resources toward southern Italy. While the staggering devastation and extreme suffering no doubt motivated officials to contribute so much aid, the largesse they demonstrated was also a function of US-Italian relations. As both an increasingly powerful nation in Europe and the source of several million immigrants to the United States in recent decades, the kingdom of Italy ranked highly in the Roosevelt administration's political calculus. For many US officials operating in this context, providing generous aid to Italy appeared nothing short of a political imperative.

More specifically, a trio of intertwined domestic and foreign policy concerns spurred US officials to act. First, within the United States, members of the Italian-American community and Italophilic Americans implored the US government and ARC to mount a concerted response, calling for acts of "charity and kindness" toward relatives and friends in the disaster zone.[23] These appeals persisted for months after the disaster, with many Italian Americans petitioning President Roosevelt or Secretary of State Philander Knox directly on their compatriots' behalf.[24] Fielding these entreaties, Roosevelt, Knox, and other US officials undoubtedly felt domestic pressure to respond.

Second, US government officials hoped humanitarian assistance would help nurture cordial diplomatic relations with Italy. The arrival of large numbers of Italians to the United States in recent years had been met in some corners with nativism, xenophobia, and calls for immigration restriction. A substantial response to the disaster stood to placate diplomatic tensions surrounding these issues, showing the kingdom of Italy that Americans did, in fact, care about the nation and its citizens. Third and last, the fact that several other European countries were providing aid made demonstrating American sympathy all the more politically expedient. Staging an equally generous humanitarian response offered the United States a way to showcase its parity with the great powers of Europe, one of the outgoing Roosevelt administration's chief foreign policy concerns.

Together, these varied motivations compelled US officials to carry out an extensive relief effort for Italy, a humanitarian operation that appeared all the more extraordinary on the ground.

☆   ☆   ☆

As the Roosevelt administration and ARC coordinated the official US response from Washington, the US ambassador to Italy, Lloyd Griscom, took charge of American relief efforts in southern Italy. Griscom represented not only the State Department but also the ARC; lacking their own personnel in Italy, ARC leaders designated the ambassador as the organization's official agent, just as they had in Chile in 1906.[25] In this capacity, Griscom assumed responsibility for disbursing Red Cross funds and administering both ARC- and congressionally financed relief activities. In contrast to his military counterparts in Jamaica two years earlier, the ambassador made sure to consult with relevant Italian authorities about his plans.

Among the ambassador's first steps was to dispatch some of his colleagues to the scene of the disaster. The earthquake had destroyed the US consulate in Messina, killing the American consul and his wife when it collapsed. On January 1, Griscom sent the US vice consul in Milan, Bayard Cutting, to Messina, accompanied by the embassy's military attaché and its interpreter. The ambassador charged these three men with locating the body of the deceased consul, establishing a new consulate, and reporting on local relief needs. American Red Cross leaders wired Cutting an additional $10,000 for this mission, designating him an ARC agent as well.[26]

Back in Rome, ARC funds were pouring into the American embassy. Initially, Griscom planned to give this money directly to the Italian Red Cross. Within a week of the earthquake, he transmitted $320,000 to the ARC's sister society.[27] Yet abruptly, Griscom lost confidence in the Italian Red Cross, and indeed in most other Italian relief agencies, save for those run by Italy's Queen Elena. In explaining his sudden about-face, the ambassador revealed deep-seated prejudices toward the individuals he was charged to assist. As Griscom confided to ARC leaders, "The relief offered by the Italian organizations is so painfully slow that it would never be tolerated for an instant in an Anglo-Saxon country." The Italians, he continued, "do not know how to spend such large sums and they are overwhelmed with the embarrassment of riches."[28]

In a pattern hardly unique to Italy, Griscom and his fellow US officials, dubious of local capabilities, determined that Americans must retain control of both US governmental and ARC funds. To assist him, the ambassador convened several prominent US citizens living in Rome, among them two financiers, a lawyer, a writer, and the embassy's naval attaché, Reginald Belknap. Calling their group the American Relief Committee, these men began planning US aid activities for southern Italy. Deciding "to invite the cooperation of a number of ladies" in Rome, Griscom also appointed a Ladies' Auxiliary Committee, naming his wife, Elizabeth, as its chair.[29]

After some deliberation, the members of the American Relief Committee agreed to use the ARC's funds to outfit a ship with relief supplies and send it to Messina. This ship was intended to supplement the efforts of the six US Navy vessels Roosevelt ordered to the disaster zone. Both the State Department and ARC readily endorsed this plan, as did the king of Italy, the Italian prime minister, and the president of the Italian Red Cross. Approval secured, Griscom used $100,000 in ARC funds to charter the German ship

Bayern, then equipped it with food, clothing, and medical supplies. Griscom put Belknap, his naval attaché, in command of the *Bayern*, while he and two other members of the American Relief Committee joined its crew. Griscom also employed a Roman physician and eighteen nurses to travel with them to provide medical care to survivors.[30]

On January 7, the *Bayern* and its passengers departed the port of Civitavecchia, just outside Rome. They arrived in Messina the following afternoon, where they joined Bayard Cutting and the two men who accompanied him to Sicily a week prior. These State Department representatives were not the only US officials converging on southern Italy, however. The *Scorpion* had arrived from Constantinople on January 3, and its crew, working in cooperation with Italian and British authorities, had begun distributing food and medical supplies.[31] Over the next several days, they were joined by Rear Admiral Sperry and the ships of the Atlantic Fleet. By mid-January, hundreds of US Navy personnel, several American embassy and consular officials, and hundreds of thousands of dollars' worth of relief supplies had converged on Sicily and Calabria. The US government—through its representatives in the State Department, the US military, and the ARC—had established a significant footprint in the Strait of Messina region.

During these first few weeks of January, US diplomats, consuls, sailors, and their associates engaged in a variety of humanitarian activities. Throughout this initial phase, their efforts focused mainly on relieving immediate suffering. In Messina and Reggio Calabria, Griscom and other embassy and consular officials oversaw the distribution of the food and other material aid the American ships had brought.[32] Naval forces, meanwhile, worked alongside their counterparts from other countries in distributing relief supplies, providing medical care, clearing debris, rescuing survivors, and recovering corpses trapped beneath the rubble—including those of the deceased consul and his wife.[33] American officials also traveled to Catania, Syracuse, Palermo, Naples, and other nearby towns and villages, where they distributed supplies among the thousands of refugees who fled the disaster zone.[34] Farther away in Rome, Elizabeth Griscom and the Ladies' Auxiliary Committee worked to supply lodging, food, and clothing to refugees congregating in Italy's capital.[35]

As they reflected on these emergency relief efforts, US officials felt considerable pride in what they were accomplishing. In Griscom's eyes, he and his American colleagues had "done much more than anybody else" to address Italian suffering.[36] Griscom and his associates also voiced compassion for Italian disaster survivors. They bemoaned the "widespread physical suffering and dangerous discomfort" they witnessed, expressing pity for the "desti-tute families" and the people "brutalized by the hopeless magnitude and horror."[37]

At the same time, however, Griscom and his associates held conflicting feelings about the men and women they assisted, with many of these same US officials displaying contempt toward Italian relief recipients. Cutting, for one, complained that "the refugees belong chiefly to the very lowest classes and are very hard to deal with," grumbling that most "spent the whole day in simple idling."[38] The new US consul in Catania, spouting contemporary anti-immigrant rhetoric, equated Italian disaster survivors with the "many Sicilians now in my country who are a menace to the public peace."[39] Amer-ican officials voiced particular concern over a presumed Italian predisposi-tion toward graft and dependency. As the secretary of the American Relief Committee remarked, "It was difficult to distinguish . . . the imposter from the deserving refugee who desired to begin again an industrious life."[40]

Guided by racialized and classist preconceptions of southern Italians and by theories of scientific charity circulating in the United States, Griscom and his colleagues employed several tactics to ensure that aid reached only "deserving cases." First, mirroring practices employed in San Francisco and Chile in 1906, they made it a policy to "investigate the actual needs" of Italians who applied for American assistance. Such scrutiny, as Elizabeth Griscom reasoned, was "essential in performing a real charity without pauperizing the individual recipients."[41] Second, hoping to place disaster survivors "beyond the need of charity"—and piloting a practice that would become common in future foreign disaster assistance efforts—US officials sought opportunities "to set the refugees to work."[42]

During the early weeks of January, the American Relief Committee con-sidered several different employment schemes, including hiring women to sew clothes for other refugees and hiring men to make shoes or rebuild roads and buildings. Although they ultimately discarded many of these ideas, deter-mining them too costly or impractical to administer, US officials employed small numbers of Italians in labor projects, compensating them with ARC funds.[43] Through such efforts, they boasted, they gave Italian disaster ref-ugees "an opportunity of earning their living and prevented [them] from becoming beggars on the streets."[44] In the minds of many US officials, this outcome was as important as relieving suffering itself. "Work," as Cutting proclaimed, "should be their salvation."[45]

Griscom and his associates may have viewed their methods of disaster relief as a godsend for earthquake survivors, but the reactions to them on the part of Italian men and women were more mixed. Many Italian citizens and government officials voiced appreciation for the aid the American Relief Committee and US Navy provided. Everywhere he went, Cutting reported, he was "greeted with friendliness, and much readiness to accept our help."[46] Sperry, likewise, noted that residents of Reggio, Scylla, Catania, and Syracuse had "thankfully accepted" the supplies he delivered.[47] Belknap, too, was gratified to receive the "heartfelt thanks" of local officials, while Griscom was "bombarded by grateful letters from the various Prefects, Mayors and Colonels along the shores of Calabria and Sicily."[48]

Yet even as Italians publicly praised US relief work, resentments stirred. By mid-January, some Italian government and military officials, though initially welcoming of US assistance, were beginning to voice frustrations. Some criticized American-led relief activities that had been organized without approval and in "direct opposition to [Italian] government policy."[49] Others worried that the massive scale of US relief efforts reflected poorly on the Italian government, implying that local authorities were incapable of responding to the crisis. Such perceptions, Griscom carped, were making Italian officials increasingly "sensitive and jealous of foreign relief."[50] Equally aware of the shifting local mood, Sperry recognized that stationing US battleships in Italian waters was becoming "very objectionable" to the Italian navy.[51]

For many Italians, finally, frustrations undoubtedly stemmed from the poor behaviors of US officials themselves. In addition to the condescending attitudes Griscom and some of his associates displayed, many US Navy sailors involved in relief work behaved in less than honorable ways. During their time in southern Italy, dozens of US sailors were punished for offenses they committed while on shore leave, including smuggling liquor, fighting, and drunkenness.[52] Many more undoubtedly engaged in these and other vices while ashore yet managed to avoid getting caught.

While Italian disaster survivors may have appreciated the aid US diplomats and sailors brought with them to the Strait of Messina region, in short, some of them surely breathed a sigh of relief as those Americans began to depart.

By mid-January 1909, three weeks after catastrophe first gripped southern Italy, the immediate crisis in the region began to subside. Italian authorities and charitable organizations, together with their foreign counterparts, had succeeded in addressing survivors' most pressing needs. With the situation better stabilized, US officials began winding down their emergency relief efforts. By the time the sixth and final US Navy ship, the *Celtic*, reached Italy from New York on January 19, three of the Atlantic Fleet's four vessels had already departed. Rear Admiral Sperry and his flagship, the *Connecticut*, left the following morning. Griscom, Cutting, Belknap, and other US embassy and consular officials had also recently returned to Rome. At this point, Griscom considered suspending US aid efforts entirely. Although he was confident that US officials did "real good and satisfied real needs," the ambassador thought it best to "get out of the relief business before the inevitable bickerings and ill feeling spring up."[53]

But Griscom would not have his way. The ARC still had hundreds of thousands of dollars in its coffers, while the Roosevelt administration had expended just $300,000 of the $800,000 congressional appropriation. Two navy ships, the *Scorpion* and the *Celtic*, remained in Italian waters, their crews on hand to provide assistance for the foreseeable future.[54] In Washington, government officials and ARC leaders, certain of the "need for long continued assistance to the sufferers in Italy," intended to prolong the aid operation until the resources they had committed were exhausted.[55] Despite their frustrations and criticisms, Italian authorities were hard-pressed to rebuff continued American assistance, given the enormous challenges of recovery and reconstruction ahead.

And so, rather than wrapping up in mid-January, the US disaster assistance operation in southern Italy endured—for nearly five more months, in fact. From this point on, however, the scope and purpose of the humanitarian mission changed appreciably. In its initial phase, the assistance operation had largely resembled US relief efforts in other disaster-stricken countries (albeit on a grander scale), with activities focused principally on addressing survivors' basic needs. While never entirely abandoning such relief work, Griscom and his associates redirected the bulk of their remaining resources toward a slate of major recovery and rebuilding projects, each intended to leave a permanent mark on Italian soil.

The decision to undertake long-term aid projects in southern Italy marked a departure from earlier patterns and precedents. It signaled a willingness, on the part of the US government and ARC, to experiment with novel, more comprehensive, and more far-reaching forms of US foreign disaster assistance. The choice to take on such projects, however, was also something of a fluke. Not planned from the outset, these new activities came to pass mainly because US officials found themselves with an unanticipated surplus of cash to spend. Despite his initial reservations, Griscom quickly came around to supporting this novel aid trajectory. "The relief work must necessarily be on a large scale," the ambassador now explained to ARC leaders. Recovery assistance "is needed," he conceded, "and will be for some time to come in the near future. I think it would be impossible," he concluded, "for us to contribute too much."[56]

The first clear indication of this shift from relief to rebuilding came on January 16, when the Roosevelt administration announced that the remaining $500,000 of the congressional appropriation would be used to build permanent housing for disaster survivors, a project the New York Times deemed "an innovation in international relief measures."[57] ARC leaders, in turn, agreed to contribute $167,000 toward this venture.[58] Under this plan, the White House, State Department, and ARC agreed to purchase lumber and building supplies in the United States and to employ seven American expert carpenters, transporting the men and materials to southern Italy on US Navy vessels. Once the supplies arrived, Griscom was to hire Italian laborers to construct cottages for those made homeless by the disaster, their work supervised by the American carpenters, embassy officials, and navy officers.[59]

This recovery program thus had twin goals: to help rebuild southern Italy while providing employment to hundreds of Italian disaster survivors. Although it relied on Italian labor, the project depended on American planning, oversight, material, and capital for its execution. In the eyes of Griscom and his associates, it would be an American venture from start to finish.

It was not the only one. As Griscom waited for the building supplies and carpenters to arrive from the United States, a process that took several weeks, he was presented with an opportunity to sponsor an entirely different recovery project. In late January, Bruno Chimirri, an Italian member of Parliament, asked the ambassador for money to endow a model farm for children who lost their parents in the disaster. As Chimirri envisioned it, this orphanage would "make citizens of abandoned children" by training them to be farmworkers, their vocational education grounded in "scientific agricultural instruction."[60]

Griscom saw great value in this project, not only for southern Italian children, but also—and from his perspective, more importantly—as a way to leave a lasting American influence on Italian affairs. As he told ARC leaders when pitching the plan, "The enterprise is to be distinctly American, and the town to be called by an American name. It seems at first sight to be a very good idea."[61] Griscom used similar arguments to sell the concept to the State Department. The institution, he explained, promised to serve "as a perpetual monument in Italy to American generosity" and to "permanently to uplift the agricultural population of the stricken districts."[62]

Griscom's pitch worked; ARC leaders authorized him to commit $250,000 to the venture, for use as a permanent endowment fund. Named the American Red Cross Orphanage, the institution was to be established near the Calabrian town of Palmi, on land donated by the Italian government. On February 16, Griscom formally presented the funds to the queen of Italy and other Italian government authorities, formally launching the first of the US-sponsored recovery projects.[63]

One week later, on February 23, a steamer carrying lumber, building supplies, and two American expert carpenters reached Messina from the United States, the first of six such shipments. With its arrival, the Roosevelt administration's scheme to erect permanent housing in southern Italy began in earnest. To command this operation, Griscom sent his naval attaché, Reginald Belknap, back to Messina.[64] On arrival, Belknap formed an American construction party to help superintend the efforts. It included the officers and enlisted men of the *Scorpion* and *Celtic*, the seven American carpenters from the United States, and several American civilians who lived in the area. Belknap also hired more than 600 Italian men to perform the actual labor of building the cottages.[65] Fulfilling a goal of many US officials involved in the aid operation, Belknap then "set the refugees to work."[66]

In Messina, Belknap established a camp as headquarters for the construction projects, hoisting the American flag above it. More than just a center of operations, the camp functioned as an exhibition of American humanitarian relief. Lauding the American camp as a "model of cleanliness, order and industry," Griscom boasted to ARC leaders that it "served as a wholesome example to the stricken people of Messina."[67] As such words suggest, US officials conceived of the housebuilding project as something grander than mere aid. Like the model farm for Italian orphans, they viewed it as a means to permanently uplift Italian disaster survivors and a spectacle of American benevolence. When they departed Messina, US officials hoped to leave behind not only American buildings but also American values.

With the work crews assembled and the American camp organized, housebuilding finally commenced.[68] Italian laborers, supervised by US officials and carpenters, built an average of 100 cottages per week. Each contained two rooms, a kitchen, and a brick fireplace, and was intended to house one refugee family, with recipients selected by local committees. Italian work crews eventually erected more than 1,000 cottages in Messina, 500 in both Reggio Calabria and the Palmi district of Calabria (near the American Red Cross Orphanage), 135 in Naples, and several hundred more in other towns and villages throughout Sicily and Calabria. American officials also agreed

**Some of the thousands of cottages constructed in southern
Italy after the 1908 earthquake and tsunami, financed
by the US Congress and the American Red Cross.**
Image courtesy of Keystone View Company, Library of
Congress Prints and Photographs Division.

to build 75 cottages at the Villagio Regina Elena, a model village for disaster survivors established by Italy's queen.[69]

In addition to the cottages, Belknap and his work crews embarked on several larger construction projects, intended to contribute to the region's rehabilitation while, at the same time, creating memorials of American aid that were "permanent in character."[70] At the Villagio Regina Elena, they began building a full hospital composed of three wards, an operating room, an emergency room, and an outpatient dispensary. The queen christened it the Ospedale Elizabeth Griscom, in honor of the ambassador's wife. In Messina, at the Italian government's request, Belknap's party erected two schoolhouses, a church, and two community centers.[71]

But their "crowning achievement," as Belknap put it, was a two-story, seventy-five room hotel in Messina.[72] Although some critics questioned the need for such a project, Belknap insisted that the hotel was an essential key to the region's economic recovery. "Many people are only waiting for an opportunity to gain a foothold here and pick up their business," he explained to Griscom, "and the absence of a hotel is a serious handicap to the revival of the city."[73] In mid-March, with approval from Italy's Ministry for Public Works and the full support of ARC leaders, Belknap's work crews began work on the Grand Hotel Regina Elena, named for Italy's queen.[74]

Observing all this work in person, notably, was the ARC's national director, Ernest Bicknell. Given the scale and complexity the humanitarian operation had assumed, ARC leaders made the unusual decision to send Bicknell to Italy to observe how the organization's funds were being used. He set sail from the United States on February 17, arriving in Rome eleven days later.[75]

More than just a tour of inspection, Bicknell's trip to Italy functioned as a formal and symbolic diplomatic visit. On arrival, Bicknell spent several days in the capital, meeting with various Italian government officials, members of the Italian Red Cross, and Queen Elena.[76] Bicknell then headed to Messina. He traveled around Sicily and Calabria for more than two weeks, meeting with local Italian government and military officials, observing the building operations, and making recommendations for future work. Finally, he headed back to Rome and, ultimately, to Washington, where he shared his impressions with both ARC leaders and State Department officials.[77] Much like the rebuilding projects he came to observe, Bicknell's tightly orchestrated visit had functioned as a visible testament to American concern for Italy.

During and following Bicknell's sojourn, Belknap, his American construction party, and the Italian laborers they employed continued to make steady progress on their assorted building projects. Like many of his American associates, Belknap expressed conflicting feelings about the Sicilians and Calabrians with whom he worked. On the whole, Belknap reported harmonious relations between US officials and their Italian workforce, a fact he attributed largely to his own system of hiring and compensation. "Men on our pay roll," he explained to Griscom, "were paid off every day, and we gave better wages than anywhere else; we also exacted more work, and promptly discharged every man that proved unsatisfactory." Using these methods, Belknap noted, he had assembled "a good force of men, who have a fair idea of how to work, understand that shirking will not be tolerated, and that pay is certain, regular, and good."[78] Belknap also enjoyed good relations with most Italian government authorities, whom he commended for their "cordial attitude towards us" and "their earnest devotion to their own duties."[79]

Though generally positive, Belknap did express some frustrations over his interactions with Italians. He criticized one Italian man for the "rotten work" he performed on the hotel's foundation and another for failing to deliver building materials punctually. He also voiced concerns with local "agitators" who, he sensed, were trying to provoke unrest among his labor crews.[80] Perhaps most significantly, Belknap became involved in a protracted dispute with Italian authorities over a parcel of land they had supposedly furnished for his hotel project but that the Railways Administration claimed as its own. Although the Italian minister of public works ultimately resolved the disagreement in Belknap's favor, the controversy left a bitter taste in the mouths of Italian and American officials, underscoring for both sides the tensions in their humanitarian relationship.[81]

In spite of these issues, construction work in southern Italy proceeded relatively smoothly and mostly without incident. In early May, Italian families began occupying the first of the completed cottages, with more and more moving in over the next few weeks. As construction efforts gradually drew toward completion, on June 11, Belknap declared the humanitarian operation complete.[82] At 5:00 that afternoon, Belknap lowered the American flag at his headquarters in Messina. He relinquished control over the cottages and other buildings to Italian government authorities, then contributed what remained of the ARC's funds to Queen Elena for her earthquake charities.[83] Italian government and military authorities honored Belknap and his American associates with a celebratory feast, served in the dining room of the Grand Hotel Regina Elena, before seeing them off in their departing ships.[84] Five and half months after a momentous earthquake and tsunami devastated the Strait of Messina region, the US government's role in relief and recovery largely came to an end.

✪ ✪ ✪

As they took stock of what the US government, US Navy, and ARC had accomplished, American officials felt pleased with the contributions they made to southern Italy's relief and recovery. Perhaps more importantly, they believed they had accrued considerable goodwill as a result. As the aid operation drew to a close, Italian authorities presented US officials with numerous accolades. Prime Minister Giovanni Giolitti thanked Griscom and his associates for rendering "a great service . . . which has aroused keen satisfaction and profound gratitude among the inhabitants of that city," while Messina's mayor bestowed honorary citizenship on Belknap and his American construction party.[85] The Italian Red Cross, meanwhile, presented its American sister society with a gold medal and diploma as a token of appreciation.[86]

Many Italian citizens also expressed their gratitude, sometimes in surprising ways. On June 27, 1909, the *New York Times* announced the birth of the "the first baby born in a new house in Messina." In honor of the American assistance they had received, the boy's parents named him (or so the *Times* claimed) "Theodore Roosevelt Lloyd Belknap Palmieri."[87] Whether or not this story was apocryphal, Italian men and women did offer words of thanks in more conventional ways, continuing to do so well after the aid operation concluded. In January 1910, more than a year after the onset of the catastrophe, the citizens of Reggio wrote the US embassy to express their "deep gratitude for the generous work which the noble American Nation accomplished in favor of the people stricken by the earthquake."[88]

For all the positive words and feelings they generated, however, American actions in the Strait of Messina region were not beyond reproach. Throughout the half-year humanitarian operation, Griscom and other US officials routinely regarded Italian citizens and authorities with condescension and even disdain. Distrustful of Italian capabilities, they remained determined to maintain control over the American resources they distributed and the aid projects they sponsored. Confident in their transformative potential of their aid, moreover, US officials endeavored to leave a lasting American mark on the region.

Inevitably, such mentalities and policies bred resentment and anger among some Italians, tempering what goodwill they may have felt toward the United States. During and after the humanitarian operation, the State Department and ARC fielded complaints about the perceived injustices of the US aid program. Some Italians criticized the unfair distribution of assistance, observing that while their compatriots received American aid, they had "been passed by and left to struggle . . . wholly unaided in any way."[89] Others questioned whether the choices US officials made were really in the best interests of disaster survivors. In Messina, for example, where many remained homeless more than a year after the earthquake, local residents proposed to convert the new hotel into a home for the elderly and infirm. But the suggestion never made much headway. The reason, as the prefect explained to the US embassy, was maddeningly simple: "The Americans insisted that it should be used as a hotel."[90]

Still others simply came away disappointed. Scores of Italians in the devastated region made inquiries about using ARC funds to finance their passage to the United States, where they hoped to join relatives and start their lives over. The State Department refused even to consider these requests, however, insisting that such a plan was "not practicable or consistent with US immigration laws."[91] Helping disaster survivors in their own nation was one thing; welcoming disaster-driven migrants onto American soil was another matter entirely.

Regardless of what Italians thought of US relief and recovery efforts at the time, the material legacies of the humanitarian operation endured long after it ended, in the American-sponsored cottages and other buildings that now dotted southern Italy's built environment. These structures functioned for years—just as US officials had intended—as perpetual monuments to US humanitarian assistance. Although the largest of these projects remained unfinished when the American Relief Committee dissolved in June 1909, they were ultimately completed under Italian direction. In 1910, the Ospedale Elizabeth Griscom opened its doors to patients.[92] It took another three years for the American Red Cross Orphanage at Palmi to commence operations. Despite the delays, US officials were no less appreciative of its function as a site of memory. Speaking at its dedication ceremony, the US embassy's military attaché declared the orphanage a "memorial of the sympathy America felt for Italy in her hours of distress" and a "memorial of America's friendship for Italy."[93]

And in 1917, when the ARC's Ernest Bicknell returned to Messina—this time as part of the ARC's First World War–era Commission to Europe—he not only dined at the American hotel but also found most of the American-sponsored cottages intact and inhabited. The houses lined streets bearing the names Viale Roosevelt, Viale Taft, Viale Belknap, and, to his pleasant surprise, Viale Bicknell.[94]

4

Cementing the Three Pillars

1909–1916

In February 1915, the Panama-Pacific International Exposition opened its doors to visitors. Held in San Francisco, this world's fair was designed to celebrate the city's renaissance following the great earthquake of 1906. More ambitiously, its planners envisioned the exposition as a testament to the United States' new standing on the world stage. In a telling nod to both these goals, the federal government's official pavilion included a massive 4,300-square-foot exhibit showcasing "the humanitarian arm of the United States Government," the American Red Cross.[1]

While the ARC's domestic activities featured prominently in the exhibition, the organization's international humanitarian activities were its real focal point. Visitors encountered a large bas-relief map of the world, with markers identifying each place the ARC had sent war or disaster aid in recent years. At the center of the exhibit, they interacted with a 360-degree panoramic display of Messina, viewing the destruction the earthquake wrought in 1908 and the American cottages built in its aftermath. From there, attendees came upon two scale models of ARC-sponsored projects in China. One portrayed a camp for famine sufferers, while the other demonstrated a proposed river conservancy scheme, intended to prevent future flooding along the Huai (Hwai) River.[2]

In the 1910s, as this exhibit portrayed, US disaster assistance was becoming increasingly global in its reach. In the decade following the Messina earthquake and tsunami, the US government, military, and American Red Cross organized responses to catastrophes as far afield as Asia, Central and South America, Europe, and the Near East. None of these relief operations matched the enormous scale of US assistance efforts in southern Italy. Even so, their sheer number attests that bilateral disaster aid had become a customary part of US foreign affairs, solidifying a process set in motion during the previous decade.

Like the dollar diplomacy and gunboat diplomacy for which the era is more commonly known, catastrophic diplomacy represented an important instrument of US foreign policy throughout both William Taft's and Woodrow Wilson's presidencies. For both administrations, a guiding foreign policy principle was that the United States must play a greater role in world affairs.[3] Contributing disaster relief to other nations represented one means to achieve this overarching diplomatic goal, an opportunity for the US government to exercise its influence and burnish its image abroad. These outcomes proved particularly important in regions that were of major strategic interest to the United States during these years: Central America and the Caribbean, Western Europe, China and Japan, and the Ottoman Empire. By responding to catastrophes in these parts of the world, members of the Taft and Wilson administrations grasped, the US government could project its political, military, and humanitarian power—all in one fell swoop.

Rather than focusing on any specific humanitarian operations, this chapter examines the broader mechanisms of US foreign disaster aid in the years before the United States entered the Great War, with an eye toward these broader foreign policy objectives. Between 1909 and 1916, the three-pillared system of US foreign disaster assistance became fully entrenched. Across these years, the State Department played an increasingly central and consistent role in the US government's system of foreign disaster assistance, assuming the primary responsibility for coordinating official responses to catastrophes abroad. The steady growth of the ARC, meanwhile, made that voluntary organization an ever more essential partner to the government.

From Washington, and from military bases in both the mainland United States and its overseas territories, the Departments of War and Navy lent crucial logistical support to their operations.

Of particular and increasing significance in these years was the humanitarian work that State Department personnel—diplomats and consuls—performed overseas. As representatives of the United States abroad, these individuals were well positioned to provide direct, hands-on aid to other nations when disaster struck. As they had in both Chile and Italy during the previous decade, the State Department and ARC regularly charged US ambassadors, envoys, foreign ministers, and consuls with administering disaster relief efforts in their host countries. Though few had any training or experience in this field, these government officials served, for better or worse, as some of the most consequential figures in the foreign disaster aid system. In assuming these duties, diplomats and consuls were regularly forced to grapple with their own sense of humanitarian and moral responsibility and to square it with the diplomatic and strategic objectives their aid was expected to achieve.

Though they were by their very nature somewhat improvisational, US responses to international catastrophes began adhering to more predictable patterns and routines during the Taft and Wilson presidencies. Firmly cemented by the eve of US entry into the First World War, this system of foreign disaster aid was to guide the nation's official responses to global catastrophes across the twentieth century.

❂   ❂   ❂

By mid-1909, as US relief operations wrapped up in the Strait of Messina region, the ARC's position as the government's chosen partner in foreign disaster relief appeared secure. Following the society's reorganization in 1905, ARC leaders had joined the Roosevelt administration in responding to catastrophes in South America, the Caribbean, and Europe, serving as a key source of funding and expert guidance for these aid operations. During William Taft's presidency and Woodrow Wilson's first term in the White House, the ARC grew larger and more financially stable. It also became ever more indispensable, in the eyes of government officials, to the nation's international disaster relief efforts.

In a very literal sense, the ARC was part of the architecture of US foreign policy, for its national offices were housed in the State, War, and Navy Building, just west of the White House. They remained there from 1905 until 1917, when the ARC relocated to its own, permanent headquarters building, just a short block away. In the heart of the nation's capital, the ARC's civilian leadership collaborated closely with some of the highest-ranking personnel of the State, War, and Navy Departments to organize the nation's foreign disaster aid operations, thus working within the US foreign policy architecture in its more figurative sense as well. Taft's first assistant secretary of state, Huntington Wilson, served as chair of the ARC's International Relief Board. So did Robert Lansing—first in his capacity as counselor to Woodrow Wilson's State Department and then, beginning in 1915, as secretary of state. Serving alongside them as vice chairs were Taft's assistant secretary of the navy, Beekman Winthrop, and his successor under Wilson, Franklin D. Roosevelt. Many other acting and retired federal government officials served terms on the International Relief Board during the 1910s, among them former secretary of war and state Elihu Root, former ambassador to Italy Lloyd Griscom, and retired general George Davis.[4]

Together, these powerful government officials worked hand in hand with the ARC's civilian leadership to plan the organization's responses to international catastrophes. Their involvement reflected not only the state's growing influence over this voluntary organization but also an ongoing effort to place the ARC under both professional and masculine leadership.

In addition to participating in the ARC's day-to-day decision-making, members of the Taft and Wilson administrations took concerted steps to strengthen the ARC, to improve the effectiveness of its foreign disaster relief work, and to bolster its relationship with the federal government. The State Department played a particularly active role in these efforts. To promote "efficiency in international relief," State Department leadership encouraged all ambassadors, envoys, foreign ministers, and consuls to "join our Red Cross" and to form ARC chapters in their host countries.[5] Facilitating the ARC's work overseas, the State Department provided many services gratis to the organization, including wiring funds abroad on the organization's behalf and transmitting communications between the ARC's Washington offices and disaster-stricken countries. At home, officials such as Huntington Wilson, the assistant secretary of state, urged citizens to donate exclusively to the ARC when disaster struck. Emphasizing its unique status among aid organizations, the ARC, he explained, "receives through the Department of State accurate information of conditions at the scene of disaster which is not available to the general public and may be of a confidential nature." Such privileged information, together with the organization's considerable experience, Wilson stressed, made the ARC "the most suitable, efficient, and reliable channel through which such contributions can pass or be administered."[6]

Reinforcing the State Department's promotional efforts, the White House supported the ARC and its disaster relief work in multiple ways. Like Roosevelt before them, Presidents Taft and Wilson issued public fundraising appeals on the ARC's behalf following many foreign catastrophes.[7] During his term in the White House, Taft agreed to serve as the ARC's honorary president, inaugurating a tradition that Woodrow Wilson and other future presidents subsequently followed.[8] Both presidents also continued the practice, mandated by the ARC's 1905 charter, of appointing the chair and one-third of the ARC's governing central committee, selecting these representatives from the federal government's executive departments. Through these various measures, the White House further cemented the ARC's position, as Taft put it in 1911, as "the official volunteer aid department of the United States."[9]

Buoyed by the government's active support—and buttressing the United States' expanding role in global affairs—ARC leaders redoubled the organization's commitment to foreign disaster aid. In 1909, they adopted a "new policy of assisting in a larger range of instances" overseas.[10] Over the next eight years, acting on this pledge, the ARC's leadership made monetary gifts to a growing number of disaster-stricken countries, drawing from both specific fundraising appeals and a slowly growing pool of general relief funds. They contributed this aid to survivors of floods in China, Colombia, France, Mexico, and Serbia, as well as to victims of earthquakes and volcanic eruptions in Costa Rica, Japan, Portugal, and Turkey, among other sites.[11]

Although the ARC contributed aid to many countries beset by disaster, the level of funding it provided for any particular crisis varied widely. Several catastrophes in these years prompted a substantial response from the organization. When the Seine overflowed its banks in January 1910, producing catastrophic flooding in Paris and the towns and villages surrounding it, the ARC's International Relief Board sent $45,000 in assistance to France.[12] The ARC's leaders also made a sizable financial contribution to China—some $70,000—in the wake of the Yangzi (Yangtze) River floods of 1911, then sent an additional $169,000 for ongoing food aid in the region the next year.[13] The decision to prioritize aid to these two countries reflected several guiding concerns, but two were paramount: the humanitarian needs of disaster survivors and the foreign policy objectives of the US government.

In contrast to these larger expenditures, most ARC foreign disaster relief appropriations were more limited, amounting to just a few thousand or even a few hundred dollars each. Here again, a combination of humanitarian imperatives and diplomatic goals steered ARC leaders' decisions about how much aid they should provide. In some instances, however, ARC leaders wished to contribute more than these token sums yet found themselves unable to do so. The whims of American donors largely determined, and often constrained, the amount of aid they could allocate. The impulse to donate money to foreign disaster relief campaigns stemmed from a wide variety of factors: diasporic politics, existing cultural ties with the affected country, media coverage, a catastrophe's timing, and perceptions—often deeply racialized—about whether the affected were "deserving" of aid.

No matter the motivations that led individual Americans to donate (or not), the ARC's reliance on private giving had critical repercussions for US foreign policy planning. Because the organization depended wholly on voluntary contributions for its operations, government officials were effectively hamstrung when fundraising efforts fell short, incapable of providing disaster assistance to other countries when they desired to do so. Although ARC leaders and their government partners attempted to privilege humanitarian need and US strategic interests in their foreign aid calculations, a lack of sufficient funding often undercut those priorities.

Such financial setbacks cropped up several times during the early 1910s, underscoring the challenges of relying on private donors to bankroll foreign disaster aid. When a disastrous hurricane struck Cuba in 1909, for instance, the ARC's national director worried that a formal relief appeal "would fall on unresponsive ears" and advised against launching a fundraising campaign unless the State Department specifically requested it.[14] In 1914, ARC leaders apologized to the State Department for failing to contribute much money toward Chinese flood relief that year, despite US diplomats' urgent pleas for assistance. While stressing that they were "exceedingly regretful . . . that we are not able to assist the people of China," ARC leaders explained that meager donations prevented them from financing US relief efforts as generously as they or their government partners wished.[15]

Desirous of making larger foreign disaster relief contributions on a more consistent basis, ARC leaders took steps during Taft's presidency to put their organization on more secure financial footing. Specifically, they started working toward three key goals: amassing a general relief fund, building a permanent endowment, and increasing the organization's national membership base.[16] Despite their efforts, progress in each of these areas initially proved sluggish. While ARC leaders made some gradual strides before 1916, it took a very different sort of crisis—US entry into the First World War in 1917—to finally put the organization on secure financial footing.

For all its importance to American catastrophic diplomacy, the American Red Cross represented only one pillar in the system of US foreign disaster assistance. Government officials also responded to international disasters through the state's own channels, whether or not they had ARC funds at their disposal.

Within the federal government bureaucracy, authority for directing the government's responses to international catastrophes was vested principally in the State Department. In the days and weeks after disasters struck abroad, Secretaries of State Philander Knox, William Jennings Bryan, Robert Lansing, and their subordinates communicated closely with both foreign government officials and US diplomats and consuls in the affected region. In any given crisis, they typically sent and received dozens—sometimes hundreds—of telegrams, diplomatic notes, memoranda, and other correspondence. Through these communications, State Department officials issued formal expressions of sympathy and extended official offers of aid. They sorted out problems related to the procurement, transport, and distribution of ARC-funded supplies and other US humanitarian assistance. And they engaged in discussions, often confidential or classified, about the political, economic, and social implications of disasters in other countries. For State Department personnel, coordinating relief efforts for earthquakes, hurricanes, floods, and other calamities in other countries was all in a day's work.

Although the State Department led Washington's responses to foreign disasters, its personnel could typically rely on other parts of the federal government to supplement those efforts. The White House normally issued its own messages of sympathy to victims of foreign catastrophes. Throughout their presidencies, Taft and Wilson sent many such communiqués to heads of state and other high-ranking diplomats. During the 1910 floods in Paris, for example, Taft wrote the president of the French Republic to express the "sympathetic distress . . . [of] the people and Government of the United States," while in 1915, after an earthquake struck Avezzano, in the Abruzzo region of Italy, Wilson wrote to the king to voice his "sincere sympathy . . . in this moment of widespread suffering and national grief."[17] Such communications by American heads of state were more than mere niceties; they represented an unwritten yet critical diplomatic protocol. Taft, Wilson, and their advisers therefore took great care in selecting their words and crafting their statements, keen to demonstrate the US government's concern to foreign leaders and publics alike.

The War and Navy Departments also played a key role in US foreign disaster aid, primarily by dispatching supplies, equipment, and personnel to assist disaster-stricken countries. When a major earthquake struck Cartago, Costa Rica, in 1910, for example, Congress authorized the donation of tents and blankets from US military stores to assist quake survivors, with the value of these supplies to be reimbursed by the ARC.[18] The secretaries of war and navy then ordered US Army and Marine Corps personnel in the Panama Canal Zone to provide this material aid from their stockpiles. Members of the Isthmian Canal Commission, a division of the War Department, subsequently transported the relief supplies to Cartago.[19] The following year, the navy secretary sent the Asiatic Fleet's commander-in-chief on a mission up the Yangzi River, dispatching him there to inspect damages caused by recent floods, to discuss the situation with Chinese officials, and to report on local relief needs.[20] And in 1912, the Navy Department ordered the crew of the USS *Scorpion*—the station ship of the US embassy in Constantinople and a veteran of US relief efforts in Messina—to the Dardanelles Strait of Turkey, to assist survivors of a major earthquake in the region.[21] Although the War and Navy Departments did not become involved in every official US relief operation, in instances such as these they provided crucial logistical support.

From the US capital, the State Department, ARC, and War and Navy Departments thus worked together to organize and coordinate the Taft and Wilson administration's official responses to foreign catastrophes. Yet, based in Washington, these government officials and ARC leaders did not travel to the scene of catastrophes themselves. Instead, they relied on US personnel who were already on the ground, or at least in the vicinity, to administer the American response. Although US military leaders occasionally sent US troops and sailors to other nations to assist with disaster relief efforts, far more frequently they relied on another core group of government agents to administer American foreign disaster relief operations: the personnel of the US diplomatic and consular corps. Representing the US government, the ARC, and their respective interests, these public servants played a central, hands-on role in American catastrophic diplomacy.

⭐ ⭐ ⭐

By the early 1910s, the US government had forty-eight embassies and legations, more than 300 consular posts, and several hundred consular agencies throughout the world.[22] Along with the territories in the Caribbean and Pacific the US government had recently annexed, these diplomatic and consular posts constituted core components of the United States' expanding humanitarian geography.

The diplomatic and consular officials stationed in these missions, together with their military and naval attachés, served as critical cogs in the US foreign disaster aid system. State Department guidelines tasked these officials with responsibility for reporting "the occurrence abroad of any disasters of unusual magnitude" and indicating "where international relief may be appropriate," so ARC leaders and their government partners could together plan the official response from Washington.[23] Once US funds and relief supplies arrived on the scene, the State Department and ARC normally delegated to ambassadors, envoys, foreign ministers, or consuls the duties of distributing this assistance and coordinating other humanitarian activities. Collectively, these policies—first reporting disasters and requesting aid, and then administering assistance efforts on the ground—gave US diplomats and consuls tremendous sway over the direction of US international disaster aid operations.

Diplomats and consuls clearly took the first part of their charge—reporting disasters and requesting aid—to heart, for the diplomatic pouch regularly arrived in Washington filled with news about recent catastrophes and appeals for assistance. In their communications, many US diplomats and consuls seemed genuinely sympathetic to the suffering they witnessed. Describing the destruction caused by the 1910 earthquake in Cartago, Costa Rica, for instance, the US chargé d'affaires lamented the "hunger and suffering among the poor" that resulted.[24] Observing floods in southeast China the following year, the US consul in Swatow (Shantou) likewise expressed grief over the "appalling" loss of life, the many survivors "with starvation staring them in the face," and conditions that appeared "beyond hope."[25] Given the widespread death, dislocation, and destruction that catastrophes wrought, many diplomats and consuls contended, the US government and ARC had a moral obligation to provide assistance. The US ambassador to Mexico expressed this conviction clearly after major floods swept through Monterrey in 1909. "The ruin and destitution," he insisted, "justifies help from our Government."[26]

But if feelings of compassion, sympathy, and duty partly compelled US diplomats and consuls to request aid from Washington, their motives for doing so were never purely altruistic. Most also saw disaster aid as a means to promote US foreign policy interests in their host countries, making this case repeatedly to Washington in their appeals for help.

Time and again, diplomats and consuls stressed the public relations benefits to be derived from disaster aid, recommending that the government and ARC could burnish the United States' image in other countries by assisting survivors of catastrophes. Following the floods in Monterrey, Mexico, in 1909, for instance, Consul Philip Hanna informed the State Department that

England, Germany, and Canada had already made relief contributions. "A more general effort on the part of our generous people seems greatly needed," Hanna advised, to show Mexico that "she is closer to us in every respect than to any country."[27] Several years later, a US consul in Colombia employed much the same logic when requesting aid for victims of flooding along the country's Magdalena River, reasoning "that the effect produced here would be excellent."[28] The American consul at Nanking (Nanjing) made a similar argument in requesting American flood aid for China. Providing assistance presented a "great opportunity for the United States not only to benefit the Chinese people," he argued, "but also to strengthen our prestige with the Government and people of this vast nation."[29]

The notion that providing aid would boost the United States' standing in disaster-stricken countries may have been self-serving, yet experience repeatedly confirmed for US officials the diplomatic advantages accrued from humanitarian aid. Diplomats and consuls routinely wrote the State Department to share messages of gratitude and praise from aid recipients. The disaster assistance the government and ARC sent to Monterrey, Philip Hanna reported, had left "Mexicans more fully convinced than ever that Americans are their friends."[30] Another US consul in Mexico concurred, telling the State Department, "The prestige of the United States has been greatly enhanced."[31] Writing from Serbia in 1911, a US consul was similarly gratified to notify the State Department that Serbian officials expressed "great appreciation and thanks" for the money the ARC sent for flood victims in the Morava Valley.[32] As such reports suggested, disaster aid could and did improve the United States' standing in the eyes of local populations, at least in the short term.

If disaster aid had the potential to improve diplomatic relations, the reverse was also true: a failure to provide aid risked having deleterious political effects in afflicted countries. Filling their diplomatic missives with dire warnings of widespread hunger, illness, lawlessness, and unrest, US diplomats and consuls stressed that generous American assistance was essential to thwart these problems.

Experiences in China offer an illustrative example. Throughout the early 1910s, US consular officials in that country reported on a broad suite of threats, both real and perceived. First, many voiced concerns about the destabilizing effects of hunger and disease, conditions greatly aggravated by disaster. After the Yangzi River flooded in 1911, the US vice-consul at Nanking warned that "dysentery, malaria and typhoid fever are becoming quite prevalent."[33] Compounding these problems, he advised, "There seems little doubt that the floods are the precursor of one of the greatest famines China has ever known."[34] Three years later, US officials in Shantung (Shandong) province sounded similar alarms when a typhoon and floods crippled the region's agricultural output, destroying most of the planned harvest. "This loss of crops," the US vice-consul cautioned, "is a very serious matter indeed, and will probably cause a famine during the approaching winter."[35]

While some worried about illness and starvation, other US consuls in China appeared more concerned with a possible uptick in criminality and illegal behaviors. Following the 1911 Yangzi River floods, the US consul in Newchang (Yingkou) advised the State Department that "Manchurian bandits" were "operating with unusual activity" due to the catastrophe, while his colleague in Swatow reported that "thieves have been travelling from house to house in sampans relieving the wretched survivors of whatever they possess."[36] After flooding in Kwangtung (Guangdong) province four years later, the US consul general in China, Fleming Cheshire, issued a similar warning, advising that "brigandage has greatly increased."[37]

If hunger, epidemics, and lawlessness were not threatening enough on their own, many US consuls in China also feared that these issues would intensify political and economic instability. Cheshire made this point to the State Department in 1914. Flooding in the country's southern provinces, the consul general warned, was "having a most disastrous effect on the import trade," rendering it "unsafe really to transport foreign imports into the interior." The problem, as Cheshire saw it, was not limited to economic matters. Stating his concerns bluntly, the consul general cautioned that the periodic floods, together with the secondary effects they produced, were certain to "have a very bad effect on the future good order" of China.[38]

American consular officials in China thus identified a broad constellation of security threats associated with the country's recurring floods. And they were not alone. From many countries, US diplomats and consuls reported similar concerns arising in the wake of disasters. Yet these officials also identified a solution—or at least a salve—for the issues they observed. By providing humanitarian assistance, they insisted, the federal government and ARC could ameliorate the political, economic, and social problems that natural hazards engendered. In an era that found the United States increasingly involved in other nations' affairs, as these officials grasped, disaster relief represented an expedient instrument of foreign policy, a seemingly innocuous way for the government to influence the internal affairs of other countries.

From missions and posts across the early twentieth-century world, US diplomats and consuls regularly reported disasters to Washington and urged the government and ARC to send humanitarian aid. Yet these tasks were only part of a diplomat's or consul's responsibilities. These State Department personnel did more than sound the alarm and advocate for aid; they also served as the United States' first responders in many disaster scenarios. The State Department and ARC typically charged these officials with the duty for organizing and administering US relief operations on the ground in their host countries. With considerable discretion over how to distribute the ARC's funds and whatever additional assistance the US government or military might contribute, US diplomatic and consular officials were in a position to exert substantial influence over the direction of US foreign assistance efforts—and, by extension, over US relations with disaster-stricken nations.

A diplomat's or consul's humanitarian duties commenced the moment a catastrophe began. In the harrowing hours and days that followed a disastrous earthquake, tropical storm, or flood, as they waited for their communications to reach Washington and for American aid to arrive, these officials performed various tasks. Some confined themselves to the embassy, legation, or consulate, assisting American citizens who lived in the region and answering inquiries from concerned relatives, business enterprises, and other interested parties. Other diplomats and consuls paid visits to local government authorities and charitable agencies, meeting with them to discuss conditions and to ascertain whether they desired or needed outside assistance. A few ventured into disaster-stricken towns and cities to assist with locally organized relief efforts, pitching in to distribute food, water, blankets, and other forms of relief.

As ARC funds, US military supplies, and other assistance arrived on the scene, the responsibilities of diplomats and consuls shifted course. Now, their objective was determining how to put this American aid to the most effective use. In some cases, these officials simply turned over the money and supplies they received to the country's Red Cross Society or, if no such branch existed, to local authorities and charitable organizations. They did so, however, only when they judged those entities suitably "organized, responsible and efficient," as one US ambassador put it, to conduct relief operations on their own.[39]

Somewhat predictably, diplomats and consuls tended to reserve such judgments for countries they considered peers to the United States. Such was the case following the 1909 earthquake in Benavente, Portugal, and in France after the 1910 floods in Paris. In both countries, the heads of the US diplomatic missions promptly surrendered control of ARC financial contributions to the Portuguese and French Red Cross Societies, respectively.[40] Following the earthquake in Kagoshima prefecture four years later, likewise, the US ambassador turned over the $5,000 ARC contribution he received to Japan's Foreign Office, entrusting this money to the government's official relief committee.[41]

Outside Western Europe and Japan, by contrast, US diplomatic and consular officials tended not to forfeit control over American relief contributions so readily. Instead, they sought to retain a tight grip on the ARC's funds and whatever material supplies arrived. Although they typically consulted with local government officials, charitable organizations, and political and social elites about where and how to direct this American aid, in the end, diplomatic and consular representatives allocated American assistance as they saw fit.

This insistence on maintaining authority over aid distributions reflected an abiding distrust, common among many US diplomats and consuls during this era, toward the individuals they assisted. These suspicions were informed by their broader racial, cultural, and class prejudices. They revealed a strong sense of paternalism, if not outright arrogance, that many US officials shared—not only abroad, it bears noting, but also domestically. In the decade after the 1906 San Francisco earthquake, state and local government officials, ARC disaster specialists, and military personnel responded to multiple catastrophes within the United States and its territories, including the 1909 Cherry Mine disaster in Illinois, the 1911 eruption of Taal Volcano in Batangas, Philippines, the 1913 Ohio River floods, and Massachusetts' Great Salem fire of 1914. Guided by the principles of the scientific charity movement, these officials exercised considerable authority over relief and recovery work, convinced that they knew what was best for disaster survivors.[42]

These same mentalities extended well beyond the borders of the US empire. Confident in the superiority of their own methods and philosophies of disaster assistance, and keen to ensure disaster aid served US foreign policy objectives, many diplomats and consuls felt they had no choice but to hold the purse strings.

Expressions of racial, cultural, and nationalist chauvinism surfaced in several US foreign relief efforts during these years. After the 1909 floods in Monterrey, Mexico, US Consul General Philip Hanna used the $7,000 the ARC sent to purchase corn, beans, flour, medicines, and household goods for disaster survivors.[43] Hanna and his American associates assumed from the outset that Mexican flood victims would be prone to graft, observing, "As a rule the town people are only too willing to receive aid and great care is necessary to avoid abuses." They therefore made it a policy to distribute all relief supplies personally, to ensure that "there is no waste and that the proper parties who are in need receive them."[44] The US chargé d'affaires in Costa Rica took an equally disdainful view of earthquake survivors he assisted in 1910. Describing Costa Ricans as "like animals," he told the ARC that "they sell the clothes and blankets that are given them in order to buy drink." To circumvent these alleged vices, he conducted personal interviews with relief applicants and required references from the area's "best families" before doling out any aid.[45] Two years later, following the Dardanelles earthquake in 1912, the US ambassador to Turkey groused that if "the work of relief" were left to the Turkish people, it "would be much retarded and very badly carried out through lack of organization and coordination."[46] Only by supervising the US relief effort, the ambassador told the State Department, could he hope to prevent this outcome.

In addition to deciding what types of aid to provide disaster survivors, some US diplomats and consuls elected to withhold certain assistance, a choice equally informed by their class and racial prejudices. For example, one of the embassy officials responsible for US relief efforts after the Dardanelles earthquake, Naval Attaché Frank Upham, recommended against providing American tents to homeless disaster victims. While admitting that "the people in the stricken district . . . could well make use of tents as temporary shelter," Upham informed ARC officials, "it has not appeared to me that such [assistance] should be afforded." The reason, he explained, was that "while enjoying temporary shelter, the able-bodied inhabitants neglect to provide themselves with more suitable shelter as the cold weather approaches."[47] In Upham's estimation, providing housing risked undermining the work ethic and initiative of Turkish quake survivors, a fate worse than the disaster itself.

As the foregoing examples attest, US diplomatic and consular officials enjoyed considerable discretion over how—or whether—to allocate ARC funds and other US disaster assistance. Such policies afforded them substantial clout and sway over many would-be recipients of American aid. Even as the US government and ARC began standardizing their disaster relief practices in the early twentieth century, these individual State Department personnel had significant latitude in shaping the nation's response to particular disasters. To a striking degree, the outcome any relief operation remained subject to the personalities and presuppositions of whatever diplomat or consul happened to be on the scene.

But if diplomats and consuls were entrusted with the formal authority to make consequential decisions about relief, rarely did they act alone. Although the State Department and ARC delegated ambassadors, envoys, foreign ministers, and consuls to lead the nation's official foreign disaster aid operations, these government agents regularly collaborated with private American citizens living in their host countries to fulfill this responsibility. These sorts of collaborations were perhaps nowhere more evident than in China, where a sizable American missionary presence meant that US diplomats and consuls could count on many willing partners in disaster relief. US officials in China routinely turned to American missionaries to deliver relief supplies or provide medical care to disaster survivors, often deputizing them to act "in the name of the American Red Cross Society."[48] In their reports to the US consuls who appointed them, these men and women recounted making "personal investigations . . . as to the actual conditions of each village" and "distributing the Red Cross rice tickets" to families affected by the floods.[49] As they traveled to remote towns and villages to deliver ARC relief, many missionaries also "distributed Christian literature" or "preached the Gospel," seizing the chance to offer spiritual as well as material assistance.[50]

Humanitarian partnerships between US officials and private American citizens occurred outside China as well. Following the 1910 earthquake in Cartago, Costa Rica, for instance, the US chargé d'affaires gave $2,000 in ARC funds to an American who headed medical charities in the area. He then used these funds to establish "a temporary hospital for the wounded."[51] After the 1912 earthquake in Turkey, the US ambassador in Constantinople used ARC funds to send two American missionary physicians "to any place the Turkish Government might designate." The two doctors spent several days in the Dardanelles region, treating patients for fractures, contusions, and other injuries, before turning the care of their patients over to the Turkish Red Crescent Society.[52] And in 1915, when a powerful earthquake struck Avezzano, Italy, the US ambassador tapped "several charitable Americans" living in Rome to help him distribute food, clothing, and other supplies among disaster victims and to establish an emergency hospital.[53]

Although many US officials embraced their roles as temporary relief workers and aid administrators, some diplomats and consuls resented the humanitarian responsibilities the State Department and ARC now entrusted to them. In 1910, one ambassador complained to the secretary of state about the disaster assistance work with which he was tasked. Arguing that "the work of the American Red Cross abroad . . . should be conducted by its own agents," he urged the secretary of state to reconsider whether these duties were appropriate for diplomats like himself.[54] Such critiques, however, largely fell on deaf ears. The ARC, still a relatively small organization in these years, employed few paid staff; dispatching these personnel to the scene of a foreign disaster would have represented a time-consuming, expensive, and impractical proposition. American diplomats and consuls, by contrast, were already in place, ready to represent the government and ARC the moment a disaster suddenly struck.

And so, despite the misgivings some diplomats and consuls voiced, State Department and ARC leaders saw little choice but to depend on them as humanitarian agents. The vast majority of diplomats and consuls, in turn, understood that administering disaster aid might one day be part of the job. Taking on this role with increasing frequency during the Taft and Wilson years, these public servants were to remain key figures in the US foreign disaster aid system in the years ahead.

❂   ❂   ❂

By the eve of US entry into the First World War, foreign disaster assistance had become a fixture of US foreign relations. In the fifteen years after Mount Pelée's catastrophic eruption in Martinique, US officials responded to dozens of sudden disasters across Europe, Asia, and the Americas, spending millions of dollars to assist untold numbers of survivors. Working with three presidential administrations and Congress, the pillars of US foreign disaster assistance—the State Department, the US Armed Forces, and their principal partner in the voluntary sector, the American Red Cross—developed a basic set of procedures and expectations to guide the nation's official international disaster aid operations. They also came to appreciate, and indeed embrace, the diplomatic possibilities of this humanitarian assistance.

But if a defined system of US foreign disaster assistance was recognizable by late 1916, much remained improvised, uncertain, even haphazard. A variety of contingent factors determined the scope and scale of any particular relief operation, including the strategic aims of US policymakers, the capriciousness of American donors, the proximity of US military installations to sites of disaster, and the inclinations of diplomats and consuls responsible for administering the humanitarian response. Together, these issues constrained both the humanitarian and political potential of US international disaster aid.

Over the next three decades, building on these early twentieth-century foundations, US foreign disaster assistance operations became increasingly routinized and ever more commonplace. In the years spanning US involvement in the First and Second World Wars, as the chapters in part II trace, US officials worked to tackle the problems and unpredictability associated with earlier aid efforts. In these same decades, US officials also experimented with bold new approaches to disaster relief, recovery, and reconstruction in other nations. As they did so, they further tested both the possibilities and the limits of American catastrophic diplomacy.

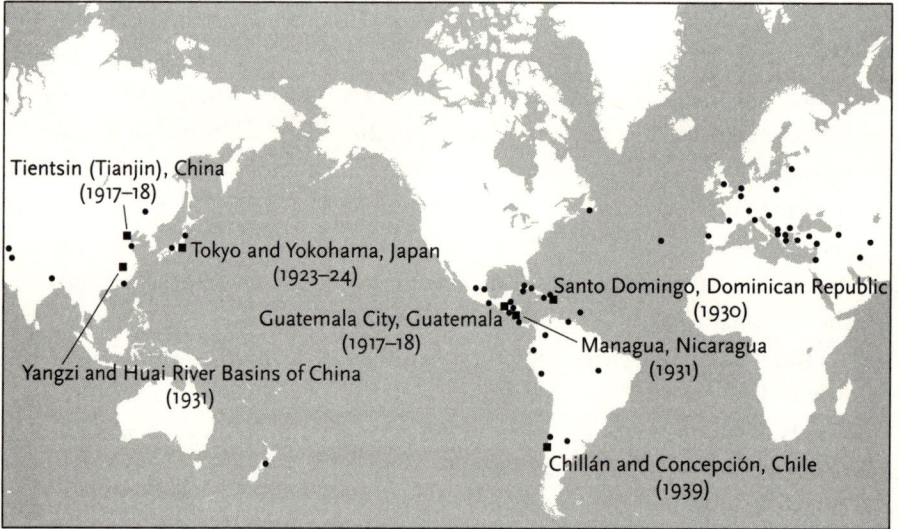

Tientsin (Tianjin), China
(1917–18)

Tokyo and Yokohama, Japan
(1923–24)

Santo Domingo, Dominican Republic
(1930)

Guatemala City, Guatemala
(1917–18)

Managua, Nicaragua
(1931)

Yangzi and Huai River Basins of China
(1931)

Chillán and Concepción, Chile
(1939)

**Recipients of official US foreign disaster
assistance referenced in part II, 1917–47**

PART TWO

Routines of Relief and the "Development" of Disaster Aid

5

Floods, Earthquakes, and the Great War

1917–1918

By 1917, the US government had honed its approach to foreign disaster aid considerably, with the State Department and its agents, the American Red Cross, and the US Armed Forces each playing a distinct part in the nation's global relief operations. The United States' entry into the First World War in early April that year had profound implications for this humanitarian system. By the signing of the Armistice in November 1918, each of the three pillars of US foreign disaster aid had undergone a momentous transformation, with lasting consequences for the nation's postwar catastrophic diplomacy. Admittedly, it was difficult to see the effects of these changes in real time; only in hindsight did these two years mark a perceptible historical turning point. Yet from this point forward, the methods and practices of US foreign disaster began evolving in new directions, in several discernible ways.

Over the next three decades, the number of US disaster relief operations continued to multiply. As they did so, those efforts became ever more routinized, guided by a consistent set of norms and procedures. Even as their aid activities grew more standardized in many respects, US officials simultaneously experimented with novel humanitarian techniques and technologies: sending expert advisers to the scene, selling surplus foodstuffs to affected countries, and using aircraft to deliver aid more quickly and across far greater distances. Mirroring the response to the Messina earthquake and tsunami in 1908–9, US officials also engaged in a growing number of long-term, comprehensive relief and recovery projects in other nations—far-reaching disaster aid operations that blurred the conceptual lines between humanitarian relief and the era's nascent development assistance initiatives.

Although these trends became steadily more apparent across the 1920s and 1930s, some initial signs of change appeared as early as the Great War era itself, the focus of this chapter. During 1917 and 1918, the US government, US military, and ARC undertook a pair of major humanitarian operations, for survivors of flooding in China and earthquakes in Guatemala. Comparable in their substantial cost, lengthy duration, and extensive scale, these episodes ranked among the largest US foreign disaster assistance efforts to date. In both countries, US officials remained involved in disaster response for months after catastrophe struck, administering an extensive and experimental array of relief and recovery projects. Their objective was not only to rebuild ruined cities and regions; embracing a developmentalist mindset, they aspired to reform Chinese and Guatemalan people and society in the process.

The timing of these episodes bears emphasis. Even as the United States waged war in Europe, the US government, military, and ARC elected to carry out sweeping disaster aid programs in both Asia and Central America. Though humanitarian motivations partly compelled US officials to stage such substantial responses, their decision to do so in the midst of an unparalleled global conflict also revealed the important position catastrophic diplomacy had achieved in US foreign policy calculations by the late 1910s. In both China and Guatemala, US officials saw a clear diplomatic and strategic rationale for providing disaster assistance, viewing it as a means to demonstrate the United States' commitments to wartime allies while helping restore order and stability in the two nations. Although the Great War upended many aspects of US foreign policy, it failed to disrupt either the practice or the political logic of US foreign disaster aid.

Lasting nineteen months, from April 1917 to November 1918, the period of US military involvement in the Great War wholly transformed the US foreign policy establishment—and, by extension, each of the three pillars of US foreign disaster aid.

For the government's official humanitarian auxiliary, the American Red Cross, the war years ushered in a period of major reorganization and explosive growth. This process started even before US entry into the conflict, between 1914 and early 1917, as ARC leaders began recruiting new members, building the organization's endowment, and raising funds for the war effort. These trends accelerated dramatically after the United States declared war. Over the next nineteen months, bolstered by the Wilson administration's strong and active support, ARC staff and volunteers carried out a mammoth program of military and civilian assistance in the United States and throughout Allied Europe, Russia, and the Near East. To finance these activities, ARC leaders launched an extraordinarily successful fundraising campaign, bringing in more than $400 million by war's end. At the same time, they built the ARC's permanent endowment to $2.5 million. The ARC's membership also soared, growing from just 286,000 dues-paying members in 1916 to 22 million adult and 11 million youth members by the Armistice in 1918.[1] American involvement in the Great War, in short, brought long-sought financial stability to the ARC while cementing its position, more firmly than ever, as the government's principal instrument of humanitarian assistance.

For the other pillars of foreign disaster aid—the US Armed Forces and the State Department—wartime mobilization proved equally consequential. The military ballooned between early 1917 and late 1918, as policymakers increased the size and capabilities of the armed forces to unprecedented levels. On the eve of its entry into the conflict, the United States had a force of 100,000 regular army troops, 60,000 navy officers and enlisted men, and 10,000 marines. Over the next two years, the strength of the army grew to more than 4 million, while the navy and marine corps reached 600,000. In that time, the number of active US Navy ships nearly tripled, from 245 vessels in late 1916 to 774 in November 1918, giving the United States "a navy second to none."[2]

The demands associated with wartime diplomacy and postwar planning presented novel challenges to the State Department as well. To accommodate its many new responsibilities, the State Department and the diplomatic and consular corps expanded dramatically, adding both permanent and temporary staff. The experience of war, however, convinced many US policymakers it was necessary to do more than just grow. What was really needed, many came to believe, was to thoroughly reform and restructure the State Department and the foreign services. Conceived during the war years, this project began in earnest once the conflict drew to a close. It culminated in 1924 with the passage of the Rogers Act, legislation that created the modern, unified US Foreign Service. Together, these changes resulted in a more professionalized, merit-based system for appointing American diplomats and consuls in the postwar years.[3]

American involvement in the Great War, in short, completely restructured the core institutions of US foreign relations and, by extension, of US foreign disaster assistance. Admittedly, many of the momentous foreign policy shifts occurring between 1917 and 1918 proved temporary. As the United States returned to a peacetime footing, many Americans sought to rein in their commitments abroad or retreat from Wilsonian internationalism.

Some developments, however, proved far more resilient. The Great War was a watershed for US foreign policy, permanently altering how the State Department, the War and Navy Departments, and the American Red Cross operated. More than this, it forever transformed the United States' role in international affairs. The United States exited the war as a creditor nation and an economic powerhouse, able to wield its financial and cultural influence— as well as its new diplomatic, military, and humanitarian power—in unprecedented ways. In the postwar years, these same developments inexorably informed the US government's evolving approach to international disaster assistance.

But what of foreign disaster aid during the war years? Throughout 1917 and 1918, the full implications of these broader structural changes were not immediately apparent. During the conflict, the primary humanitarian concerns of both the Wilson administration and ARC lay in aiding US troops and Allied soldiers and civilians, not victims of "natural" catastrophes. And yet, even as war raged across Europe, Russia, and the Near East, earthquakes, tropical storms, floods, and other natural hazards continued to trigger disasters throughout the world. Although ARC leaders acknowledged that they were "committed by public opinion . . . principally to war relief," they could still count on requests for aid to "come with certainty through our State Department" when major catastrophes occurred abroad.[4] For the Wilson administration, foreign disasters remained a pressing diplomatic concern.

Even as they concentrated most of their humanitarian energies on the Great War, then, ARC leaders and their government partners continued to aid survivors of foreign disasters during these years, from victims of an earthquake in El Salvador in June 1917 to those of a hurricane in the West Indies in October 1918.[5] By far the largest of their wartime relief efforts came in response to a pair of immense catastrophes in East Asia and Central America. The first of these disasters began in September 1917, when torrential rain and river flooding ravaged China's Chihli (Zhili) province. The second commenced a few months later, after a series of powerful earthquakes struck Guatemala, causing extensive damage to the nation's capital. Together, these two disasters upended the lives of more than a million people. Although they caused comparatively few deaths, by any other measure the events in China and Guatemala constituted extraordinary humanitarian crises, among the most devastating disasters either country had experienced in decades.

In spite of their extensive wartime commitments in Europe, the US government, military, and ARC responded actively to both crises, carrying out comprehensive programs of humanitarian assistance in each nation. Though considerably less costly than the US aid efforts for southern Italy nine years prior, they nevertheless ranked as some of the most expensive foreign disaster relief operations the US government and ARC had yet undertaken. They were also among the longest in duration and most far-reaching in scope. For nearly six months in Guatemala and almost a full year in China, American relief committees—composed of diplomats and consuls, army officers and troops, ARC personnel, and private citizens—administered a wide array of relief and recovery projects. In the process, they tested novel methods of disaster aid, some of them intended to leave a lasting influence on local populations. These twin disaster relief operations, much like the United States' concurrent participation in the Great War, constituted significant manifestations of American political, military, and humanitarian power in the early twentieth-century world.

❂   ❂   ❂

In late September 1917, following several weeks of unusually heavy rainfall, a dike burst on the Hai River near Tientsin (Tianjin), China. The country's second largest port city, Tientsin was also the site of eight foreign concessions, extraterritorial enclaves occupied and administered by multiple European powers, Japan, and the United States. Within a few days, the floodwaters had coursed through Tientsin and much of China's northeast Chihli province, leaving 500,000 people homeless and adversely affecting twice that many. In Tientsin itself, the floods displaced 100,000 people and left much of the city underwater, creating an urgent need for food, clothing, and shelter.[6]

By the time of this cataclysm, the United States had a long history of diplomatic, philanthropic, financial, and military involvement in Tientsin, as well as in China more broadly. In recent years, the US government and American voluntary sector had provided significant levels of disaster aid to China. Since its reorganization in 1905, the ARC had sent more than $650,000 to the country to assist survivors of floods, famine, epidemics, and other humanitarian crises.[7] American missionary societies also provided substantial aid to China, and the country had long been a focal point of their overseas activities. Although the US government had effectively barred immigration from the country since 1882, under the racially motivated Chinese Exclusion Act, the xenophobic and discriminatory impulses of some Americans coexisted with the humanitarian and paternalistic impulses of others.

In the decade leading up to the floods, with the collapse of the Qing dynasty and the formation and subsequent fragmentation of the Republic of China, US interests in China had only intensified.[8] With China's entry into the Great War in August 1917—less than two months before the floods occurred—the country had also become an official ally of the United States. For US government officials, ensuring order and stability in the country now appeared more important than ever. Unfolding at this critical moment in Sino-American relations, US disaster assistance efforts in Chihli province represented one of the many ways US officials intervened in, and attempted to influence, Chinese affairs during the early twentieth century.

Initially, the US response to the Hai River disaster followed a customary path. Adhering to the well-trod routine, the State Department took the lead in initiating these efforts. Following usual protocol, the US consul general in Tientsin, Paul R. Josselyn, and the US minister to China, Paul S. Reinsch, reported the catastrophe to Washington and requested American aid.[9] ARC leaders, in consultation with the State Department, subsequently wired $125,000 to the American legation in Peking (Beijing).[10]

What distinguished the US response to the Tientsin floods of 1917 from previous relief efforts in China was the central role the US Army played in executing it. As the highest-ranking US diplomat in China, Reinsch was officially responsible for overseeing American humanitarian operations in Tientsin. Consumed with wartime demands and other obligations in Peking, however, he delegated authority for "the distribution of all American relief" to a US Army officer already on the scene: Lieutenant Colonel Edward

Sigerfoos, commander of the Fifteenth US Infantry Regiment.[11] Since 1912, the Fifteenth Infantry had been garrisoned in Tientsin, stationed there to protect US and other foreign interests in the city.[12] To its roughly 1,000 officers and enlisted men now fell a new mission: assisting survivors of the Hai River floods.

Although Reinsch's decision gave the Fifteenth Infantry substantial power over American relief efforts, the military's authority did not go entirely unchecked. To maintain some level of civilian control over the humanitarian operation—a condition ARC leaders in Washington demanded—Reinsch formed an American Red Cross Flood Relief Committee to work in partnership with Sigerfoos and his troops. At the ARC's recommendation, Reinsch appointed Roger Greene, the resident director of the Rockefeller Foundation's China Medical Board, to serve alongside Sigerfoos as the committee's vice chair. He tapped seven other American citizens in Tientsin to fill out its ranks, among them a physician, a businessman, and several Protestant missionaries. To ensure a voice for the State Department in the US response, Reinsch named Paul Josselyn, the US consul general to China, as a member of the committee.[13] Working together under the banner of the ARC, representatives of the US Army, State Department, and American philanthropic and private sectors administered the official US response to Tientsin's cataclysmic floods.

In early October 1917, their respective roles now assigned, Sigerfoos, Greene, Josselyn, and their American associates began discussing how to use the funds at their disposal. By this point, more than a week after the disaster commenced, the Chinese Red Cross, the Tientsin police, and the Chinese central government had already launched a sizable humanitarian response. Chinese authorities organized food and clothing distributions and provided other forms of relief to survivors. They also established several large camps for the tens of thousands of refugees who congregated in Tientsin.[14] Residents of the Japanese, French, and British concessions, too, started organizing their own aid work.[15]

Relief activities thus proliferated in Tientsin, yet as they surveyed the scene, US officials expressed apprehension with these existing efforts—particularly those under Chinese direction. As Josselyn told the State Department, echoing the rhetoric of the era's scientific charity movement, the existing relief work appeared "to lack system and efficiency. Money and efforts," he added, were "ill-applied [and] not calculated to get [the] best results."[16] Particularly troubling to Josselyn and his American colleagues were the refugee camps Chinese authorities administered. Judging these camps overcrowded and unsanitary, Josselyn disparaged them as "a menace to the health of Tientsin" and a "menace to the community."[17] Equally critical, Greene concluded that "more sustained and far-reaching measures would be needed than could easily be undertaken by the Government and other Chinese agencies."[18]

Determined to remedy the problems they perceived, Sigerfoos, Greene, and Josselyn resolved to undertake a pair of major aid projects, both intended to make a "sustained and far-reaching" contribution to the region's recovery. With this decision, they moved the US response out of the realm of emergency relief and into more comprehensive recovery and reconstruction efforts, designed with the region's longer-term development in mind.

The first of these projects was a camp for flood survivors, built and administered by the Fifteenth Infantry. In mid-October, Sigerfoos and his colleagues presented a tentative proposal for an American-run camp to the Chinese government's director of flood relief, Hsiung Hsi-ling (Xiong Xiling).[19] From the start, US officials conceived of this camp as a model for China. Assuming that "our experiment proves a success," Greene imagined, it promised to "materially influence Chinese policy in the conduct of similar camps, thus rendering a great service to the whole community at Tientsin."[20] To their satisfaction, Chinese officials not only gave permission to move ahead with the project but also provided a site for it: a large, vacant lot in the ex-German concession, which the Chinese government had reoccupied after its recent declaration of war on Germany.[21]

Having secured formal approval and a location for the camp, US officials commissioned Henry Hussey, a Chicago architect who was in China designing buildings for the Rockefeller Foundation's Peking Union Medical College, to design it. By involving a renowned figure like Hussey in the design of a temporary refugee camp, US officials clearly sought to send a message, highlighting both American humanitarian expertise and the United States' commitment to China. Hussey soon drew up plans for a camp containing 800 temporary houses, capable of sheltering roughly 4,000 Chinese men, women, and children.[22]

With his blueprints in hand, soldiers of the Fifteenth Infantry set to work constructing houses and other supplemental buildings. On November 13, 1917, they officially opened the American Red Cross Flood Relief Camp to residents. Sigerfoos appointed one of his officers, Lieutenant Colonel Charles Morrow, as the camp's director, and assigned several additional officers and enlisted men to oversee its day-to-day operations. The Fifteenth Infantry ran the camp until late March 1918, when Morrow and Sigerfoos, citing improved conditions, closed the site and declared its mission complete.[23]

During the four and a half months they administered the flood-relief camp, the soldiers of the Fifteenth Infantry closely governed the nearly 4,000 Chinese residents who inhabited it, keeping tabs on their activities, behaviors, and bodies.[24] Morrow and his men expected all camp residents to submit to a strict regime of medical and hygienic discipline. Flood refugees who desired entry first had to undergo a physical examination, performed by an American physician. Those with infectious diseases were either sent to a clinic for treatment or denied entry. Those who were admitted received a medical identification card, a series of vaccinations, a bath, and a set of clean clothes.

Once inside the camp, residents were expected to receive regular health check-ups and additional inoculations, performed by American missionary doctors and nurses. Morrow also required them to follow a long list of sanitary and hygienic rules. These included such tasks as cleaning their homes, urinating and defecating solely in latrines, and bathing once per week. To ensure adherence to these regulations, Morrow and his subordinates held a formal inspection of the camp each morning. They also appointed several trusted Chinese men to police their compatriots and report any violations. Those who broke the rules risked losing rations or, for greater offenses, expulsion from the camp.

In addition to subjecting camp residents to medical and hygienic protocols, Morrow and his men compelled Chinese men and women to occupy their time in prescribed ways. They required all able-bodied adults to engage in some form of productive labor in order to receive daily food and fuel rations. Some residents left the camp each day for jobs in Tientsin, but Morrow also established workrooms inside the camp's walls for the unemployed, where residents manufactured clothing, shoes, sleeping mats, and other material goods. While adults labored, the camp's youth residents were encouraged to attend school. Morrow designated several buildings as classrooms and hired two instructors to teach the camp's boys and girls. Although attendance at the school remained voluntary, engaging in exercise did not. Morrow required all boys between the ages of ten and eighteen to participate in ninety minutes of calisthenics and marching each day, a drill routine supervised by officers of the Fifteenth Infantry.

Through their administration of the American Red Cross Flood Relief Camp, US Army troops thus surveilled and tightly regulated the bodies and behaviors of thousands of Chinese flood survivors. Their actions were guided by a deep-seated classist and racialized mistrust of the very individuals they assisted. With their oversight, US officials hoped to shape Chinese behaviors in the longer term, leaving a lasting appreciation for hygiene, discipline, and hard work. They sought, in effect, to rebuild the people of Tientsin along with their city.

Of course, it was relatively easy to exercise such controls within the closed confines of a refugee camp. A second US disaster aid project—a large-scale employment and public works program for Chihli province—put US soldiers in a position of authority over Chinese civilians beyond the camp's walls.

Much like their counterparts in the Strait of Messina region in 1909, Sigerfoos, Greene, Josselyn, and their associates regarded the gainful employment of flood survivors as an essential component of Tientsin's recovery. Rather than limiting their efforts to providing food and shelter, US officials agreed, they should "devis[e] means by which the destitute people of Tientsin and [its] vicinity may be provided with a livelihood by being given work."[25] Only in this way, Greene argued, could he and his colleagues assist the region's recovery "without pauperizing the people." Hiring aid recipients to construct a significant public works project, he continued, would be an especially prudent use of the ARC's investments. Because such a "large sum of money was to be spent by our committee," he explained, it should "be given in return for productive work of some kind . . . on work that was permanently useful."[26] The decision to undertake this experimental scheme, in short, stemmed from US officials' suspicions about traditional forms of charity, their classist and racially motivated aspirations to improve Chinese civilians' work ethic, and their desire to leave a lasting testament to US assistance in the region.

In late October 1917, guided by these impulses and objectives, the members of the ARC Flood Relief Committee formulated plans for a large-scale road-building project, financed with the ARC's remaining funds. They proposed to hire several thousand Chinese men, recruited from the worst-flooded areas of Chihli province, to construct a new, modern highway between Peking and the nearby district of Tungchow (Tongzhou). From the US legation, Paul Reinsch expressed his full support for the proposal, telling his superiors in Washington that "by initiating joint relief and improvement work, [the] Red Cross would be setting a fine example."[27] State Department officials and ARC leaders concurred, deeming the project a "splendid example of intelligent relief measures."[28] Chinese government officials, too, were evidently persuaded. In addition to giving their consent, Hsiung and his associates agreed to pay half the workers' wages and to cover the costs of machinery, materials, and the salaries of two American engineers.[29]

With the approval of both US and Chinese officials, the ARC Flood Relief Committee moved forward with the execution of the highway-building scheme. Sigerfoos placed this project, like the relief camp, under the command of a Fifteenth Infantry officer, deputizing Captain R. T. McDonnell to oversee the highway's construction. Employing some 4,000 Chinese laborers under US military command, this roadbuilding program represented an enormous undertaking, a large-scale experiment with manual labor as a form of disaster aid.[30] During the nine months that they administered the roadbuilding project, McDonnell and his associates supervised Chinese flood refugees as they cleared land, broke up granite slabs for paving material, and built the new highway. Breaking ground on the Peking-Tungchow Highway in early December 1917, the laborers continued its construction until August 31, 1918, almost a year after the onset of the catastrophe. Compared with most contemporaneous US disaster assistance operations, which typically lasted no more than a few weeks, the project was exceptional for its longevity.

Several key details of the roadbuilding project reflected the prejudiced assumptions of its organizers. In return for their labor, all workers received housing, basic provisions, and a salary of $6.50 per month. McDonnell and his men did not give these wages to Chinese laborers directly, however, or in full. All workers were first required to cover the costs of the monthly grain and coal rations allotted to them. Once those expenses were met, US officials remitted the balance of the wages—roughly $3.00 per month—not to the workers themselves but to their wives and families. Presuming that Chinese men would otherwise fritter their earnings away on alcohol, prostitution, and other vices, McDonnell and his associates kept tight control over the flow of wages and, by extension, over the Chinese workers who earned them.

Like their counterparts in the American Red Cross Flood Relief Camp, McDonnell and his men understood their mission as something more than relieving Chinese flood victims. In their minds, administering the highway-building program presented a unique opportunity to promote—and if necessary, to compel by military authority—desired behaviors among Chinese civilians. Intended as much to transform behavior as to build infrastructure, the Chihli roadbuilding project went well beyond the traditional goals of humanitarian aid. It functioned as an analogue to other contemporary forms of development assistance.

With the completion of the Peking-Tungchow Highway in August 1918, US flood relief efforts in Chihli province officially concluded. As they looked back over what they accomplished during the past eleven months, the US officials who oversaw this operation believed that they had done considerable good—not only in humanitarian terms but also as measured by the mark they left on Chinese relief recipients. Calling the refugee camp "eminently successful" and "a model for relief work in China," US minister Paul Reinsch commended the US troops who ran it. He praised the Fifteenth Infantry for "keeping the adults employed in useful labor," for "schooling the children in mental and physical branches of education," and for teaching all camp residents "not only to keep clean, but to like and desire cleanliness." Further extolling the camp's transformative potential, Reinsch crowed, "It is certain that the 3,700 people who found refuge in this camp will look upon their stay there as an experience which has changed their outlook on life."[31]

American officials expressed similar enthusiasm for the roadbuilding project, calling it a "very satisfactory . . . experiment" in humanitarian aid.[32] As a form of disaster assistance, Greene asserted, the employment of refugees had proven "much more satisfactory than camp maintenance or grain distribution." In addition to promoting the value of work and encouraging relief recipients' self-sufficiency, Greene and his colleagues observed, "there was an important work done in return, which will be of use to the public for many years."[33] Long after the US aid operation ended, Greene and his American associates hoped, its legacies would endure, etched into the mindsets and behaviors of Chinese relief recipients, and memorialized on the new stretch of the Peking-Tungchow Highway.

☆   ☆   ☆

From late 1917 to mid-1918, as Fifteenth US Infantry soldiers and their associates administered flood relief projects in Tientsin and its environs, US officials halfway around the world in Guatemala carried out a similarly "sustained and far-reaching" program of disaster assistance. This parallel humanitarian episode began on December 25, 1917, when a major earthquake struck Guatemala City, leveling hundreds of buildings. Over the next two weeks, dozens of additional shocks occurred, adding to the devastation. By early January, Guatemala's capital lay in ruins. Although relatively few people died as a result of the earthquakes, more than 100,000 lost their homes.[34]

Guatemala, like China, was a nation shaped by a long and often fraught legacy of US economic and political incursions. By the time the Guatemala

City earthquakes began in late 1917, American business interests—chief among them the United Fruit Company—had amassed enormous control over Guatemalan land, infrastructure, capital, and labor. They secured this power thanks to the active support of the US government and the complicity of Guatemalan politicians and elites, chief among them Guatemala's authoritarian president, Manuel Estrada Cabrera. Although Guatemala initially remained neutral in the Great War, the Wilson administration had for months been pressuring the country to declare war on Germany, which it eventually did in April 1918.[35]

Much like their counterparts in Chihli province, US officials in Guatemala City saw in the disaster an opportunity to exercise additional influence, not only over the relief and recovery process, but within Guatemalan society and politics more broadly. Through the provision of disaster aid, they believed, they could continue wooing a potential wartime ally while helping to preserve order and stability in a nation critical to US economic interests.

Guided by these foreign policy objectives, the US government's official response to the catastrophe in Guatemala City commenced the day after the Christmas Day earthquake. Leading it initially was the US chargé d'affaires in Guatemala, Walter Thurston, a young clerk presiding over the American legation while the US minister took a leave of absence. On the morning of December 26, following established procedures, Thurston reported the disaster to the State Department. He met several times that day with Guatemalan president Manuel Estrada Cabrera, assuring him that the "[US] Government desires to be of any possible assistance to her sister republic" and offering his help in "organizing and maintaining relief work."[36] Thurston had little time to act on this pledge, however, for the situation abruptly worsened. Severe shocks occurred again on December 27 and 29, adding to the already appalling devastation. Facing an escalating humanitarian crisis, Estrada Cabrera declared martial law and accepted the US government's aid offer, setting American relief efforts in motion.[37]

At first, the US response to the Guatemala City earthquakes followed the routine course of many contemporary US foreign disaster relief operations, remaining fairly limited in scale and cost. From Washington, the chief of naval operations instructed a nearby cruiser, the USS *Cincinnati*, to head to Guatemala and present the Wilson administration's sympathies to Estrada Cabrera.[38] Leaders of the ARC, meanwhile, wired $5,000 to Thurston and made preparations to ship food and medical supplies from New Orleans. As they had done with Reinsch in China, they directed Thurston to organize a relief committee composed of individuals he judged "competent to advise and assist in control of [the] Red Cross contribution and of its relief work."[39] Assuming the role of chair of this ARC Relief Committee, Thurston appointed eight other US citizens to assist him, including the general manager of the International Railways of Central America and several others with stakes in Guatemala's coffee plantations and railroads.[40]

But before Thurston and his ARC Relief Committee could accomplish much more, the situation worsened. Major aftershocks occurred again during the nights of January 3 and 4, 1918. As one observer described it, they "finished everything."[41]

The disaster in Guatemala City was now a catastrophe of major proportions, and Thurston, in turn, was growing increasingly uneasy. On January 5, the chargé d'affaires reported to the State Department that "conditions are becoming alarming." Thurston expressed deep reservations about the capacity of either Guatemalan citizens or the Guatemalan government to respond effectively to the crisis. Warning his superiors that the "native relief committee" was "unanimously condemned for grafting and inefficiency," Thurston exhorted the State Department and ARC to send reinforcements, calling for a "commission of able Americans capable of handling the situation."[42]

His appeal worked. Responding favorably to Thurston's requests, US government and military officials and ARC leaders expanded the humanitarian operation in Guatemala. The escalation began on January 7, when a seven-man American relief party left the US-controlled Canal Zone on a steamer bound for Guatemala. The members of this Panama Commission included the chair of the ARC's Canal Zone chapter and six American employees of the quasi-governmental Panama Canal and Panama Railroad Company. They brought with them 130 tons of food, medical equipment, blankets, and other supplies, an allocation authorized by the War Department and financed with a new infusion of ARC funds. Arriving in Guatemala City five days later, the party joined Thurston and his existing ARC Relief Committee in their work.[43] Augmenting the resources at their disposal, the secretary of war agreed to loan 4,000 US Army tents to Guatemala. From the US naval base on Key West, a navy cargo ship transported the tents to Puerto Barrios, Guatemala's main Caribbean seaport, arriving on January 12.[44]

In Guatemala, as in China, US officials' xenophobic assumptions about the people they assisted had an early and direct bearing on the American response. From the start, Thurston was adamant that he and his American colleagues maintain full control over the relief operation. "It is absolutely essential that we strictly supervise the distribution of all this work," he told the State Department. Otherwise, he explained, "money, medicines, food and shelter will all be wasted through maladministration on the part of native committees, and, I regret to say, graft." Meeting again with Estrada Cabrera, Thurston insisted that the president give the ARC Relief Committee "absolute authority to handle the relief supplies and administration of general relief in our discretion."[45]

Somewhat to Thurston's surprise, Estrada Cabrera not only agreed to these demands but also offered something more. Promising to give the chargé d'affaires and his associates "full and unquestioned powers" and "supreme control of the city," Estrada Cabrera declared "the military, the police, and the existing relief organizations all to be subordinate" to the ARC Relief Committee.[46] Elated with this outcome, Thurston informed the State Department that "the President seemed to realize quite thoroughly the necessity of placing the whole situation in the hands of people who would impartially and drastically handle it."[47] Although Thurston included several Guatemalans (including Estrada Cabrera) as honorary members of the ARC Relief Committee, he and other US citizens secured a remarkable degree of power over Guatemala City's relief and recovery efforts. As Thurston desired, the disaster aid operation was firmly in American hands.

Though this decision may have surprised Thurston, it was a logical outgrowth of Estrada Cabrera's approach to governance. After assuming power in 1898, as one historian has written, Estrada Cabrera "established a political system in which he alone made and executed public policy."[48] He built and sustained his authoritarian rule, moreover, largely by cozying up to US interests. In exchange for granting concessions to American capitalists and investors, the Guatemalan president demanded a hefty cut of their proceeds. He also expected US officials to turn a blind eye toward his corruption and brutality. By delegating his authority to Thurston and his American associates, Estrada Cabrera stood to benefit in two ways. This move enabled him to perform fealty to US officials friendly to his regime while freeing his own government from the costs and burden of administering a major relief effort.

During the next several weeks, with Estrada Cabrera's blessings, Thurston and his expanded ARC Relief Committee began organizing a broad slate of humanitarian activities in Guatemala City. American officials oversaw the distribution of the many tons of food, medical supplies, and other material assistance arriving from the Canal Zone, Key West, and New Orleans. They also established an outpatient dispensary and began digging latrines throughout the city, with an eye toward addressing pressing medical and sanitary concerns.[49] They started clearing the ruins of Guatemala City's now-demolished General Hospital, planning to construct a new, temporary hospital in its place. And finally, they made preparations for an American-led refugee camp, intended to house earthquake survivors in the 4,000 tents the War Department sent.[50] To finance these efforts, ARC leaders wired additional funds to the American legation in Guatemala City, bringing their total contributions to $200,000.[51]

By late January, a month after the onset of the disaster, the ARC Relief Committee had laid the foundations for several ambitious assistance projects, which promised to keep them involved in Guatemala City's recovery for the foreseeable future.

Responsibility for seeing these complex, longer-term efforts through to completion, however, was about to pass into new hands. In a move that was then atypical—but that became more commonplace during the 1920s and 1930s—ARC leaders decided to send an "experienced relief administrator" to Guatemala to take over the US assistance operation.[52] Their willingness to do so was "especially gratifying to the [State] Department," the acting secretary of state told ARC leaders, "because of the proof which Guatemala has given of its cooperation with the United States in the war. The Department," he noted, "attaches great importance to the aid thus furnished by the Red Cross."[53]

For this dual humanitarian and diplomatic mission, ARC leaders selected John J. O'Connor, the director of the ARC's Central Division, who was normally responsible for the organization's activities in the midwestern United States. To assist him, the ARC also sent to Guatemala an American sanitary engineer, Edward Stuart, and O'Connor's wife, Louise, who was a social worker. On January 18, their party sailed from Mobile, Alabama, on a United Fruit Company ship, reaching Guatemala City five days later. Although some members of the original ARC Relief Committee remained involved in assistance activities, the US aid operation was now in the hands of an experienced ARC disaster relief specialist.

From the moment he arrived, O'Connor insisted on keeping aid efforts under US control for he, like Thurston, was highly dubious of locally led assistance efforts. O'Connor abruptly dismissed the Guatemalan Red Cross as "a poor thing at best," calling its personnel "to a large extent ineffective, full of words and promises."[54] He had little better to say for other Guatemalan charitable agencies. Much to O'Connor's satisfaction, Estrada Cabrera reiterated his support for the US-led operation. Moreover, he invited O'Connor and his associates to provide "expert advisory assistance" to the Guatemalan government and Guatemalan Red Cross on sanitary engineering, public health, and related subjects.[55] Enjoying Estrada Cabrera's continued blessings, O'Connor and his colleagues began steering Guatemala City's recovery efforts in new directions.

Remaining in Guatemala for more than four months, O'Connor transformed what was already a major US relief operation into an even more comprehensive program of recovery and development assistance. As Estrada Cabrera requested, O'Connor and his associates started counseling the president, other government authorities, and Guatemalan Red Cross officials on a variety of matters. Observing that "there is no food card system" in place and concerned with the potential for abuse, O'Connor persuaded Guatemalan authorities to adopt a more regimented approach to the distribution of rations. Adopting a procedure that had become common in the United States and elsewhere since the 1906 San Francisco earthquake, they required would-be recipients to register with the government and verifiably prove their need for aid.[56]

Having dealt with the matter of food distributions, O'Connor set his sights on more far-reaching changes. Straying far afield from emergency relief, and drawing inspiration from contemporary health and welfare reforms in the United States, O'Connor set out to improve Guatemala City's existing health and sanitation infrastructure. To achieve this goal, O'Connor and his American colleagues urged Estrada Cabrera to adopt many new laws and policies. Some of their recommendations were for structural reforms, such as chlorinating the city's water supply, improving the sewage system, inspecting the milk supply, and establishing public delousing stations "under military control." Others were measures intended to reform individual behaviors and cultural customs, such as making vaccination for typhoid and smallpox compulsory, forbidding above-ground burials, and requiring the "imprisonment of any person found guilty of soil pollution."[57]

Besides doling out advice, O'Connor used the ARC funds at his disposal to finance several large-scale aid projects. Perhaps the most visible of these activities was the American tent camp for earthquake refugees, a project that shared many parallels with the Fifteenth Infantry's camp in Tientsin. Initiated by O'Connor's predecessors, Camp Manuel Estrada Cabrera became fully operational shortly after his arrival, housing 1,200 families by early February.[58] To direct the camp, O'Connor appointed Herbert Apfel, an American businessman living in Guatemala City.

Much like his military counterparts in China, Apfel closely policed the behaviors of camp inhabitants, endeavoring to restrict aid only to those he deemed worthy of assistance. Like officials in Tientsin, he viewed his camp as a reformatory institution. Its purpose was not only to relieve disaster

**Camp Manuel Estrada Cabrera, established and
administered by the American Red Cross Relief
Committee in Guatemala after the 1918 earthquake.**
Image courtesy of American National Red Cross photograph
collection, Library of Congress Prints and Photographs Division.

survivors' suffering but to transform their attitudes and behaviors, particularly surrounding labor and public health. In these respects, not only was Apfel imitating his colleagues in Chihli Province, but he was also mirroring the actions and ideologies of camp administrators in San Francisco and other sites of disaster *within* the United States and its territories. The parallels in aid projects across these far-flung sites attest to the global reach of the period's scientific charity movement and the methods and philosophies of disaster assistance it inspired.

The surveillance and control at Camp Manuel Estrada Cabrera started with the admissions process. Concerned that people "scarcely deserving relief" might be "admitted indiscriminately," Apfel required any Guatemalan who desired entry "to have the endorsement of two citizens" and to receive vaccinations against smallpox and typhoid.[59] Once admitted, residents were expected to adhere to a slew of regulations. Apfel imposed a 9:00 p.m. curfew and required all camp inhabitants to follow rules related to drawing water, using latrines, disposing of refuse, and other hygienic matters. He expected all men, unless they were elderly or infirm, to serve as laborers in the camp construction department. In exchange for performing useful manual labor, these men received rations for their families and two hot meals per day. Those who failed to meet expectations were denied food aid. Apfel also refused rations to camp residents who could not show a certificate proving they had received second and third rounds of typhoid vaccination. While withholding rations represented one tool to encourage compliance, Apfel also relied on sanitary police squads to conduct daily inspections of tents and to report all violations of rules, illnesses, and deaths.[60]

Further extending American surveillance over camp residents, Apfel tasked Louise O'Connor with conducting a comprehensive survey of camp inhabitants. Working with a team of volunteers, she collected a full social and medical history of every resident. O'Connor and her associates cataloged a variety of problems, including widespread infectious diseases and other maladies, improper feeding of babies and young children, and high infant mortality rates. O'Connor subsequently presented her findings to Estrada Cabrera, together with a set of recommendations for reform. Improvements could be "easily accomplished," she advised the president, with "practical demonstrations and simple lectures" on public health, childrearing, and other social welfare topics.[61] Such educational programs stood to benefit not only camp residents, O'Connor stressed, but the citizens of Guatemala City as a whole.

Outside the camp, O'Connor deputized other US citizens to direct additional facets of the Guatemalan recovery effort. To administer activities related to medicine and public health, he turned to Alvin Struse, a physician for the Rockefeller Foundation's International Health Board living in Guatemala. Like many of his American associates, Struse was highly critical of the public health situation in Guatemala City. "There has been no effort at system, nor has there been any towards organization," he complained to ARC leaders, "and the work done so far is far below what it should have been."[62]

Hoping to overcome the problems he perceived, Struse implemented multiple health-related projects. On the former site of Guatemala City's General Hospital, he supervised the building and subsequent operation of a temporary hospital and outpatient dispensary. Housed in forty-one US Army tents, these institutions together provided care to roughly 4,000 patients over the next few months. Struse also established a specialized children's dispensary, which provided free milk to Guatemalan youth and offered demonstrations to "teach the mothers proper methods to prepare and administer food." Additionally, extending his reach to the community at large, Struse organized a citywide immunization campaign against smallpox and typhoid, using vaccines donated by his employer, the Rockefeller Foundation.[63]

To oversee sanitation issues, O'Connor appointed Edward Stuart, the engineer who accompanied him to Guatemala. Echoing his other American colleagues, Stuart lamented the situation in Guatemala, calling "the sanitary conditions ... appalling." He was especially disgusted to observe that "people urinate and defecate at will upon the surface of the ground."[64] Endeavoring to halt this behavior, he hired Guatemalan laborers to construct dozens of latrines throughout the city while urging Estrada Cabrera to make their use mandatory. Stuart also set his sights on providing safe water for Guatemala City's residents. To accomplish this task, he first conducted a sanitary survey of the entire city. He then drew up plans for a new municipal water supply system. Specifically, Stuart advised the Guatemalan government to purchase $40,000 worth of water piping to upgrade the city's water supply. Hoping to incentivize this initiative, ARC leaders pledged to provide a chlorinating plant for the new system. Persuaded by Stuart's recommendations and the ARC's offer, Estrada Cabrera accepted this proposal, allowing the project to commence.[65]

With each of these recovery projects, O'Connor and his associates moved well beyond the bounds of emergency relief work. Though an earthquake brought them to Guatemala City, US officials aimed to do more than minister to acute suffering; they strove to implement comprehensive hygienic, infrastructural, and behavioral reforms. Together, O'Connor and his colleagues intended to leave a permanent mark in Guatemala City, a testament to the United States' involvement and expertise that would remain visible for years to come.

Even as they administered this far-reaching program of disaster aid, by March 1918—three months into their aid mission—US officials began discussing an eventual exit strategy. Their desire to withdraw was motivated, in part, by a conviction that Guatemalans should assume the primary responsibility for their own recovery, a belief rooted in prevailing ideologies of self-help and self-sufficiency.[66] As the ARC's director of civilian relief put it, US officials must not take on "obligations which it is the prime responsibility of the Guatemalan Government to discharge. Neither must we stay too long," he continued, "for to do so is to weaken the resolution and the self-dependence of those who should assume the continuing burden of restoration of the city and rehabilitation of its people."[67]

O'Connor and his associates also wanted to conclude their involvement for a second reason: they were growing increasingly frustrated with the slow pace of recovery, a problem they attributed largely to Guatemalan indolence and incompetence. William Hayne Leavell—who had by now resumed his post as US minister at the American legation—expressed these feelings quite clearly. "With the leisurely ways and the inadequate methods of these people," Leavell told the State Department, "it will take months to clear away the debris found everywhere in the city." Certain that abuse was rampant, Leavell further criticized Guatemalans for "making gain of the generous help that has been sent from the United States."[68] Like Leavell, O'Connor lamented the "lethargy, delay and hesitation" he felt characterized the Guatemalan government's response. "While putting all the pressure we can and giving all the assistance we can," he stressed, US officials must plan on "seizing the first favorable opportunity of withdrawing."[69]

Yet for all their talk of departing imminently, it was not until early April that members of the relief committee decided that "the American Red Cross has done all that should be done" and that they could "safely withdraw, leaving to the Government and to the Municipality the obligations to look after the people."[70] At this point, US officials finally took some concrete steps to terminate the US aid program. O'Connor sold off what remained of the ARC's food supplies, netting a surplus of $22,000.[71] Struse, meanwhile, "turned our hospital work . . . into the hands of the Guatemalan [medical] profession," while Apfel notified residents of Camp Manuel Estrada Cabrera that it was poised to close.[72] A few weeks later, Stuart and his crew completed the installation of Guatemala City's new water supply system.[73]

American involvement in Guatemala City's recovery was *still* not finished, however. Even as they were shutting down operations, O'Connor and his colleagues simultaneously initiated two entirely new aid projects, each intended to endure long after the US operation ended. First, determined to leave "some memorial gift of the Red Cross—some work that has been needed for years, that is constructive and permanent," O'Connor used the $22,000 he made selling surplus foodstuffs to purchase equipment for the city's new General Hospital.[74] By outfitting the hospital, as Struse told ARC leaders, he and his colleagues would "leave a permanent stamp" on Guatemala City, a functional monument to their humanitarian work.[75] Second, acknowledging that many Guatemalans remained homeless, O'Connor established a second, smaller refugee camp, capable of housing 400 of the "poorest families" from the original facility. The Guatemalan government, in turn, pledged to administer the camp once the American relief committee disbanded.[76] Although both projects served an undeniable need for Guatemalan earthquake survivors, the desire to make a diplomatic impact—to leave a lasting testament to American aid and beneficence—was never far from US officials' minds.

Eventually, more than five months after it started, the US relief operation in Guatemala City concluded. In late May, Estrada Cabrera hosted a lavish farewell banquet and awards ceremony for the ARC Relief Committee at his residence, La Palma.[77] Two weeks later, declaring, "All our work is closed," O'Connor officially disbanded the relief committee. He, his wife Louise, and Stuart then returned to the United States, while the remaining members of the relief committee resumed their normal responsibilities in Guatemala. By the time it formally concluded in June 1918, the US disaster relief operation in Guatemala was valued at nearly $400,000, a substantial figure for its time.[78]

Reflecting on the results of the assistance effort, O'Connor frankly admitted that "Guatemala is still a city of ruins," conceding, "Five months are too short a time in which to rebuild a city that . . . is as completely destroyed as this capital." Still, he found reason for optimism, confident that "more progress has been made than impatient eyes and anxious minds can discern." Highlighting the American relief committee's accomplishments, O'Connor boasted, "We taught thousands the value of cleanliness and the benefits of open air life," all while giving "proof of the friendliness of the United States Government."[79] Echoing his counterparts in Chihli province, O'Connor believed that he and his American associates had brought tangible, lasting benefits to Guatemala City and US-Guatemalan relations. In their minds, American disaster aid had achieved something much greater than alleviating immediate suffering. It had permanently uplifted Guatemalan people and society.

★ ★ ★

From September 1917 through August 1918, as most Americans trained their attention on the Great War in Europe, US officials in East Asia and Central America focused on crises of a different sort, carrying out humanitarian responses to momentous catastrophes in China and Guatemala. By the time US relief efforts in Chihli province and Guatemala City formally concluded, the Great War was nearing its end. Just a few months later, in November 1918, the armistice brought the conflict's hostilities to a close.

In the months that followed, the US government, US military, and ARC gradually reverted to a peacetime footing. Yet they did not, indeed could not, simply return to the prewar status quo. While waging war, each of these institutions of US foreign policy had undergone a fundamental transformation. The system of US foreign disaster assistance, built on these three pillars of US foreign relations, could not help but evolve with them.

Across the next three decades, many of the patterns and precedents that had defined US foreign relief operations since the early twentieth century were to endure. Despite calls for American retrenchment from world affairs, a global Great Depression, and a Second World War, US foreign disaster aid operations continued apace, becoming a common, even routine facet of the nation's foreign affairs. At the same time, however, the mechanisms of US international disaster aid changed in some appreciable ways, adapting to new geopolitical contexts, technological innovations, and shifting philosophies of humanitarian aid. Increasingly, American officials tested novel techniques and methods of disaster aid, paralleling the sorts of recovery, reconstruction, and proto-development projects they had piloted in Chihli province, Guatemala City, and the Strait of Messina region. Throughout it all, US foreign disaster aid remained inextricably linked to broader foreign policy objectives and agendas. As they navigated a new era, government officials continued to explore the potentials and pitfalls of American catastrophic diplomacy.

6

The Possibilities and Limits of Catastrophic Diplomacy in Japan

1919–1924

On September 1, 1923, an exceptionally powerful earthquake occurred in the Kantō region of Japan, its epicenter near the bustling metropolises of Tokyo and Yokohama. The seismic event and ensuing fires destroyed Yokohama and much of Tokyo, along with hundreds of smaller towns and villages. More than 100,000 people perished, while another 43,000 were declared missing. An estimated 113,000 more suffered injuries, many of them quite grave. The disaster left 2 million people homeless and reduced numerous hospitals, schools, businesses, and other buildings to rubble. To this day, the Great Kantō earthquake remains one of the deadliest and most destructive catastrophes ever to occur.[1]

When the earthquake struck, nearly five years had elapsed since the conclusion of the Great War. In that time, the US government and ARC concentrated their humanitarian energies principally on the relief and re-covery of war-torn Europe, Russia, and the Near East, abstaining from the sorts of expansive disaster aid operations they had undertaken in China and Guatemala. In 1919, the US government made a major foray into postwar foreign aid when it established the American Relief Administration. Over the next four years, funded by congressional appropriations and private do-nations, its personnel distributed more than 4 million tons of foodstuffs throughout twenty-three war-torn countries, helping feed millions of peo-ple. Hundreds of ARC personnel also remained in Europe through 1922, administering a broad-based relief and recovery program for the Great War's survivors.[2] While carrying out these postwar relief activities, US officials learned valuable lessons and established precedents that would shape their responses to many future humanitarian crises—including not only war and armed violence but also "natural" disasters.

Even as they devoted most of their humanitarian attention to postwar assistance, the US government and ARC never entirely ignored catastrophes caused by natural hazards, ensuring foreign disaster aid remained a routine part of US international affairs. Between 1919 and 1923, ARC personnel in Europe responded to several such crises, including those triggered by earth-quakes in Tepelenë and Elbasan, Albania, and the Tuscany province of Italy. Staff of the ARC's new Insular and Foreign Division, meanwhile, dispatched disaster aid to other parts of the world, including El Salvador, the Dominican Republic, Persia, and China.[3] As had usually been the case before the war, this assistance generally took the form of modest financial contributions or limited quantities of material supplies, designed to assist other countries in the short term while, as one US consul put it, "encouraging a kindlier feeling toward Americans."[4]

And then came the Great Kantō earthquake. In the days, weeks, and months after the onset of this incredible catastrophe, the US State Depart-ment, US Army and Navy, and ARC organized and administered an immense official response, fueled by the unparalleled donations and impassioned support of the American public. This humanitarian operation became the costliest foreign disaster aid operation the US government and ARC had ever undertaken; indeed, as historian J. Charles Schencking has calculated, "disaster relief to Japan stands unrivalled to this day."[5] Valued at nearly $20 million, the aid operation involved scores of US diplomats and consuls, army soldiers and navy sailors, military physicians and Red Cross nurses. As in Chihli province and Guatemala City, the US response moved well beyond emergency relief to include long-term recovery aid and more permanent forms of reconstruction assistance, designed to leave a lasting mark on the Japanese landscape. American officials participated in these assorted humanitarian efforts for months after the onset of the disaster, remaining involved well into 1924.

While the sheer enormity of the disaster in Tokyo and Yokohama was partly responsible for triggering such a monumental American humanitarian response, both its incredible magnitude and its specific form must also be understood in another way: as a function of US-Japanese relations in the early 1920s.[6] Occurring five years after the Great War ended and less than two years after the Washington Naval Conference, the Great Kantō earthquake came at a time of simmering tensions between the two nations. Since 1919, US officials had grown deeply concerned by the Japanese government's increasing militarism and its imperial ambitions in China and the Pacific. Japanese authorities, for their part, were aggrieved by the US government's refusal to recognize Japan as an equal power in diplomatic and military affairs. Rising, xenophobic demands in the United States to bar Japanese immigration and restrict the rights of Japanese Americans compounded this issue, revealing to many Japanese the depths of American racism.[7]

In this heated milieu, US officials recognized the potential diplomatic benefits to be gained not only from assisting Japan but also by treating the Japanese government and its citizens as peers in the humanitarian field. Conducted in the right way, they anticipated, a US relief operation stood to demonstrate the United States' respect for Japan and to highlight the racial, cultural, and political parity between the two nations. In stark contrast to places like Guatemala City, Chihli province, and the Strait of Messina region—where Americans had insisted on maintaining tight control over the aid they contributed—US diplomatic and military officials in this case ceded much of their authority to the Japanese government and its relief agencies.

In the short term, this approach appeared to work, for US actions generated considerable goodwill throughout Japan. Yet, whatever Japanese-American amity resulted from this episode ultimately proved ephemeral, and by some measures wholly illusory. Temporarily subdued, deep-seated political animosities and racial prejudices quickly resurfaced, tearing apart the fragile comity that emerged in the disaster's wake and squandering the diplomatic gains US assistance had achieved. In the early 1920s, the US response to the Great Kantō earthquake laid bare the possibilities and the limits of catastrophic diplomacy.

✪ ✪ ✪

Ambassador Cyrus Woods was closing the embassy's offices for the weekend when the earthquake began, at just before noon on September 1. The violent tremors lasted just five minutes, but "the whole afternoon," Woods recounted, "was one of terror."[8] Fleeing the crumbling embassy for the relatively safer streets of Tokyo, Woods and his military attaché, Lieutenant Colonel Charles Burnett, stood alongside the city's hundreds of thousands of surviving residents, looking on as building after building collapsed and as raging fires enveloped much of what remained.[9] As dire as the situation felt for Woods and Burnett, the situation proved far worse for their colleagues in Yokohama. There, both the US consul and vice-consul lost their lives as the powerful shocks and ensuing inferno razed the city.[10]

Woods and his colleagues were not the only US citizens affected. The Kantō region was home to a sizable American community in 1923, which included Protestant missionaries, employees of corporations such as General Electric and Standard Oil, and their families. In total, seventy-eight US citizens died as a result of the earthquake, while roughly 1,200 survived as refugees.[11] Tens of thousands of other disaster victims also hailed from outside Japan, chiefly from Korea (then a Japanese colony), China, and the Soviet Union. This cosmopolitan cast of characters joined the more than 2 million Japanese men and women affected by the Great Kantō earthquake, a catastrophe, as Woods put it, "without precedent in history."[12]

Though still reeling from the devastation, Japanese government officials responded swiftly and vigorously. The following day, the imperial government established an Earthquake Relief Bureau to oversee the provision of food, water, medical care, and shelter. Declaring martial law, Japanese authorities deployed tens of thousands of soldiers and sailors to the region, charging them with restoring order and assisting relief efforts. In the coming weeks and months, the Japanese government spearheaded a vast recovery and reconstruction program.[13]

Japanese officials did not act in isolation. A major international response to the crisis coalesced, with dozens of nations contributing to the collective relief effort. But the leading foreign donor, by leaps and bounds, was the United States.[14]

The US response to the Great Kantō earthquake began taking form in the first days of September, as US officials throughout Japan, within the United States, and in the nearby US territory of the Philippines simultaneously organized different facets of the relief operation. Hewing to established protocols, Ambassador Woods reported the catastrophe to his superiors at the State Department the day after the quakes, imploring the government and

ARC to send assistance.[15] He sent additional appeals to Rear Admiral Earl Anderson, the commander of the US Asiatic fleet, requesting that he proceed to Tokyo Bay with assistance, and to Leonard Wood, governor general of the Philippines, asking for food and medical supplies from Manila.

Woods next called on Japan's prime minister, Yamamoto Gonnohyōe, to advise him of the actions he had taken. Telling him "that the American people wanted the privilege of helping Japan in this great calamity," Woods received the prime minister's consent to move forward with his plans.[16] The following morning, Woods placed his military attaché, Charles Burnett, in charge of the American relief effort, tasking him with coordinating the voluntary activities of US citizens living in Japan.[17] Over the next three days, now operating out of the Imperial Hotel in Tokyo, the two men and their associates organized what limited relief measures they could as they waited for outside aid to arrive.[18]

While Woods attempted to organize a humanitarian response in Tokyo, the US consul in Kobe, Erle Dickover, administered his own relief activities. In the days after the quake, several US commercial ships in nearby waters ferried American, Chinese, Russian, and other foreign refugees to Kobe, some 200 nautical miles to the southwest of Tokyo and Yokohama. The first of these vessels, the SS *West Prospect*, reached Kobe on September 4, carrying several hundred disaster survivors. Others followed in quick succession, unloading thousands of passengers in Kobe within a few days.[19] Learning from these individuals about the full extent of the catastrophe, Dickover ordered his vice-consul to proceed to the disaster zone with a shipload of food, water, and medical supplies.[20]

Dickover then took steps to assist the swelling foreign refugee population in Kobe. In collaboration with British, French, and other foreign consular officials, he participated in efforts to establish temporary hospitals, set up improvised barracks, and organize food and clothing distributions. He also created a registry bureau for American disaster survivors and began investigating possible avenues for their eventual repatriation.[21] To assist those needing medical care, Dickover summoned to Kobe the Shanghai Unit of the American Red Cross, a chapter composed of US citizens living in that Chinese city. On arrival, this contingent of ARC doctors and nurses organized three temporary hospitals, where they treated injuries and provided other medical care.[22]

As Woods, Dickover, and other State Department personnel improvised on the ground in Japan, other components of US foreign disaster aid system swung into action, adhering to well-trod humanitarian routines. In the

United States and its territories, government and military officials and ARC leaders quickly got to work. Calvin Coolidge, who had assumed the presidency a month earlier upon Warren Harding's sudden death, was determined to provide the government's assistance to Japan. He found himself hamstrung, however, by the fact that Congress was not in session and thus could not appropriate funds or authorize any extraordinary emergency actions. Eager to take what steps he could "in the absence of special legislative authority," Coolidge was forced to look elsewhere.[23]

Like several of his predecessors, he turned to the US military. On September 3, Coolidge instructed the War and Navy Departments to direct supplies, men, and ships to the disaster zone. By this point, however, the US Asiatic Fleet was *already* steaming toward the Kantō region, having received Woods's distress calls well before they reached Washington; with his orders, Coolidge thus gave this mission his retroactive approval.[24] Meanwhile, the US Army made its own contributions to the recovery efforts. On September 5, the *Somme*, an army transport ship, departed San Francisco for Japan, carrying a large cargo of medical supplies, several dozen members of the Army Medical Department, and thirteen field hospitals.[25] Over the next four days, two other US Army transports, the *Merritt* and the *Meigs*, sailed from Manila, carrying dozens of additional army officers, troops, physicians, nurses, and several tons of supplies to the Kantō region.[26] To command the relief operation, the secretary of war detailed Brigadier General Frank McCoy, an assistant to Governor General Leonard Wood in the Philippines, directing him to proceed to the disaster zone.[27]

The third pillar of US foreign disaster aid—the American Red Cross— was equally involved in organizing the nation's official response to the catastrophe. A block away from the State, War, and Navy Building, at the ARC's new national headquarters, discussions about how to finance and administer the relief operation began soon after news of the disaster reached Washington. On September 3, the ARC's vice chair convened an emergency meeting with several high-ranking government officials, including Commerce Secretary Herbert Hoover, the assistant treasury secretary, and Japan's ambassador to the United States.[28] They agreed to allocate an initial $100,000 from the ARC's reserves for Japanese relief, plus an additional $10,000 to assist US citizens in the region.[29]

Hoping to augment these funds, ARC leaders looked to the government for help. They first asked Coolidge to make a special appeal on the ARC's behalf. The president readily complied. In an official proclamation, he called on US citizens to assist "the people of the friendly nation of Japan" by sending

Sailors of USS *Blackhawk*, loading relief supplies at Tsingtao (Qingdao), China, for shipment to Tokyo after the 1923 earthquake and fire.
Image courtesy of Eugene R. O'Brien Collection,
Naval History and Heritage Command.

"all contributions, clearly designated," directly to the American Red Cross.[30] Taking additional steps to ensure the "the utmost coordination and effectiveness" of the ARC's campaign, State Department and White House officials subsequently sent telegrams to hundreds of religious leaders, civic officials, and voluntary organizations throughout the country, requesting that they channel all donations they received through the ARC.[31] Further facilitating the ARC's work, Coolidge ordered the army and navy to put their ships and communication facilities at the organization's disposal.[32] He also asked the US Shipping Board, a federal agency established during the Great War to regulate commercial shipping and operate a merchant fleet, to make its steamers available for transporting ARC relief supplies to Japan.[33]

Much to the satisfaction of ARC leaders and US government officials, these combined appeals and promotional efforts spurred a wildly successful relief campaign. Donors swiftly surpassed the ARC's $5 million fundraising target, going on to contribute a record-setting $11 million over the next several weeks.[34] "Not since the war days," the ARC's vice chair observed, had there been a "response to the Red Cross appeal as the earthquake in Japan has produced."[35]

While the desire to assist both Japanese and American survivors of this immense calamity undoubtedly compelled officials and citizens to act, those planning the relief effort were also keenly aware of the diplomatic stakes involved, seeing in this humanitarian crisis a propitious chance to improve Japanese-American relations. "I felt that we had a great opportunity," Woods later recounted to the State Department, "to break down the suspicion and antagonism against the United States existing in the minds of many Japanese."[36] Spinning this in a more positive way to Japanese ambassador Hanihara Masanao, Secretary of State Charles Evan Hughes explained that the catastrophe "afforded an opportunity to the people of the United States to evidence their friendly and sympathetic feeling for the people of Japan."[37]

High-minded words and lofty rhetoric, of course, could only go so far. If they were to succeed in enhancing the United States' image in Japan—an image tarnished by the government's perceived snubs of the Japanese empire and by widespread American xenophobia and violence toward peoples of Japanese descent—US officials would have to translate their plans and promises into action.

✪ ✪ ✪

In the aftermath of the Great Kantō earthquake, the Coolidge administration and ARC laid the foundations for an extensive humanitarian operation in Japan. Four days after the disaster commenced, the American aid they dispatched started arriving in the Kantō region. Downed telegraph wires and limited radio service made communications between Tokyo and Washington slow and erratic, but on September 5, Ambassador Woods received the State Department's instructions to draw an initial $110,000 in ARC funds and to appoint a committee of trusted US citizens to help administer it.[38] Arriving concurrently with this financial assistance was military support. The first vessel of the Asiatic Fleet's Thirty-Eighth Destroyer Division, the USS *Stewart*, docked in Yokohama's harbor on the morning of September 5, the first foreign battleship to reach the disaster zone.[39]

From there, the US presence in the region swelled rapidly. Over the next few days, more than a dozen US Navy destroyers appeared on the scene, accompanied by several smaller supply and transport ships. They carried not only large cargoes of food, fresh water, and other relief supplies, but also hundreds of US Navy officers and sailors. Several high-ranking military officers arrived in the Kantō region, too, among them Rear Admiral Earl Anderson and Brigadier General Frank McCoy—who, at the instructions of the ARC and War Department, became director general of the American Red Cross Relief Committee.[40] State Department personnel came as well, including the US consul-at-large from Shanghai, Nelson T. Johnson, and the vice-consul dispatched from Kobe.[41] Meanwhile, scores of additional army officers, medical corpsmen, and Red Cross nurses were en route from Manila and San Francisco, due to arrive later in the month.[42] From the US Pacific Coast, and from nearby Kobe, Nagasaki, and Shanghai, more than a dozen US Shipping Board vessels were also heading toward the Kantō region. In their holds were thousands of additional tons of supplies, purchased by the ARC and valued at $3 million.[43]

Within a week of the earthquake, vast quantities of US monetary and material aid, soldiers and sailors, and military and commercial ships had converged on the Kantō region, with still more on the way. Such an immense, highly visible response seemed likely to make just the sort of diplomatic splash US officials desired. And indeed, as Woods reported to the State Department, Japanese officials and civilians appeared "profoundly touched" by the assistance pouring in from the United States.[44]

Yet as Woods and his associates realized, the sudden influx of American aid and military personnel also presented something of a diplomatic dilemma. Several officials sensed an "attitude of suspicion" among their Japanese counterparts, who appeared concerned as to just "what we were going to do."[45] As the embassy's naval attaché reasoned perceptively, Japanese authorities "feared that we would come in the usual wholesale American way, land forces, take over and repair railroads, administer relief and in fact run practically all relief work and as much of the country as possible."[46]

Woods and his colleagues were acutely aware of the risks that landing hundreds of US soldiers and sailors on Japanese soil might entail, recognizing this significant presence of military force could easily foment tensions. As Woods cautioned, a "small incident could turn present happy relations into hostility." If the US government and ARC hoped "to preserve [the] beneficial effect of this action," the ambassador stressed, "it is essential that no friction between Americans and Japanese subsequently develop."[47]

Hoping to circumvent any such problems, Woods took proactive steps to mitigate tensions. First, he issued an official notice to Americans in the region, declaring it "incumbent on all foreigners . . . to maintain a constantly

Headquarters of the American Red Cross Relief Committee in Tokyo, Japan, following the 1923 earthquake and fire.
Image courtesy of American National Red Cross photograph collection, Library of Congress Prints and Photographs Division.

friendly attitude towards all Japanese."[48] Second, the ambassador recommended to the State Department and ARC that all American "relief supplies should be delivered to Japanese authorities on arrival and no American organization should attempt distribution."[49] Further, he counseled, "it would be extremely inadvisable to have any relief expeditions sent from the United States for the purpose of functioning ashore."[50]

Such a hands-off approach represented a major shift in policy. It ran directly counter to the patterns characterizing many recent US foreign disaster relief operations, where ARC leaders and US government officials insisted on maintaining tight control over American-financed relief activities. In light of the tense relations between the United States and Japan, however, Woods maintained that a strategy of restraint was in the US government's best interests politically, affording US officials a chance to show Japanese citizens they regarded them as equals. Favoring this diplomatic objective over whatever potential advantages a US-controlled relief operation might afford, Woods stressed it was crucial to respect Japanese sovereignty and authority.

To Woods's satisfaction, the relevant parties in Washington concurred with his recommendations, agreeing to place their "full confidence in [the] administrative effectiveness of Japanese relief agencies."[51] With the assent of the State, War, and Navy Departments and the ARC, the American Relief Committee adopted Woods's recommendations as official policy.[52] Their stated objective, Anderson explained to his superiors, was "to assist the Japanese in every way possible, to be extremely courteous, to offer them all of our facilities, but not to force upon them in any way our services."[53] Summarizing the policy more concisely, Woods affirmed, "American help stops at shore."[54]

Guided by these principles, McCoy, Woods, Anderson, and their subordinates worked together to assist their Japanese counterparts in responding to the catastrophe. At the Japanese government's request, the crews and ships of the Asiatic Fleet performed a variety of tasks, including evacuating refugees and providing emergency electricity and communications services.[55] While the navy provided logistical assistance, US officials delivered the monetary and material aid they received to Japanese authorities. After turning over the ARC's initial $100,000 to the Japanese Red Cross, Woods transmitted the next infusion of ARC funds—this time to the tune of $1 million—to the Japanese government's Earthquake Relief Bureau.[56] Anderson, meanwhile, ordered the navy sailors under his command and the crews of US Shipping Board vessels to unload relief supplies from their ships then hand them over to Japanese authorities. When the *Merritt* and *Meigs* arrived from the Philippines, bringing to the region more than 150 army officers and enlisted men and millions of dollars' worth of army supplies, McCoy similarly instructed these troops to transfer their cargoes to the Japanese government, stressing, "No armed forces are to be landed in Japan[,] no armed men are to be landed or sent ashore and no arms are to be taken ashore."[57]

This collective American assistance made a tangible contribution to the larger emergency response, helping the Japanese Relief Bureau provide for the basic needs of earthquake survivors more efficiently and effectively. As importantly—at least in US officials' minds—the outpouring of aid appeared to make a positive impression among earthquake survivors. "The Japanese people are, without exception, overwhelmed by the genuine sense of gratitude," the military attaché of the Japanese embassy told the US secretary of war.[58] Japan's Prince Tokugawa Iesato, likewise, declared, "Nothing could be more gratifying and more touching than the promptness and efficiency with which the whole people of the United States have spontaneously responded to the call of human brotherliness."[59] Thanking the White House and State

Department for the "spontaneous and prompt measures" the US government and ARC took, Japanese prime minister Yamamoto affirmed the response had made "a profound impression" in Japan, "drawing still closer the bond of friendship and trust between the two countries."[60]

American officials, in turn, felt confident that their efforts had effected a profound change in Japanese attitudes toward the United States. Anderson, for one, reported that the Asiatic Fleet had "done much to lessen their suspicion by doing away with any possible national ill-will."[61] Concurring with these observations, Woods advised the State Department that he and his associates had "played a tremendous part not only in performing a great work of humanity but in demonstrating to the Japanese a sincerity of friendship that will have far-reaching effects."[62]

By mid-September, as the emergency began to subside, so did Japan's need and desire for extensive American aid. On September 17, Japanese authorities informed McCoy that no additional food supplies were necessary and requested that the ARC cancel all future relief shipments.[63] The following day, Japanese rear admiral Kobayashi Seizō told Anderson that he and his men could depart. While offering his "sincerest and heartfelt thanks to the US Navy for the prompt and invaluable assistance," Kobayashi explained that "as the relief works are getting pretty well in hand, the Department does not think it proper to ask further assistance of the US Asiatic Fleet."[64]

Just as they appreciated the diplomatic importance of surrendering American supplies and oversight to Japanese authorities, Woods and other US officials saw value in complying promptly with the Japanese government's requests. As the acting secretary of state explained to ARC leaders, withdrawing now "would greatly increase the confidence of the Japanese in our disinterested motives" and, he hoped, "have a most beneficial effect upon our relations with Japan."[65]

And so, roughly three weeks after the earthquake, US officials followed the Japanese government's wishes and brought the bulk of their emergency aid operations to a close. At the State Department's request, ARC leaders canceled their outstanding orders for relief supplies. The US Shipping Board issued a general order to the many ships ferrying relief supplies and refugees across the Pacific, instructing their crews to resume normal operations.[66] At this juncture, most of the Asiatic Fleet departed as well, save for three ships that remained in Tokyo Bay to provide ongoing logistical assistance.[67] Recognizing the symbolic value of this punctual exit, Woods observed that the navy "arrived at the psychological time and it left at the psychological time, insofar as the necessities and sentiment of the Japanese are concerned."[68]

And yet, although US officials acceded to Japanese requests and wrapped up most relief activities, US involvement in the Kantō region had not ended entirely. Rather, much as it had in Italy in 1909 and Guatemala in 1918, the US aid operation took on a new form. Beginning in mid-September and continuing for the next several months, US officials embarked on several large-scale recovery and reconstruction activities in Tokyo and Yokohama. While intended to assist Japanese disaster survivors in the longer-term, these projects were also designed as lasting testaments to US humanitarian assistance and US amity.

Japanese authorities, it bears noting, *invited* this prolonged American participation in the region's recovery. When they asked the Asiatic Fleet to withdraw, Japanese government officials simultaneously requested that US Army personnel stay on to assist with the restoration of hospital services.[69] The earthquake had crippled the region's medical facilities, destroying two-thirds of Tokyo's hospitals and leaving Yokohama without a single functioning healthcare institution. Anxious to address this problem, the Japanese government invited McCoy to remain in the Kantō region to supervise the construction of temporary medical facilities.[70] As part of their respective relief shipments, the US Army and ARC had delivered medical and surgical equipment, tents and cots, and field hospitals to Japan. They had also dispatched dozens of medical professionals from Manila and San Francisco, including officers and enlisted men of the US Army Medical Department, US Army nurses, and nurses of the ARC's Philippines chapter. These men and women were to extend the American humanitarian operation in Japan for another month—this time on the ground in Tokyo and Yokohama.

Throughout September and into mid-October, working together under McCoy's command, teams of American and Filipino medical professionals installed nearly 100 temporary medical facilities throughout the devastated region. These projects included an enormous 1,000-bed base hospital in Tokyo, a 432-bed evacuation hospital in Yokohama, roughly a dozen smaller field hospitals, and eighty-two outpatient dispensaries.[71] Once these facilities were constructed, McCoy handed them over to the Japanese government. On October 13, he turned over the last major project, the 1,000-bed Bei-Hi (American-Philippine) Hospital in Tokyo, to Japanese authorities. Though the ARC agreed to provide supplies and equipment to these hospitals and medical facilities for an additional six months, the institutions themselves were now under local control.[72]

To commemorate this transfer of authority, Woods and McCoy joined Japan's foreign minister, the Japanese ministers of war and navy, and other high-ranking Japanese political and military leaders in a formal ceremony, an event laden with diplomatic symbolism. Together, the men lowered the American flag flying over the hospital, leaving only the Japanese flag still raised above it.[73] In a speech thanking the United States for "this visible token of sympathy and friendship," Japanese officials proclaimed the Bei-Hi Hospital "a fitting monument to the spirit of humanity and universal brotherhood which is the soul of the American nation."[74] In a further symbolic gesture, Japanese Red Cross officials erected a large sign at the hospital's entrance, identifying it as a "gift of the United States" and reminding all who entered of "the sympathy and friendship thus shown us by the American people."[75]

In mid-October, with these hospital projects completed, the US Army's involvement in Tokyo and Yokohama concluded. On October 14, army and ARC medical personnel departed, heading back to Manila and, for some, the mainland United States.[76] McCoy left Japan five days later, sailing for the Philippines to resume his normal responsibilities. At this juncture, Ambassador Cyrus Woods, too, decided that it was time for a much-needed vacation. Toward the end of the month, he left for an extended leave of absence in the United States.[77]

American involvement in the recovery effort, however, was *still* not fully complete. One final project remained. In early November, ARC leaders in Washington notified the US embassy in Tokyo that $4.5 million remained in the ARC's relief fund for Japan.[78] Reasoning that the US public gave this money to help Japanese earthquake survivors, ARC leaders felt obligated to spend the funds in Japan, as donors intended. How to use this money, though, remained an open question.

In Tokyo, the US embassy's chargé d'affaires, Jefferson Caffery, took up the matter with Japanese government authorities and Japanese Red Cross leaders, asking how they preferred to see the money spent. Although some Japanese officials hoped to use the funds to purchase additional supplies, such as winter clothes and lumber for rebuilding homes, they eventually agreed on a grander project: the construction and maintenance of a new charity relief hospital in Tokyo. Welcoming the opportunity to erect a "permanent relief memorial" in Japan, ARC leaders and US government officials quickly gave their assent, authorizing Caffery to move forward with the project.[79]

With this move, the US disaster assistance operation in Japan turned a decisive corner, transitioning from short-term relief and recovery aid into a large-scale rebuilding project, designed to endure for decades to come.

While Japanese authorities had their own reasons for accepting this gift, US officials intended the charity hospital in Tokyo to serve as nothing less than a perpetual, functional monument to American aid. As the ARC's vice chair crowed, "The new hospital will be a permanent credit to the United States and the hundreds of thousands of Americans who contributed."[80]

During the next several months, Caffery and his military attaché worked closely with Japanese government officials and Japanese Red Cross leaders in Tokyo, and with ARC leaders in Washington, to draw up plans for the institution. Ultimately, they agreed to allocate one-third of the remaining ARC funds, or $1.5 million, to the "construction of a practical, up-to-date earthquake proof hospital."[81] Leaders of the ARC committed another third of the funds to endow the new building, to be named the Doai (Fraternity) Memorial Hospital. They transferred the final $1.5 million to the Japanese government for "direct social service work among [the] destitute refugee population."[82] In March 1924, having reached a final agreement on these plans, Caffery handed over the ARC's cheque to Japanese authorities.[83] Though construction of the Doai Memorial Hospital did not commence for another year, the ARC and US government had secured a durable and visible monument to their humanitarian efforts in Japan.[84]

With plans for the permanent hospital set in motion, the official US assistance operation for Japan finally concluded. On June 30, 1924, ten months after the Great Kantō earthquake, the ARC closed out its Japanese Relief Account. By this point, the organization had collected and spent more than $11 million on this humanitarian operation, the largest expenditure it had ever made for a foreign "natural" disaster. Although a small portion of these funds went toward US citizens living in Japan, the vast majority were reserved for the relief and recovery of Japanese citizens.[85] The US military, too, had committed unprecedented resources to this foreign disaster relief operation. Both the army and navy deployed hundreds of soldiers, sailors, officers, and medical personnel to the Kantō region.[86] The War Department contributed $6 million worth of supplies from army stores, and the Navy Department another $2 million; Congress eventually reimbursed the military for these expenses, through a belated relief appropriation in 1925.[87] Woods, Caffery, Burnett, and other diplomatic and consular officials, meanwhile, devoted countless hours to planning and overseeing American relief activities in Tokyo, Yokohama, and beyond.

All told, the US government and ARC provided more than $20 million worth of cash and supplies to Japan, more than all other nations combined.[88]

The episode represented, if nothing else, an extraordinary humanitarian undertaking.

✪ ✪ ✪

But of course, it did represent something more. More than a major disaster relief operation, the episode marked a critical chapter in US-Japanese relations. As US officials first embarked on this massive relief effort, in early September 1923, they understood it as a golden opportunity—a chance not only to help Japanese earthquake survivors but also to dramatically improve relations between Japan and the United States. Had they succeeded in their efforts "to break down the suspicion and antagonism against the United States," as Woods and his colleagues hoped?[89] Or was a single US disaster assistance operation insufficient to overcome long-standing grievances, prejudices, and diplomatic tensions between the two nations?

For a moment in late 1923, it appeared that the outpouring of US assistance had indeed transformed US-Japanese relations in positive and concrete ways. Throughout the monthslong relief effort, Japanese authorities heaped considerable praise on the actions Americans took in and for the Kantō region, lauding the "nation-wide understanding and heartfelt sympathy" US citizens showed toward Japan and "the admirable work which the American relief mission accomplished."[90] Beyond expressing their immediate gratitude, many Japanese officials predicted that American aid would have enduring political consequences. "The zealous relief work" Americans had conducted, Japan's minister of foreign affairs told Woods, had "left a lasting impression" throughout the country. Making a similar point, Ambassador Hanihara told Secretary of State Hughes that Americans left "behind them a glow of admiration and gratitude that will live for years to come."[91]

It was not only Japanese government officials who voiced appreciation for US aid; many ordinary Japanese citizens did as well. Newspaper editorialists celebrated the "benevolence and charity of the United States . . . and the magnanimity of Americans," declaring that US aid had "accomplished a success far greater than that accomplished by a stroke of diplomacy for the Japan-American friendship."[92] In letters to Woods and his associates, residents of Tokyo and Yokohama offered their "heartfelt thanks" for American aid and applauded the "manifestation of humanity and philanthropy for which America stands so valiantly."[93] Others demonstrated their gratitude in action rather than in words. When Woods departed Tokyo to begin his leave of absence, the US naval attaché observed, "The ovation which he received from the Japanese people . . . was unprecedented in the history of Japan."[94]

On the receiving end of these accolades, the US diplomatic and military officials who executed the relief effort expressed considerable pride at what they had accomplished, extolling the beneficial results—both humanitarian and political—they felt their aid had achieved. The US humanitarian response had "so touched the hearts of the Japanese," McCoy informed his superiors in the War Department, "that Americans long-resident in Japan note a distinctly new attitude toward all Americans on the part of all classes of the Japanese."[95] Echoing his colleague, Woods told the State Department that wherever he went, "I hear the opinion expressed that our countries at last understand each other, and that we are united in ties of friendship more strongly than any paper treaty could possibly establish."[96]

As late as January 1924, US officials continued to celebrate the diplomatic volte-face. "Never in the history of Japanese-American relations," observed the US embassy's military attaché, "has there been such an era of gratitude and friendly feeling as the Japanese feel now for America."[97] Writing that same month, an ARC publicist claimed, "America's action in this crisis had done more to strengthen the friendship between Japan and America than all the treaties and Washington conferences could ever have achieved."[98]

And yet, for all the celebratory rhetoric, for all the optimistic words, aspirations, and fanfare about the future of US-Japanese relations, the newfound friendship between the two nations was in the end neither as deep nor as lasting as many American and Japanese observers hoped.[99] To be sure, many of the positive emotions and sentiments that Japanese and American citizens felt for one another were undoubtedly genuine and sincere. But beneath the surface goodwill and harmony, a pair of major diplomatic fault lines remained. Together, they belied—and ultimately undermined—newfound feelings of Japanese-American amity.

The first of these points of fracture lay in the attitudes and assumptions many US officials held behind closed doors. Despite their public assertions of respect and goodwill, many of the individuals who carried out the American relief effort in Japan harbored deep-seated racial, cultural, and nationalist prejudices toward Japanese people, views they freely expressed in personal correspondence. As their more private missives revealed, US officials' outward declarations of goodwill toward Japan proved in some cases superficial and in others disingenuous, with lofty rhetoric masking underlying feelings of superiority and animus.

Erle Dickover, the US consul in Kobe, was one of these officials. In early October 1923, having concluded his relief work for American refugees who evacuated to Kobe, Dickover redirected his energies toward assisting US businessmen who lost property in the disaster. Over the next few months, he lobbied the ARC and State Department to provide financial aid to these individuals, only to have his requests shot down repeatedly.[100] Dickover grew "bitterly disappointed" by these multiple rejections, incensed "that the United States should be willing to raise a huge sum . . . to help an alien people" while refusing "to help out its own nationals." Although Japanese authorities informed US businessmen they could apply for aid through Japanese recovery agencies, Dickover deemed this option "utterly repugnant and impossible." Taking aid from Japan, he lectured the State Department, "would greatly weaken the prestige of the white man in the Orient." In times of crisis, he stressed, "a small community of white people, set down amidst hordes of people of an alien race, must inevitably cling together."[101]

Dickover's superiors, to their credit, condemned the consul for his "querulous tone" and his "most unfortunate and undignified" words, remaining steadfast in their opposition to his proposal.[102] Even so, as his disparaging remarks illustrate, the empathy US officials widely professed was sometimes woefully insincere.

Dickover, moreover, was hardly alone in holding such derogatory views. Many of his American associates conveyed their own criticisms about the Japanese men and women they assisted, regularly deploying orientalist tropes and racist stereotypes within these critiques. Authors of several US Navy intelligence reports, for instance, identified a litany of "weaknesses" inherent in "the Japanese character"—traits, they claimed, that hindered the effectiveness of the Japanese government's relief efforts. In one such report, the US embassy's naval attaché disparaged Japanese officials for their "harmful secretiveness" and their "distrust of all foreign help," arguing that these qualities led to "maddening slowness and ineffectiveness" in the Japanese government's disaster response. The only explanation for such behaviors, he concluded, "must be based on their psychology."[103]

The commander of the Forty-Fifth Destroyer Division offered a similar analysis. Explaining that the Japanese had adopted "Western ideas" only recently, he judged it only logical they should "revert to their natural state of civilization" as soon as the earthquake struck. Japanese men and women, he explained, were "capable of being stripped of this civilization in whole or in part by any catastrophic event." As a result of this "reversion to type," he reasoned, Japanese officials responded to the disaster with a "fatalistic stupor,

...an incomprehensible unresponsiveness, [and] a startling lack of initiative." Worse still, so far as US-Japanese relations were concerned, they displayed "an utter inability to analyze foreign motives or to accept graciously and spontaneously the unqualified offers of immediate foreign aid."[104]

Like these navy officers, members of the US Army frequently pointed out flaws they believed innate to Japanese people. Upon arrival in Japan, for instance, the chief of the US Army medical corps warned his men of "the extreme danger of contracting venereal disease," cautioning, "These diseases, especially Syphilis, are supposedly more virulent when contracted from oriental sources."[105] An army lieutenant, assigned to build temporary hospitals in Yokohama, directed his indignation toward the Japanese laborers who worked alongside him. "They seem to lack any initiative whatsoever" and "frequent rests were indulged in to smoke," he complained, concluding, "The Japanese laborer is a willing beast of burden, but requires an energetic leader, and must be watched."[106] Another army officer, meanwhile, criticized the Japanese military for failing to "jump into this relief work with the promptness, energy and initiative" he expected. Such shortcomings, he reasoned, were due to "the standpoint of the oriental psychology which is different from ours particularly in the matter of sentiment and fatalistic conceptions."[107] The kind words and flattery US officials expressed to Japanese faces, in short, conflicted markedly with the caustic and insulting remarks they made behind Japanese backs.

Compounding these entrenched racial and cultural prejudices was a second diplomatic fracture point: even the sincerest feelings of gratitude, appreciation, and compassion that arose in the aftermath of the catastrophe ultimately proved fleeting, quickly overshadowed by subsequent US policies and actions. Although Woods and his colleagues initially expressed "no doubt" that their aid would "have a marked and lasting beneficial effect on the relations between the United States and Japan," such confident projections steadily gave way in the year or so after the earthquake, replaced by increasingly pessimistic assessments about the future of Japanese-American relations.[108]

Already by late 1923, chargé d'affaires Jefferson Caffery noticed a change of heart. Although Japanese officials remained "really appreciative of America's efforts," he reported to the State Department, some Japanese politicians seemed to feel that "this gratitude and the resultant favorable sentiment to America may be carried too far." The reason, Caffery explained, was that anti-Japanese policies in the United States—including both the "California land question and [the] immigration question"—had once again surfaced in the news. Renewed anger over these issues in Japan, he warned, was rapidly undermining the better relations that he and his colleagues had so recently nurtured.[109]

Initially, Caffery held out hope that the Doai Memorial Hospital project might "counteract [the] extraordinary recent change of Japanese popular sentiment toward the United States."[110] Yet from that point forward, the diplomatic situation only deteriorated further. Writing to the State Department in early 1924, Caffery reported that "agitation over the recent developments in the Japanese question in the United States is continually increasing."[111] Specifically, he cited "recent Supreme Court decisions" that upheld discriminatory laws barring Japanese immigrants from owning agricultural land. These rulings, coupled with proposed congressional legislation "aiming at Japanese exclusion," were fueling a swelling wave of Japanese animus. Were it not for the goodwill US officials generated through disaster assistance, Caffery advised, there would likely be "a much more severe outbreak of feeling over the news . . . than is actually the case."[112]

Try as they might, Caffery and his colleagues could do little to reverse the rising tide of discontent over blatantly anti-Japanese rhetoric, policies, and actions in the United States. Just a few months later, in May 1924, the passage of the Johnson-Reed Immigration Act—legislation effectively barring all further immigration from Japan—struck another decisive blow to US-Japanese relations, nullifying what remained of the goodwill US officials had achieved the previous year.

Reflecting on the abrupt swing in Japanese public opinion in September 1924, at the one-year anniversary of the Great Kantō earthquake, the head of the new ARC chapter in Tokyo, D. H. Blake, expressed deep regret. Moreover, he chastised US policymakers for squandering such an auspicious diplomatic opportunity. The "truly great work of the American people," he declared, "was inspired solely by sympathy for a friendly nation in the hour of need, and by humanitarian principles which have always appealed to our countrymen." Now, he lamented, the positive feelings American aid generated were "greatly discounted and to a large extent rendered ineffective because of the action of our last Congress in passing the Japanese Exclusion Act." Although Blake held out hope that "time will heal the breach that has been made in the friendly relations of the two peoples," the buoyant forecasts that US officials made just one year earlier were fast becoming a distant memory.[113]

Perhaps more than any previous foreign disaster assistance effort, the US response to the Great Kantō earthquake demonstrated both the enormous potential and the immense challenges of American catastrophic diplomacy. As the government moved on from the Great War and into the postwar era, foreign disaster assistance remained a potentially valuable tool of US foreign relations. Yet as events in Japan starkly revealed, its beneficial influence and effects could only go so far.

7

The Sun Never Sets

1924–1931

On March 4, 1929, a man best known for his humanitarian pursuits became president of the United States and commander-in-chief of its armed forces. In accordance with established custom, he also assumed the honorary presidency of the American Red Cross.

By the time he arrived in the White House, Herbert Hoover had earned an international reputation for organizing large-scale relief efforts at home and abroad, assisting victims of both conflict and disaster. During the Great War era, he headed two major aid agencies: the wartime Commission for Relief in Belgium and the postwar American Relief Administration. In the 1920s, while secretary of commerce, Hoover also helped lead the ARC. Serving as a member of its governing central committee throughout the decade, he joined in planning the organization's responses to dozens of domestic and foreign crises, including the 1923 earthquake in Japan. Hoover burnished his humanitarian credentials further still in 1927, when he became director of relief efforts for the Great Mississippi floods, a catastrophe of historic proportions within the United States. Although Hoover and his actions were not without critics, his career demonstrates an important truth: by the 1920s, American politics, American diplomacy, and American humanitarianism had become intimately entwined.

The 1920s and early 1930s are popularly remembered as a period of American retrenchment from world affairs. The US government's sustained program of foreign disaster assistance during these years, however, offers ample evidence to the contrary. Throughout Calvin Coolidge's presidency, from 1924 to early 1929, the government, military, and American Red Cross regularly contributed disaster aid to other countries. Even the arrival of the Great Depression in 1929 failed to curtail these humanitarian commitments. During Herbert Hoover's time in the White House, in fact, US officials undertook several major foreign disaster assistance operations, dispatching aid abroad despite the worsening financial crisis at home. Across the 1920s and early 1930s, foreign disaster assistance remained a routine part of US foreign relations, one of the many ways the government and its partners continued engaging with the world.

Among the most notable US disaster aid efforts during these years were those following a hurricane in the Dominican Republic in 1930 and an earthquake in Nicaragua in 1931. In both countries, US officials swiftly intervened, sending to the scene not only substantial amounts of cash and relief supplies but also considerable numbers of American personnel. In each country, these diplomats, military officers, and ARC experts secured considerable powers, over not only American relief activities but the entire disaster response. Remaining involved in both nations for months after the disasters commenced, US officials experimented with a broad suite of recovery projects, which once again muddled the distinctions between humanitarian relief and development assistance. Reflecting and exacerbating hemispheric power disparities, these episodes formed part of a long history of US involvement—humanitarian and otherwise, welcome or not—in the affairs of the Caribbean Basin.

✪   ✪   ✪

As they tuned their radio dials in 1925, American listeners may have chanced upon a broadcast by an influential figure in US international affairs: Ernest Bicknell. A leading figure in the ARC since 1908, Bicknell now directed the organization's Department of Insular and Foreign Operations, a division established after the Great War to oversee ARC activities in other countries and US territories. Addressing audiences about the ARC's work in these "far places," Bicknell advanced a rather imperial sounding claim. "It may be truthfully said," he declared, "that the sun never sets on the work of the American Red Cross."[1]

This was, perhaps, a slight exaggeration. Still, the gist of Bicknell's remark was reasonably accurate. What he was describing was the routinization and global reach of US foreign aid—and, by extension, of American catastrophic diplomacy. Having taken shape in the years before the Great War, the three-pillared system of US foreign disaster assistance was firmly ensconced by the late 1920s. When catastrophes occurred in other nations, it was now customary, even expected, for the ARC and State Department to spearhead a response. Cooperating with them to conduct these efforts were US diplomats and consuls—now part of a unified agency, the US Foreign Service—and military servicemen throughout the world.

Across the latter part of the 1920s, ARC leaders, in close concert with the State Department, made financial contributions to many nations experiencing "great and sudden emergencies."[2] ARC funds went to Armenia, Colombia, and Costa Rica after earthquakes occurred in those countries in 1924. They arrived in the Netherlands and Belgium when a cyclone struck in 1925, in Mexico and the Kingdom of Slovenes, Croats, and Serbs following major floods in 1926, and in the Azores and Albania after earthquakes that same year. In 1927, ARC aid went to Mandatory Palestine for earthquake relief and Switzerland for flood assistance. And in 1928, the organization sent contributions to survivors of a hurricane in Haiti, floods in Latvia, and earthquakes in Chile, Bulgaria, Greece, and Turkey.[3]

The majority of these relief allocations were relatively small in scale. Most were in the range of $5,000, though some reached a more substantial $15,000, $20,000, or $25,000. A few catastrophes in these years, however, prompted more liberal contributions from the ARC. When a major earthquake hit central Japan in 1927, killing more than 3,000 people, ARC leaders wired $50,000 to that country.[4] They allocated twice that much to China in 1924, after severe flooding affected 10 million people in nine northern Chinese provinces.[5] They made an equivalent contribution to Cuba in 1926, appropriating $100,000 after a hurricane devastated both the capital, Havana, and the Isle of Pines.[6] Like their predecessors, members of the Coolidge administration depended on the ARC to bankroll much of the United States' official foreign disaster assistance. They regarded the ARC as the nation's official voluntary aid organization, "the appointed agent of the people."[7]

On occasion, however, hoping to deliver more than the ARC could comfortably provide, federal officials also partnered with other voluntary organizations and private citizens to send additional aid abroad. After the 1926 hurricane in Cuba, desiring to supplement the ARC's allocation with even more "substantial contributions," State Department officials invited Dwight Morrow, a partner with J. P. Morgan, to chair a Cuban Relief Committee, tasked with raising additional funds.[8] The committee ultimately brought in another $125,000 for hurricane relief, sending these funds to Havana through the ARC's channels.[9] Following the earthquakes in Bulgaria, Greece, and Turkey two years later, the US ambassador in Athens urged the philanthropy Near East Relief to contribute assistance, explaining it "would certainly cement the good relations" between those countries and the United States.[10] Although Near East Relief's leaders agreed to provide some aid, they simultaneously took the opportunity to remind government officials of the ARC's primacy in this humanitarian field. "The Red Cross," the organization's acting director observed, "is the one great organization prepared to meet emergencies such as the present Corinth disaster." Providing foreign disaster relief, he lectured US diplomats, "is really the entire responsibility of the Red Cross."[11]

Once funding was secured, the question turned to the administration of American relief. In a few instances, a high-ranking member of the ARC's staff traveled to the affected country to assist local disaster aid operations, as had happened after the Guatemala City earthquakes in 1918. Following the 1926 hurricane in Cuba, for example, ARC leaders sent their national director of disaster relief, Henry Baker, to Havana as a special representative. Invited there by President Gerardo Machado to advise the Cuban National Reconstruction Commission, Baker remained on the island for two weeks, working with Cuban officials to prepare an official rehabilitation plan.[12] Although Baker ostensibly went to Cuba to cooperate with the Machado government in its disaster response, his superiors at the ARC had privately given him a different mission: to "'take charge' of the work" in Cuba.[13] Doubting Cuban officials' ability to administer disaster assistance on their own—a suspicion the US minister to Cuba reinforced in confidential missives to the State Department— ARC leaders embraced the opportunity to steer the country's recovery.[14]

Although ARC advisers were more likely to travel abroad in the 1920s than in previous years, staffing and transportation limitations still made this practice fairly infrequent. Often, ARC leaders delegated someone else to administer funds and distribute supplies. When they judged the National Red Cross Society of a disaster-stricken country "prepared to organize and direct the relief operation with efficiency and dispatch," ARC leaders simply transferred funds to that sister organization.[15] This list of trusted partners had grown significantly since the early twentieth century, thanks to the ongoing expansion of the International Red Cross and Red Crescent Movement. By the 1920s, ARC leaders regarded the Red Cross Societies of Japan, most European nations and empires, some South American nations, and a handful of other countries as sufficiently organized, professional, and competent to administer American aid themselves.

In other instances, when they lacked confidence in either the Red Cross Society or the government of a disaster-stricken nation—as they did throughout the Caribbean, Central America, and China, especially—ARC leaders instead wired funds directly to US ambassadors, ministers, and consuls in the region. Insisting that ARC contributions "should be safeguarded" by US Foreign Service Officers and their trusted American associates, ARC leaders relied on these government officials to "serve as the outposts of the Red Cross," counting on their "cooperation in supporting and facilitating our work in foreign countries."[16] By and large, US Foreign Service personnel took the job of safeguarding the ARC's disaster relief funds seriously, viewing it as their duty to ensure ARC aid reached only deserving relief recipients. Determining who genuinely deserved US aid, however, remained as subjective a process as it ever had been, indelibly shaped by officials' cultural assumptions, class positions, and racial prejudices.

While relying on US Foreign Service Officers to administer relief, the State Department and ARC also depended on the US Armed Forces, the third pillar of US foreign disaster aid, to lend supporting assistance. During the late 1920s, military personnel responded to several disasters in the Caribbean Basin, humanitarian operations that reflected the United States' imperial geography in this region. When a destructive hurricane struck Haiti, then under US military occupation, in 1928, officers in Port-au-Prince detailed two US marines, five navy sailors, and three members of the US-trained Garde d'Haiti to the disaster zone to deliver food and medical care.[17] The forces staged an even larger relief operation in Cuba following the hurricane of October 1926. The US Navy sent a destroyer and a cruiser from its base at Guantánamo Bay, while the US Coast Guard detailed a cutter and several patrol boats from Key West. Together, these ships brought several dozen army and navy doctors, surgeons, nurses, and medical corpsmen to the Isle of Pines—an island officially under Cuban jurisdiction since 1925 but that also housed a sizable concentration of US settlers, landowners, and businesses.

In this quasi-colonial space, the US military treated injuries and distributed food and clothing among the island's Cuban and American residents.[18]

Whether it took the form of a monetary donation or a physical intervention, whether limited in scale or more extensive in scope, foreign disaster assistance represented a valuable and flexible instrument of US diplomacy. Speaking to American audiences once again in 1928, this time about "the world-wide purposes and activities of the Red Cross," Ernest Bicknell lauded foreign relief for forging "kindlier and closer relations between the countries" during the 1920s. Through disaster aid and other humanitarian assistance "we are creating bonds of friendship," he promised listeners, "which supplement and strengthen the attempts of our country to cultivate sentiments of peace with all nations."[19]

Echoing Bicknell, many government officials praised disaster assistance for its "favorable influence" on US relations with other countries.[20] Following floods in Nayarit, Mexico, in 1926, for instance, US consul William Blocker and his associates described their relief work as a means to counteract prevailing Mexican nationalism and anti-Americanism. Flood aid, they reasoned, stood to "lessen the feeling . . . against the American people who have large property interests in this state" while compelling Mexican officials to rethink their policy of "expropriating American property whenever possible."[21] State Department officials attributed comparable diplomatic power to the hurricane aid they sent to Haiti in 1928. Eager to assuage Haitian anger over the now thirteen-year US military occupation, they hoped that humanitarian assistance "would go a long way toward helping a better understanding and relationship between the Haitian population and the American people."[22]

In other instances, government officials viewed disaster aid as a means for the United States to outshine its geopolitical rivals in the global battle for hearts and minds. The State Department's motivation to offer flood relief to China in 1924, for example, stemmed in large part from the knowledge that the Soviet Red Cross had already contributed 20,000 rubles to the cause. In light of "Soviet efforts to cultivate Chinese friendship and simultaneously inflame Chinese animosity against the West," one Foreign Service Officer advised, the US government and ARC should stage their own, generous response to Chinese distress.[23]

Throughout the mid- to late 1920s, adhering to patterns now a quarter-century in the making, foreign disaster assistance remained a consistent element of US international relations. Repudiating the myth of US isolationism in the 1920s, the government, military, and ARC undertook international disaster relief operations all over the world during this decade, with a frequency unimaginable just a generation prior. Foreign disaster aid was now a common and rather conventional element of US diplomacy and statecraft. For the architects and agents of American foreign policy, it had become routine.

As the 1920s drew to a close, a trio of successive events seemed poised to transform the character of American catastrophic diplomacy in significant— yet countervailing—ways. The first, the election of Herbert Hoover in November 1928, appeared highly auspicious for the future of US foreign disaster aid. The arrival of the "Great Humanitarian" to the White House in March 1929, many contemporaries imagined, would surely be conducive to a robust US foreign disaster aid program. Frustrating this possibility, however, was a second set of pivotal events: the onset of the Great Depression in late 1929 and the beginning of a severe, prolonged period of domestic drought the following year. The financial crash and ensuing economic crisis, together with the ecological and social devastation of the Dust Bowl, threatened to rein in US international assistance considerably. Facing reduced donations from the American public and increased demands for relief domestically, ARC leaders began wondering whether they could sustain the financial commitments that had fueled humanitarian operations during the 1920s. So, too, did their partners in the Hoover administration.

But although the Great Depression would have some negative bearing on US foreign disaster aid, its consequences were not so dire as many officials feared, at least not initially. During Hoover's four years in office, the ARC continued making small foreign disaster aid contributions on a consistent basis. In 1929 alone, ARC leaders and their government partners provided aid to survivors of earthquakes in Albania, Persia, and Venezuela, floods in Latvia and Turkey, and a tsunami in Newfoundland. Between 1930 and 1932, as the Great Depression reached its nadir in the United States, the ARC allocated disaster relief to multiple countries, aiding survivors of floods in France and Turkey, hurricanes in British Honduras and Cuba, and earthquakes in Albania, Greece, Mexico, New Zealand, and Persia.[24] Even amid the most devastating economic catastrophe the United States had ever experienced, disaster assistance remained a fixture of US foreign affairs.

In these same years, moreover, US officials administered a trio of major foreign disaster operations, responding to a hurricane in the Dominican Republic in 1930, an earthquake in Nicaragua in 1931, and extensive flooding in China later that year. These three humanitarian operations shared some important commonalities. All occurred in regions of critical strategic and diplomatic interest to the United States. Each also served as a crucible for testing new humanitarian techniques, approaches that became increasingly commonplace in foreign disaster aid operations in subsequent years. And finally, these three humanitarian operations ranked as some of the most extensive and far-reaching foreign disaster relief efforts the US government, US military, and ARC had together administered. All of this, remarkably, occurred during some of the worst years of the Great Depression and Dust Bowl, suggesting just how deeply entrenched disaster aid had become in US foreign affairs.

The remainder of this chapter examines the first two of these aid efforts, offering a comparative analysis of the US responses to the 1930 hurricane in the Dominican Republic and the 1931 earthquake in Nicaragua. The following chapter opens with a discussion of the third, the US response to the 1931 Yangzi-Huai (Yangtze-Hwai) floods in China.

⭐ ⭐ ⭐

On September 3, 1930, a powerful hurricane made landfall near Santo Domingo, the capital of the Dominican Republic. The San Zenon hurricane, as contemporaries named it, left as many as 6,000 dead and injured thousands more. The storm-force winds leveled most of the city, leaving 29,000 people without shelter.[25] Less than seven months later, on March 31, 1931, a second major catastrophe visited the Caribbean Basin, this one triggered by an earthquake in Nicaragua. The quake and resulting fires destroyed 90 percent of Managua, the country's capital, rendering 40,000 homeless. More than 1,000 people lost their lives, while 7,000 more suffered serious injuries.[26]

Historic in their proportions, the 1930 San Zenon hurricane and 1931 Managua earthquake occurred on opposite corners of the Caribbean Basin, a region that was long a focal point of American diplomatic, economic, and strategic interest and interference. As crises gripped first Santo Domingo and then Managua, US officials became deeply involved, organizing major humanitarian responses in both cities. While these aid operations shared many commonalities, their contours were also shaped by each country's internal politics and by their respective relations with the United States.

Since the early twentieth century, the US government had intervened repeatedly in the affairs of the Dominican Republic. Most notably, US officials seized control of Dominican customs house revenues in 1905 and then, from 1916 to 1924, occupied and administered the country under direct military rule.[27] When American forces finally withdrew, they vested considerable power in a US Marine–trained constabulary force, the Guardia Nacional, and one of its chief officers, Rafael Trujillo, trusting them to maintain political order and economic stability. By the time of the San Zenon hurricane six years later, Trujillo had gone on to consolidate significant control over the Dominican Republic, having assumed the presidency (through questionable means) just weeks before the storm struck.[28]

While the political circumstances in Managua differed from those in Santo Domingo, many noteworthy parallels existed. As they had in the Dominican Republic, the US government and military intervened in Nicaragua multiple times during the first three decades of the twentieth century. Though a democratically elected president, José María Moncada, governed the country at the time of the disaster, Nicaragua was simultaneously under US military occupation and had been, aside from a brief hiatus, since 1912.[29] Although the Hoover administration had recently announced its intention to withdraw from the country, roughly 1,000 US Marines remained in Nicaragua in March 1931, part of a controversial, multipronged mission to maintain political order, recruit and train a Guardia Nacional, and combat antigovernment, anti-US rebels led by Augusto César Sandino.[30]

The US government thus had a long, fraught history of interfering in Dominican and Nicaraguan political, economic, and military matters. In 1930 and 1931, American officials intervened in the two nations' humanitarian affairs as well. As both a visible manifestation of US sympathy and a means for restoring order and stability, these disaster relief operations stood to advance multiple US foreign policy objectives in the Dominican Republic and Nicaragua. Assuming all went well, officials anticipated, they could also help improve the United States' tarnished image throughout the Caribbean Basin.

After the San Zenon hurricane hit the Dominican Republic, Trujillo quickly declared martial law. He then took command of the Dominican Red Cross, ousting its existing leadership, and began organizing emergency relief activities under its auspices.[31] But Trujillo did not act entirely on his own. When the US minister, Charles Curtis, offered to call for outside assistance, Trujillo quickly gave his consent, allowing in both US material aid and personnel.[32] In Managua, although the process differed, the end results were largely the same. On the day of the earthquake, both Moncada and the US minister, Matthew Hanna, were away from the city for the Easter Holy Week. In their absence, the American commander of the Guardia Nacional declared martial law, proclaiming broad powers for US Marines and Guardia forces. Returning to Managua the following morning, Moncada gave his blessings to this decision and agreed to additional US aid.[33]

In the US capital, government and military officials and ARC leaders also reacted to the two catastrophes in similar ways, adhering to customary protocol. As news of the crises reached Washington, the system of US foreign disaster assistance quickly mobilized. Following each calamity, ARC leaders convened emergency meetings with President Hoover and with members of the State, War, and Navy Departments to plan their collective responses.[34] Following the San Zenon hurricane, US officials instructed Curtis, the US minister, to draw funds from the ARC's treasury.[35] After the Managua earthquake, the ARC similarly entrusted monetary contributions to Hanna at the American legation.[36] Eventually, the ARC allocated more than $222,000 for relief and recovery in the Dominican Republic and some $128,000 for Nicaragua, among the organization's largest foreign disaster relief appropriations in recent years.[37]

Supplementing the ARC's monetary aid, government officials deployed American military personnel to the Dominican Republic and Nicaragua. Facilitating these actions were the military's extensive footprint and ongoing occupations in the region, reflecting the United States' imbricated imperial and humanitarian geographies in the Caribbean Basin. To Santo Domingo, they ordered personnel of the US Army, Marines, and Navy, sending them there from Puerto Rico, the US Virgin Islands, US-occupied Haiti, and the continental United States.[38] In Managua, officers and troops of the Second Marine Brigade, already on the scene, began organizing relief activities immediately. Joining them a few hours later was a detachment of US Army Engineers, coming from nearby Grenada. Additional navy and army detachments arrived over the next few days, detailed to Managua from US bases in the Panama Canal Zone, Guantánamo Bay, and San Diego.[39]

Expediting the arrival of military personnel and supplies to both nations was a relatively new technology: aircraft. The day after the San Zenon hurricane, a Pan-American Airways "airship" delivered two US Army officers from Puerto Rico to Santo Domingo, together with a cargo of relief supplies.[40] The commander of the US Marine Corps in Haiti, meanwhile, dispatched several marine pilots to conduct aerial surveys of the disaster zone. The commander of the US Navy Medical Corps in Haiti also flew in from Port-au-Prince to assist with medical needs on the ground. And from Norfolk and Hampton Roads, Virginia, came three navy planes, carrying thousands of pounds of vaccines and medical supplies.[41]

Seven months later, Nicaraguans witnessed a comparable display of US air power after the disaster in Managua. Within twenty-four hours of the earthquake, two planes arrived from the US naval air base at Coco Solo,

Panama, bringing a navy medical unit and supplies to the city.[42] The following day, an aircraft carrier from Guantánamo anchored off Nicaragua's Caribbean coast, from where it proceeded to launch more than a dozen relief flights.[43] Declaring it his "duty and obligation" to help, the president of Pan-American Airways loaned six transport planes to the US government at no cost, and for as long as the military needed them.[44]

In the years ahead, the use of aircraft became increasingly central to US foreign disaster relief operations. Even as navy and commercial ships continued to play a role in humanitarian response, military and civilian aircraft extended the reach of US foreign disaster assistance dramatically, enabling the government and ARC to dispatch aid and personnel more quickly and over far greater distances than before. From these foundations in the Dominican Republic and Nicaragua during the early 1930s, American air power was to fundamentally remap the United States' future humanitarian geography.[45]

Supplementing the sizable US military interventions in both countries— and reflecting a practice that had become more common in recent years—the ARC's leaders sent an adviser to the Dominican Republic and Nicaragua. For both assignments, they selected Ernest J. Swift, the ARC's assistant director of insular and foreign operations. The decision to send Swift on these twin missions reflected a conviction, shared by both the Hoover administration and ARC leaders, that the Trujillo and Moncada administrations needed an American expert to help plan and implement a competent disaster response.[46] Swift, they believed, was just the man for the job. "A person of [Swift's] extensive experience cannot only make the best out of the American contribution," the ARC's acting chair told Hoover after the San Zenon hurricane, "but possibly be of assistance as an advisor in setting up the whole relief operation."[47] After the Managua earthquake, likewise, ARC leaders promised the State, War, and Navy Departments that Swift would "make his experience available in formulating the general relief and rehabilitation measures."[48] Traveling on Pan-American planes, Swift arrived in both countries just a few days after the disasters occurred. He remained in the Dominican Republic for more than a week and in Nicaragua for nearly a month supervising relief operations.[49]

In the early 1930s, in light of prevailing discontent over the long history of US gunboat diplomacy, military occupations, and political and financial coercion in the Caribbean Basin, staging such extensive US humanitarian operations in the Dominican Republic and Nicaragua was not without political risk. In this context, the deployment of US troops, military aircraft, and American advisers to these nations had the potential to spark a profound backlash or, as the acting secretary of state put it, to "create misunderstanding in other countries."[50]

Both Trujillo and Moncada, however, proved largely receptive to the massive influx of American personnel, cash, and supplies in their nations' devastated capitals. In Santo Domingo, US officials encountered some initial resistance from Trujillo, who was reportedly "insulted" by their insistence on retaining control over ARC funds and refusal to "turn the money over to him."[51] Those early frictions quickly subsided, however. Within a few days of the hurricane, both Swift and Curtis observed, Trujillo's attitude completely transformed, with the president now giving US officials his "unstinted support."[52] In Managua, meanwhile, relations with the government proved cordial from the beginning. There, Moncada cooperated straightaway, working with Hanna, Swift, and other US officials to organize the disaster response.[53]

But Trujillo and Moncada did not merely acquiesce to the significant US presence in their countries; much like Manuel Estrada Cabrera had done in Guatemala in 1918, they also granted wide powers to US officials at the scene, delegating to these Americans tremendous authority to administer the disaster response in their respective nations. After the San Zenon hurricane, Trujillo relied heavily on Swift to devise the official relief and recovery program for Santo Domingo. He also appointed two US military officers to direct medical relief efforts and oversee the distribution of aid supplies.[54] After Swift departed to return to his normal responsibilities, Trujillo tapped Major Thomas E. Watson, a US Marine and naval attaché, to assume command of the ongoing relief efforts. As Watson explained to his superiors, just "about every day I think of some other job I can take over and of course it is all right with T[rujillo]."[55] Watson remained in this position of authority until early November 1930, two months after the hurricane, when the aid operation officially closed.[56] Praising Trujillo for entrusting "continental Americans and Porto Ricans" with these responsibilities, Curtis, the US minister, commended the president for having "the courage to give such wide powers and such firm support to foreigners."[57]

In Nicaragua, Moncada likewise ceded considerable authority to Swift and other US officials. Upon his arrival in Managua, with Moncada's assent, Swift began organizing a disaster response plan for the city. To help him oversee this work, he established a central relief committee. Chairing the committee was Hanna, the US minister, with Moncada assuming the post of honorary chairman. Filling out its ranks were three US military officers— the commander of the US Marines' Second Brigade, the American commander of the Guardia Nacional, and the head of the US Army Engineers in

Nicaragua—and one noteworthy Nicaraguan: the undersecretary of foreign affairs (and future dictator) Anastasio García Somoza.[58] After Swift returned to Washington, leadership of the relief projects he initiated fell to Hanna. Working closely with Moncada, Somoza, and other Nicaraguan officials, the US minister continued to oversee elements the American response until mid-August 1931, four and a half months after the onset of the catastrophe.[59]

Following major disasters in the Dominican Republic and Nicaragua, US diplomatic, military, and ARC officials thus planned and administered prolonged humanitarian operations, each stretching several months beyond the events that first triggered them. These successive disaster aid operations, moreover, were not limited to emergency assistance. As had occurred in Italy, Guatemala, China, and Japan, US officials also pursued far-reaching reconstruction and development projects, designed to leave a lasting mark on the two countries. In their hands, foreign disaster assistance became yet another tool for promoting political order, economic stability, and social change in the Caribbean Basin. A closer look at the specific aid activities that US officials administered in both countries—from short-term relief efforts to longer-term recovery projects—further underscores the extent and character of their involvement.

In the initial hours and days after the onset of the two disasters, US military personnel in Santo Domingo and Managua assumed a central role in the immediate response. In both cities, US Army and Navy medical units set up emergency hospitals, where they treated thousands of injured patients in just a few days' time. Military personnel also established field kitchens and stations for distributing rations and chlorinated water, providing basic sustenance to tens of thousands of people. Others performed manual labor, such as cleaning the streets, clearing away debris, and digging graves for the deceased. In Nicaragua, where raging fires compounded the damages caused by the earthquake, marines and members of the US Army Corps of Engineers worked together to extinguish the flames.[60]

Not all this work was clearly humanitarian in character. In Managua, US Marines and Guardia Nacional troops served as emergency police forces, assigned to restore order and "prevent looting on the part of irresponsible inhabitants."[61] Establishing patrols of the devastated capital, they prohibited the city's residents from entering abandoned buildings and seizing property, their authority backed with the authorization to shoot any looter on sight.[62] This emphasis on policing was hardly unique to Nicaragua. American officials implemented similar policies during prior foreign disaster relief

**A US Marine posing with bodies recovered from the ruins
of the 1931 earthquake and fire in Managua, Nicaragua.**
Image courtesy of Dr. Hugh E. Maudlin Photograph
Collection, Naval History and Heritage Command.

operations, including in Guatemala and China. They did much the same after catastrophes *within* the United States and its territories. Following the Great Mississippi Flood of 1927, for instance, white authorities in the Mississippi Delta exercised considerable police powers over the region's many African American flood survivors, justifying their actions as essential for preserving public order and stability.[63] For many US officials involved in disaster assistance efforts, whether at home or abroad, relieving victims coexisted comfortably with policing them.

Ernest Swift's arrival in each country—five days after the San Zenon hurricane and just forty-eight hours after the Managua earthquake—signaled an important turning point in both humanitarian operations. During the eight days he spent in the Dominican Republic and the more than three weeks he remained in Nicaragua, Swift worked with Trujillo and Moncada, and with US diplomatic and military officials, to devise and implement a structured program of disaster response. Under his supervision, the humanitarian operations in both countries moved steadily away from short-term, emergency aid and toward longer-term recovery, rebuilding, and development initiatives.

Initially, activities related to medical care and public health remained a key priority for Swift and his associates. In tent hospitals and makeshift clinics, US military physicians, surgeons, and nurses continued to treat sick and injured patients. Once these more acute cases subsided, however, they shifted their energies to preventing the spread of infectious diseases. To accomplish this goal, US officials in both cities undertook a variety of tasks intended to improve sanitation, including digging latrines and repairing damaged water and sewer systems. They also established inoculation clinics in the two cities, where they administered typhoid vaccinations shipped from the United States. To ensure compliance, officials employed coercive tactics, mandating that disaster survivors present a certificate of inoculation in order to receive aid.[64]

Feeding disaster survivors remained a second short-term priority, and also transformed in significant ways under Swift's watch. In Santo Domingo, US officials investigated anyone who requested food aid and then issued ration cards to those they deemed eligible, a policy intended to minimize abuse and "to avoid duplication of relief." At distribution stations throughout the city, US officials provided food to as many as 18,000 people daily, guarded by "a squad of soldiers" Trujillo dispatched "to handle the crowd."[65] In Managua, the system functioned in a similar manner. After conducting surveys to determine who qualified for food assistance, US officials distributed ration cards to ensure "the possibility of abuse was eliminated or minimized."[66] They then provided food—in the form of cooked meals, raw staples, and milk for babies—to as many as 8,000 people per day, conducting these distributions under marine supervision and with the assistance of trusted Nicaraguan authorities.[67]

Though US officials initially concentrated on feeding the hungry and addressing critical health needs, their duty—at least, as they understood it—was not to provide aid indefinitely. The more important objective, as Swift saw it, was to "encourage restoration of normal conditions."[68] Or, as Watson put it, to "get back to a normal stage."[69] Within a few weeks of both the San Zenon hurricane and the Managua earthquake, American officials therefore moved away from relief activities and toward longer-term rehabilitation projects. Launched initially by Swift, and then continued after his departure by Watson in the Dominican Republic and Hanna in Nicaragua, these initiatives kept US officials involved in recovery efforts for several months after the disasters commenced.

Of particular anxiety to the officials was the belief that providing free food to disaster survivors would foster dependency and idleness. From Santo Domingo, Watson wrote his superiors bemoaning "the evil of distributing food to the people."[70] Convinced that "the continued issue of free food would tend to keep some of the people from going to work," he concluded, "very drastic measures were necessary to force the people to understand that free food was not going to be available to them forever."[71] In Managua, Hanna similarly emphasized "the necessity of reducing and eventually discontinuing the free distribution" of food, advising Moncada that this policy was essential to avoid "the creation of a beggar class."[72] Moncada, it turned out, agreed with these sentiments. Noting, "Relief engenders bad habits among simple and humble people," the Nicaraguan president told the US minister that "the free distribution of food will bring to Nicaragua a greater misfortune than the earthquake."[73]

Guided by these fears and their philosophies of charity, US officials and their political allies made it a priority to discontinue feeding as rapidly as they deemed practical. Within just two weeks of the San Zenon hurricane, Watson reported proudly, he had barred all men from the food distribution stations in Santo Domingo. Two weeks later, he also cut off free food for women and children.[74] While conceding, "It is very probable that some worthy people suffered from this step," Watson easily rationalized his decision, reasoning, "No one was allowed to starve to death."[75] After the Managua earthquake, the US relief committee stopped serving cooked food three weeks after the earthquake, though they distributed uncooked food to those deemed eligible for a few weeks longer.[76] In place of food distribution stations, they set up four retail stores, "where essential foods were sold at reasonable prices" rather than given away gratis.[77]

As they took steps to reduce the distribution of free food, Watson, Hanna, and their associates simultaneously put labor relief programs in place. In so doing, they embraced a method of assistance US officials had used in the Strait of Messina region, in Chihli province, in the Mississippi Delta, and in other contemporary sites of catastrophe. Like their counterparts in these other places, US officials in the Dominican Republic and Nicaragua believed that creating "remunerative and useful work for unoccupied labor" was essential for at least three reasons.[78] First, it promised to speed both cities' recovery. By performing "useful work in connection with the relief," as Hanna put it, disaster survivors could provide "material assistance in the reconstruction" and help restore the city to "a normal self-supporting life."[79] Second, such an approach reinforced US officials' efforts to counteract the presumed risks of dependency and idleness, a perception that was manifestly racialized and classist. Putting the matter bluntly to his superiors, Watson explained, "These damned people won't work as long as they can eat so I'll remove the incentive to do no work."[80] Third and finally, US officials viewed labor relief programs as "vitally necessary" to inhibit crime, vice, and unrest. Warning the State Department that "the germs [of disorder] are present and are spreading and may easily become dangerous if not stamped out," Hanna insisted that "employment for those out of work . . . as a consequence of the disaster will help to prevent such disorder."[81]

American officials thus held high hopes for the positive and transformative benefits of labor relief projects. They sought to discourage laziness and instill a stronger work ethic, providing lessons in morality and a dose of paternalistic tutelage along with their aid. In practice, however, the results of their programs were uneven. In Santo Domingo, they proved something of a flop. There, Watson made "a standing offer to any man that if he wanted to work and couldn't find work that the Committee would give him work with a pick and shovel." In exchange for such recovery-related tasks as clearing streets and rebuilding roads, he promised "three meals a day and some cigarettes" as compensation. Most Dominicans turned down his offer, however, and in the end, Watson hired just thirty-five men. These men "were retained," he griped, "until some labor organizer convinced them that they should receive cash," prompting his small crew to quit.[82]

In contrast with Watson's experience, US officials in Managua came away pleased with the labor relief program they implemented. Initially, Swift and Hanna mandated that all "able bodied males without occupation will be required to give a reasonable amount of labor" in exchange for food rations.[83] Once they cut off food distributions, however, they shifted to a system of monetary compensation. Offering recruits forty cents per day, US officials succeeded in hiring a force of 1,400 men and still had "an excess of applicants." Working under the supervision of an American engineer, these laborers performed such tasks as clearing streets, hauling away debris, and repairing roads and buildings. Officials in Nicaragua also organized "gangs of women street sweepers," reporting with pride that "they proved more efficient than men."[84] Speaking for many of his colleagues, Hanna judged "the employment of labor . . . our principal and most beneficial form of relief."[85]

As another central facet of their recovery work, US officials launched several major rebuilding projects. Their ambitions were not limited to reconstructing the two cities but aimed at something bolder: to improve upon what had been there before. At the same time, they desired to create tangible monuments to US assistance. Once again, this was not the first time American officials had undertaken these sorts of far-reaching programs. Actions in Italy in 1909, in China and Guatemala during the Great War, and in Japan after the 1923 earthquake provided key precedents and models. In the Caribbean

Basin, US disaster aid once again straddled the boundaries between humanitarian relief, development assistance, and cultural propaganda.

In Santo Domingo, Watson spent the bulk of the ARC funds at his disposal on a "housebuilding scheme."[86] This project entailed the construction of 600 small homes, intended for individual families who became homeless due to the hurricane. In light of his difficulties recruiting willing workers—and adopting an approach used in the Mississippi Delta after the floods three years before—Watson turned to another, more dependable source: convict labor. With Trujillo's blessings, he conscripted imprisoned Dominicans to construct the homes and to repair hospitals, schools, and other institutions.[87] In Managua, meanwhile, Hanna and his American colleagues focused on improvements to the sanitary and medical infrastructure. Using the ARC's funds, they hired Nicaraguan laborers to repair the city's water pumps and its two largest hospitals.[88] Setting his sights beyond rebuilding, Hanna also persuaded ARC leaders to fund the construction of a brand-new hospital for leprosy patients. Noting that the conditions in Managua's current leper asylum were "not suitable for a modern pig sty," he reasoned that the new institution would serve as both a model, state-of-the-art medical facility and a perpetual memorial to US disaster assistance.[89]

If homes and hospitals functioned as concrete symbols of reconstruction and US assistance, one final recovery project was less material in nature. Three weeks after the San Zenon hurricane, President Hoover sent Eliot Wadsworth, a member of the ARC's governing central committee and former assistant secretary of the treasury, on an economic mission to Santo Domingo, naming him a special commissioner to President Trujillo. "Now that the emergency work . . . is fully organized and under way," Hoover announced, questions related to the "readjustment of San Domingo finances call for careful planning and first-hand knowledge."[90] Charged "to advise [the] Government on those subjects," Wadsworth remained in the Dominican Republic for several weeks, working with Trujillo to design a "financial and rehabilitation program" for the country.[91] Twenty-six years after Theodore Roosevelt first seized the Dominican customs houses, the Hoover administration had found, in the aftermath of disaster, an opportunity to exert its own sway on the Dominican Republic's financial affairs.

In 1930 and 1931, US officials engaged in lengthy, far-reaching relief and recovery efforts in the Dominican Republic and Nicaragua, occupying positions of tremendous authority over disaster survivors in these two countries. In the eyes of both US officials and their allies in the Trujillo and Moncada administrations, these aid operations appeared to be humanitarian and diplomatic success stories. Trujillo offered his "profound and sincere gratitude" to President Hoover, thanking him "for the efficient aid and timely cooperation and assistance."[92] Moncada, likewise, extolled the ARC for its "devotion to the relief of the sufferers and to the health and sanitation of the Capital."[93] In a concrete gesture of appreciation, he awarded the Medal of Merit to Hanna and nineteen of the US Marine Corps, Army, and Navy officers involved in the assistance operation.[94]

Returning the compliments, US officials sang the praises of the government leaders with whom they collaborated. Swift and his American associates hailed Trujillo for "bringing order out of chaos" and "for the splendid order that prevailed" under his watch.[95] In Managua, likewise, Hanna and other US officials applauded Moncada for giving them "the whole-hearted moral and material support of the Government." They singled out "the energy and activities of General Somoza" for particular praise, stressing that his "services were of incalculable value" to the relief operation.[96]

The glowing tributes that prominent officials exchanged with one another, however, were not the final word. Even as Dominican and Nicaraguan political elites cheered the "genuine spirit of Pan American fraternity" they witnessed, some residents of Santo Domingo and Managua took a far more negative view.[97] As these critics saw it, the US aid operation represented an undesirable intrusion into their nations' affairs, a humanitarian action marred by coercion, condescension, and even violence.

Though critiques of this nature surfaced in both countries, they were especially pronounced in Nicaragua. There, a vocal chorus of critics advanced an alternate narrative of US disaster assistance, an interpretation of events that differed vastly from the stuff of official reports and public accolades. As they aired their collective grievances, these individuals laid bare the complicated politics of US disaster aid in the Caribbean Basin, and indeed around the world.

Rumors of violence, destruction, and lawlessness at the hands of US Marines began circulating in Managua almost as soon as the US disaster response commenced. The precise details varied. Some claimed that "thirty-five or forty natives had been shot by Marines for looting," while others charged a single marine lieutenant with having "executed fourteen natives caught looting."[98] Still others, disparaging US forces for their "criminal" behavior, insisted that "Managua burned down" not as a consequence of the earthquake but for a more sinister reason: "because the Marines wanted it."[99] A man claiming to be a representative of Augusto Sandino accused US officials of looting crumbled houses while leaving Nicaraguan disaster victims buried in the rubble—and, in some cases, "riddled by marines' bullets." The disaster response, he proclaimed, was the most "cowardly, cruel, and shameful crime that the Yankee pirates have every committed."[100] Although they differed in specifics, the central thrust of these rumors was essentially the same: the US aid operation in Managua was plagued from the start by brutality and savagery, committed by the very individuals who claimed to be helping Nicaraguans in their moment of crisis.

Not contained to Managua or even Nicaragua, the rumors soon crossed borders and began circulating throughout other Central American countries. In El Salvador, a group of doctors and engineers who had participated in relief work reported that US soldiers commandeered all of the bread the Salvadoran government sent to Nicaragua.[101] In Mexico, one newspaper accused US Marines of looting Managua's jewelry stores. Another condemned the "Yankee forces of occupation" for having made the disaster far worse by their presence.[102] Echoing these claims, a noted Mexican labor leader charged US troops with assassinating survivors, decrying the "sack of the city of Managua by Yankee Marines."[103] A Mexican physician, who had gone to Nicaragua as a health envoy, concurred with his compatriot, averring that "the Yankees . . . conducted themselves like pretorian guards, sowing panic and terror among the inhabitants."[104]

It did not take long before these damning indictments reached US officials. Gravely concerned with this "anti-American propaganda" and its implications for the "prestige of the Americans" in the region, US diplomats in the United States, Nicaragua, and elsewhere in Central America urgently endeavored to neutralize the criticism.[105] Calling the reports "a libel" and "grossly exaggerated," Hanna implored Moncada to help quash the rumors, declaring them a "vicious, erroneous, and malintentioned attack on the American Marines."[106] In Washington, State Department officials and ARC leaders refuted the questions of probing reporters, insisting that "if the shooting of so many natives by Marines had actually occurred we would have a report of it."[107]

Determined to reap the diplomatic benefits of American disaster aid, US officials refused even to countenance the possibility these stories might contain some grain of truth.

Reading these accounts, it is tempting to wonder what *really* happened in Managua—to ask what, precisely, US Marines and other American officials did or did not do in the wake of the great earthquake. And yet, the veracity of these rumors is in many ways beside the point. The counternarratives circulating in and beyond Nicaragua during the early 1930s are important not for their reliability or accuracy but rather for what they reveal about the contested meanings of US foreign disaster aid.

Though lauded by their allies, the Americans who assumed control over disaster aid efforts in Managua were, in their critics' eyes, proverbial wolves in sheep's clothing. Their humanitarian actions served to mask—and according to some, to exacerbate—the violence US occupying forces had long committed in Nicaragua. And indeed, even as relief and recovery efforts progressed in Managua, elsewhere in the country US Marines and the Guardia forces they trained were continuing to carry out their primary mission: conducting "operations against the bandits" and hunting for Augusto Sandino.[108]

Little wonder, then, that some Nicaraguans saw the aid operation in Managua as inseparable from the hostility and criminality of the American occupation. Just as surely, the presence of US military personnel, ARC experts, and financial advisers in Santo Domingo—all deputized by the authoritarian Trujillo to oversee the nation's relief and recovery program—raised a similar, all-too-familiar specter of US intervention for many Dominicans. In the eyes of these critics, and for others like them throughout the world, US disaster assistance was not altruism. It was just another manifestation of American interference in their own affairs and governance, an excuse for the United States to exercise its police power abroad.

8

Navigating the Great Depression and Playing the Good Neighbor

1931–1939

When Franklin D. Roosevelt became president of the United States in 1933, his administration ushered in a dramatic expansion of the American welfare state. Over the next several years, the US government implemented a sweeping slate of social programs and economic reforms, known collectively as the New Deal. Roosevelt and his fellow New Dealers justified these actions, notably, by pointing to a long tradition of federal disaster relief. Since 1790, as legal scholar Michele Landis Dauber has shown, Congress had regularly provided aid to survivors of fires, floods, hurricanes, earthquakes, and other hazards within the United States, locating its power to do so under the General Welfare Clause of the Constitution. In the 1930s, New Dealers claimed authority to respond to the Great Depression in precisely the same way.

Between 1933 and 1939, they committed unparalleled levels of federal funding to assist victims of this sweeping economic and social disaster.[1]

But if domestic relief and recovery programs swelled tremendously across the 1930s, the same cannot be said for foreign disaster aid. After sustaining an active program of international disaster assistance throughout the 1920s and early 1930s, the US government and ARC reined in their foreign aid commitments during the remainder of the decade. In part, this retrenchment was a function of the ongoing economic crisis. Facing continued financial constraints at home, officials reduced the frequency and scale of their humanitarian efforts in other nations. By the latter half of the decade, as new conflicts abroad led to rising calls for nonentanglement within the United States, the political climate also became less conducive to US foreign interventions of *any* kind, disaster aid among them. There was to be no new deal, in short, for American catastrophic diplomacy.

If these factors encouraged a diminution of foreign disaster assistance during the 1930s, however, they never squelched it entirely. Across these economically and politically fraught years, members of the Herbert Hoover and Franklin D. Roosevelt administrations, together with the ARC, responded to plenty of international calamities. They simply did so on a more selective basis. While they kept in mind the humanitarian needs of disaster survivors, US officials also privileged their own diplomatic designs, economic interests, and geographical and technological limitations in choosing where to send their limited assistance.

Included among these aid operations was the US response to one of the twentieth century's most cataclysmic disasters: the central and south China floods of 1931. Despite the enormity of this catastrophe, the ARC contributed a comparatively trivial amount of aid to Chinese flood sufferers, an outcome driven as much by fiscal austerity as by institutional apathy and antagonism toward China. Motivated to supplement this aid for reasons both political and humanitarian, the Hoover administration experimented with a relatively novel form of US governmental disaster assistance: selling surplus American commodities to China to relieve hunger. Tested in the early 1930s, this new state-led approach to humanitarian aid grew increasingly commonplace in future years, eventually becoming a cornerstone of US foreign disaster assistance by the 1950s.

Between Franklin Roosevelt's presidential election in late 1932 and the Second World War's eruption in Europe in 1939, the US government and ARC continued to respond to foreign disasters on a regular basis—particularly when it appeared likely to serve the Roosevelt administration's broader foreign policy agendas. For this reason, the lion's share of US international disaster assistance flowed into the Western Hemisphere, where US officials hoped it might bolster the Roosevelt administration's Good Neighbor Policy. But smaller amounts of disaster aid still trickled into many other parts of the world, as US officials strove to maintain their earlier humanitarian reach. Even amid the domestic and global tumults of the 1930s, foreign disaster aid remained a routine part of US international relations.

✪ ✪ ✪

In late June 1931, as US earthquake relief efforts in Managua were drawing to a close, heavy rains began falling across central and southern China. Although summer was typically the rainy season in this part of the world, this year's precipitation was more extreme than usual. For several weeks, incessant rainfall deluged the region. Rivers soon breached their banks and dikes gave way, resulting in destructive flooding along several major river systems. By late July, the Yangzi, Huai, and Yellow Rivers had all overflowed, as had China's Grand Canal. According to estimates at the time, the floods inundated an incredible 34,000 square miles—double the area, one US official observed with dismay, as the Great Mississippi River floods four years prior.[2] According to recent estimates, it was actually closer to *four times* that size.[3] Cresting at more than fifty feet in some places, the floods laid waste to bustling cities, productive agricultural lands, and everything in between.

Though relatively few people died immediately—a few thousand were believed to have drowned—the human toll of the disaster was nevertheless staggering. According to contemporary estimates, the floods left 10 million people homeless while affecting 31 million in some way; more recently, some historians suggest that figure may have exceeded 50 million. Hundreds of thousands (and perhaps millions) of these survivors eventually perished due to the flood's secondary effects, including famine, epidemic disease, and exposure to the elements.[4] The catastrophe, one commenter grimly surmised, would "probably prove to be in all of its aspects the greatest tragedy in human history."[5] Certainly, it ranked high among them.[6]

The floods presented an enormous challenge for China's Nationalist government, testing its capacity to respond to an immense public emergency. On August 14, calling the episode the "most serious natural calamity in China in recent times," government authorities established a National Flood Relief Commission, charged "with supreme authority to coordinate and supervise all relief operations."[7] Headed by Minister of Finance T. V. Soong, the commission included sixty Chinese and more than thirty foreign members, roughly half of them American, among them financiers, businessmen, lawyers, missionaries, and journalists.[8] Armed with a $2 million appropriation from the Nationalist government, the commission took steps to respond to the crisis while seeking additional funds.[9] With the latter goal in mind, Soong and his commission appealed to US diplomats and consuls in China, inquiring "whether it might be possible to obtain assistance from the United States."[10]

In the minds of these US Foreign Service personnel, both the disaster itself and the Chinese government's request for aid were inextricable from broader political and foreign policy concerns, rooted in China's recent past. After becoming a republic in 1911, China subsequently experienced a period of political turmoil and fragmentation, with power devolving among regional warlords. In 1928, ending this "warlord era," Chiang Kai-shek and the Nationalist Party (Kuomintang) led a successful military campaign to reunify the country, then began governing China. National unification was in name only, however. A civil war soon broke out, pitting the Nationalists against a new rival, the Communist Party of China. By the time the floods commenced in 1931, the two sides were embroiled in a bloody struggle for control of the country. Further complicating the political situation, tensions between China and neighboring Japan had recently begun to escalate. Throughout 1931, rifts between the two nations widened, paving the way for Japan's invasion of Manchuria in mid-September. Together, these events led to decisive changes in US foreign policy toward China, geared toward supporting the Nationalist government in its multipronged fight against communism and Japanese incursion.[11]

In this highly charged political climate, US officials viewed the unfolding catastrophe with considerable apprehension, wondering what ramifications it would have for the political situation in China. Many US Foreign Service personnel feared the disaster would exacerbate communist influence and activity. "People are suffering frightfully both from floods and consequent famine," the US minister to China, Nelson T. Johnson, informed the State Department, "but the most brutal part of it all is that Communists bandits [are] taking advantage of [the] situation . . . robbing and murdering to an extent hitherto unknown."[12] While Johnson predicted an upsurge in violence, some of his colleagues worried that the floods had given Chinese communists a golden opportunity for spreading their political ideologies and influence. "The terrible hardships and suffering which the present flood will cause," the US consul general at Hankow (Hankou) cautioned, would undoubtedly "serve to assist the propagandists who are seeking to instill the ideals and principles of Soviet Russia into the minds of the people."[13] Still other officials speculated that the unrest the floods triggered would further aggravate tensions with Japan. "What I fear most," the commander of the Asiatic Fleet confided to Johnson, "is that these people will resort to banditry in order to get food without much thought of the Chinese-Japanese situation."[14]

Hoping to combat the apparent political perils and the very real human suffering the floods produced, most US Foreign Service personnel in China agreed that it was vital for the United States to respond favorably to the Nationalist government's request for assistance. Shortly after receiving Soong's appeal, Johnson wrote the State Department to convey this recommendation. "I feel that foreign aid will be necessary," he reported, "and hope that [the] Red Cross can do something in this emergency."[15] Ernest Bicknell, a veteran leader of the ARC's international division, agreed wholeheartedly with Johnson, advising his ARC colleagues that they must "respond to the needs of this disaster." Calling the catastrophe "one of the most severe and extensive which has occurred in China during a great many years," Bicknell affirmed, "This is a disaster which meets every essential condition to justify Red Cross intervention."[16]

Although high-ranking US diplomats and ARC leaders deemed Red Cross aid "necessary" and "essential," a pair of concerns made them hesitant about contributing this assistance. First, the financial situation caused by the Great Depression and Dust Bowl had seriously hampered the ARC's ability to attract donations from the public, particularly for its international activities. Well aware of this fact, the ARC's governing central committee quickly determined that a "campaign at this time for a large flood relief fund would not be feasible because of the situation in this country."[17] This left the committee with the option of contributing funds from the ARC's reserves, a strategy to which it had often resorted in past crises.

There was, however, a second and more glaring issue: many ARC leaders had grown reluctant to aid China at all. In the years since its reorganization in 1905, the ARC had allocated $2.2 million in humanitarian aid to China, making the country one of the leading recipients of the organization's foreign relief funds.[18] And then, in 1929, ARC leaders sent a commission to China to inspect how those funds were being used. The result of the commission's investigation was a damning report. Rehashing a litany of deep-rooted racial prejudices and cultural chauvinisms, it excoriated Chinese authorities for misgovernment, corruption, and abuse of American aid.[19] In late 1929, based on the commission's findings, the ARC's central committee issued an official statement on future relief to China, declaring that the ARC would no longer extend assistance to the country except in emergencies "plainly due to an act of God."[20]

Less than two years later, the conclusions and consequences of the ARC's China report continued to reverberate. Although the flood disaster was clearly attributable—at least in part—to an "act of God," many US officials insisted that providing *any* aid to China remained a fool's errand. "If the American Red Cross came into this area for relief purposes," warned the American consul general in Hankow, Walter Adams, "it would be faced with insistent demands from Chinese military authorities" who would "prevent such expenditures from attaining their objectives."[21] Emphasizing that "little can be expected from the Chinese Government authorities in the way of intelligent preventive measures or effective relief work," Adams advised, "I am not convinced that relief measures . . . are desirable at this time."[22]

Despite their concerns, ARC leaders ultimately concluded they must extend an offer of Red Cross assistance to China, reasoning that the humanitarian and political imperatives outweighed the difficulties and objections. However, the amount they chose to send—$100,000—reflected persistent ambivalence about that decision.[23] Although $100,000 represented a fairly sizable contribution in relative terms (ranking as one of the ARC's more substantial foreign disaster relief appropriations during the 1920s and 1930s), it was a mere drop in the bucket for a catastrophe affecting tens of millions of people. The figure was less than the ARC contributed to either Nicaragua or the Dominican Republic the previous year. As Chinese critics were quick to point out, it was also exponentially less than the $11 million the ARC contributed to Japan after the Great Kantō earthquake in 1923.[24] As this egregious discrepancy underscores, humanitarian need was never the sole consideration guiding US foreign aid decision-making. A host of additional factors—foreign policy goals, geography, ideological assumptions and prejudices, donor whims, and chance timing—all influenced calculations about where to send US international disaster relief.

Although US Foreign Service personnel in China inquired whether the ARC could give more, the ARC's central committee remained steadfast in its decision. Explaining that the organization was "not in [a] position and does not desire to assume an obligation with regard to the major problem of relief created by this flood," they told the State Department that $100,000 was all they were prepared to give.[25] At the State Department's request, however, ARC leaders did agree to provide one additional form of assistance: free advice to Chinese authorities, detailing "measures . . . taken in connection with the Mississippi Flood" of 1927.[26] The guidance took the form of a two-page letter, transmitted via US Foreign Service personnel in China, summarizing "the methods used by the American Red Cross in dealing with large flood disasters."[27]

Adding insult to injury, ARC leaders selected Walter Adams, the US consul general in Hankow who was so skeptical about contributing aid to China in the first place, to oversee the administration of the organization's funds. After receiving the money, Adams held onto it for several weeks while he considered how the "gift of the American Red Cross could be handled most efficiently."[28] Eventually, he decided to entrust the contribution to one of the American members of the National Flood Relief Commission, James Earl Baker. Although T. V. Soong raised objections to this move, asking why the funds were not turned over to the headquarters of the National Flood Relief Commission, Adams refused to reconsider, believing Baker would expend the funds more "efficiently and honestly" than his Chinese counterparts.[29]

Over the next few weeks, Baker used the $100,000 at his disposal to purchase rice, beans, flour, coal, and clothing for flood survivors and to assist with sanitation measures in refugee camps.[30] It remained the sole funding the ARC sent to China for one of the most devastating catastrophes in the country's—and the world's—history.

Although the ARC resolved to limit its aid to China, many members of the Hoover administration wanted to provide additional humanitarian assistance to the country, for reasons both diplomatic and moral. Soon after the floods commenced, therefore, they started considering how the government might supplement the ARC's monetary contribution. One tried-and-true option, used in many prior foreign disasters, was to lend the military's assistance to China. In mid-August, at the State Department's request, the US Navy secretary authorized the commander of the Asiatic Fleet to dispatch "for such assistance as they can render, such ships and personnel as he may feel advisable to spare from other duties."[31] Both Chinese government officials and US diplomats on the scene, however, responded that US Navy assistance would be superfluous, advising that the services it offered could be "carried on by local means."[32]

With military assistance off the table, government officials instead opted to test out an experimental form of aid, piloting a novel humanitarian technique that would one day become a fixture of US foreign disaster assistance: selling surplus agricultural crops to the Chinese government for use as food relief. Key precedents for such an action had been established during the Great War and its aftermath, when Congress charged two successive federal agencies—the US Food Administration and the American Relief Administration—with purchasing and selling American foodstuffs to war-torn Europe, Russia, and the Near East. These wartime aid programs, however, did not immediately become a model for peacetime disaster relief. In 1931, selling excess commodities to assist survivors of "natural" catastrophes was not yet a customary practice.

More than a decade after the Great War, moreover, both the context and the mechanisms for such a transaction had changed considerably. The 1920s was a devastating decade for American farmers, who struggled to stay afloat following postwar collapses in crop prices. In 1929, Congress attempted to respond to the so-called farm crisis by passing the Agricultural Marketing Act. Intended to stabilize farm prices, this legislation authorized the US government to buy and store surplus commodities, to be managed by a new agency called the Federal Farm Board.[33] The system did not function as its designers hoped, however. As a result, the Farm Board soon amassed more wheat and other agricultural products than it knew what to do with.

In 1931, as floods inundated central and southern China, Chinese officials alighted on the idea of repurposing this overabundance as humanitarian assistance. Concrete plans for this scheme began germinating in early August, when a representative of China's Huai River Commission approached the American consul general in Nanking with an informal proposal. "In view of the large surplus supply of wheat said to have been purchased by the American Government and still on hand," he asked, "might be possible to obtain a portion" for flood survivors?[34] A few days later, officials at the Chinese Ministry for Foreign Affairs followed up this inquiry with a formal request to the American legation, proposing that the Farm Board sell the Chinese government "wheat on liberal terms for the purpose of assisting in relief." The wheat "would be utilized in connection with a carefully supervised program of work," they explained, "and particularly [for the] repair and construction of dikes and river improvements designed to prevent recurrence of similar disasters in future."[35]

Taken with these ideas, especially with their dual emphasis on productive labor and long-term development, US minister Nelson Johnson forwarded the Chinese government's proposal to Washington, where it received an equally warm reception.[36] Assessing its merits, the chief of the State Department's Division of Far Eastern Affairs noted that the US government had long "desired that the friendly attitude of the United States toward China be demonstrated, if and when an opportunity arose." The request to sell surplus wheat as flood aid afforded just such an occasion, he argued, a chance "simultaneously to do something for China's benefit and to advance the interests of the United States in China." In addition to its obvious "philanthropic side," he continued, the scheme promised multiple commercial and diplomatic benefits: "to dispose of American surplus goods and services, to connect up with constructive enterprises in China, and to maintain and strengthen the asset of good will." For these reasons, he strongly recommended that the government move forward with the proposal.[37] Asked to offer his impressions, the chair of the Senate Committee on Foreign Relations, William Borah, expressed similar logic. "I should like to see the Farm Board turn over a large part of its surplus to the starving people of China," Borah asserted. "Such a course can be justified," he added, "not only as a matter of humanity but as considered from the standpoint of economics for our own."[38]

On August 19, acting on these and other recommendations, the chairman of the Federal Farm Board affirmed his agency was prepared to act favorably on Soong's request.[39] Shortly thereafter, US and Chinese government officials entered into negotiations over the wheat sales. During these discussions, however, a few sticking points arose. First, some local government officials in China raised objections to the plan, on the grounds that surplus foodstuffs *already* existed domestically, in the three eastern provinces (Manchuria). These crops "should be sent south to feed flood sufferers," they complained, reasoning "that native products are cheaper than that from America and that less time and cost will be required to transport [them]."[40] Soong brushed aside this suggestion, however, arguing that securing Manchurian grain— given mounting tensions with Japan in and over this region—would "depend too much on fortuitous circumstances."[41] And indeed, the Japanese invasion of Manchuria just a few weeks later, on September 18, proved his assessment correct.

A second and more diplomatically fraught obstacle, from the perspective of Sino-American relations, arose over the specific terms of the agreement. The Farm Board initially proposed selling the wheat to the Chinese government on credit, repayable in four years at a rate of 4.5 percent interest.[42] In response, Chinese authorities appealed for more generous terms. "While the offer made by the Farm Board is appreciated," they explained, the Nationalist government would have a difficult time repaying the loan, given its "existing financial and economic difficulties."[43] State Department officials refused to budge, however, insisting that the "terms given by the Farm Board are the most liberal which could probably be given without special authority from Congress."[44] Backing the Farm Board, a White House spokesman maintained that the original offer "amounts to very generous action on the part of our government," particularly given the state of "Chinese credit in the money markets of the world."[45]

Despite these hurdles, the deal eventually went through, with US negotiators mostly having their way.[46] Finalized in early September, the agreement stipulated that the Farm Board would sell 450,000 tons of surplus Western white wheat and wheat flour to China, to be "used by the buyer exclusively for charitable purposes in the flooded areas of China."[47] The Nationalist government, in turn, agreed to assume all transportation costs and to repay the US government in three equal installments, due by 1936, at the slightly reduced rate of 4 percent interest. American wheat would soon be on its way to China.

On September 15, two weeks after concluding its agreement with the Federal Farm Board, the Nationalist government declared the emergency period complete. The floodwaters had finally receded, and the time had come, Chinese officials determined, to transition into the recovery stage. Yet if the emergency was officially over, the catastrophe the floods produced most certainly was not. Considerable suffering persisted across central and southern China. Millions remained homeless or displaced. In crowded refugee camps, cases of cholera, dysentery, and other infectious diseases surged, sickening and killing thousands of people. The threat of famine also loomed, with millions on the brink of starvation.[48]

As troubling as the situation appeared for humanitarian reasons, equally worrisome, in US officials' estimation, were the unresolved political problems the catastrophe had exacerbated. Communists were now "working at the very roots where the suffering is most acute and people feel most the intolerable conditions of life, and the desperate need for change," the American vice-consul in Hankow advised the State Department in late September. "In their suffering," he cautioned, "the people will turn to him who says that he knows the way out of it."[49] Nelson Johnson offered a similarly bleak assessment. Writing a few weeks later, in early November 1931, the minister observed that millions of people remained "in great distress" and warned of the "political dangers" this ongoing crisis presented. "It is in such conditions," he observed, "that the leaders of Communistic propaganda find their best opportunity to spread it among the mass of people." Making matters worse, Johnson continued, the Japanese government was now pointing to these "internal political troubles ... as a confirmation of its claim that China really has no government."[50] Surveying this situation in the wake of Japan's invasion of Manchuria in mid-September, Johnson predicted considerable turmoil, both domestically and regionally, in the months ahead.

It was at this moment of intensifying concern that the first shipments of surplus US wheat reached China, fulfilling the Hoover administration's contribution to flood relief. Between mid-November 1931 and mid-May 1932, sixty-seven American steamers arrived in Chinese ports, carrying hundreds of thousands of tons of wheat and wheat flour in their holds.[51] After distributing a third of the wheat directly to disaster survivors, the National Flood Relief Commission used the remaining two-thirds to compensate laborers involved in disaster recovery projects, just as Soong originally proposed. During the first half of 1932, some 1.4 million Chinese flood refugees took part in these work-for-food programs, most of them tasked with rebuilding and refortifying levees along the flooded rivers. Together, they constructed an estimated 3,100 miles of dikes by the time the program drew to a close, on July 1, 1932.[52]

As they witnessed the arrival, distribution, and consumption of US wheat supplies in China, the program's various stakeholders expressed their thoughts about this experimental aid scheme. Many of them came away impressed with its concomitant humanitarian and political effects. Foreign Service personnel in China widely praised the program, certain that the US government had played a vital part in "curtailing the spread of communism" by feeding and employing refugees.[53] In the Nationalist government, T. V. Soong and many of his associates proved outwardly enthusiastic as well, commending the wheat-for-work relief projects for "their productive nature and tendency to check Communism."[54] The director general of the National Flood Relief Commission, the British politician Sir John Hope Simpson, likewise emphasized the "importance of the reconstruction work" while lauding "the generous attitude of the American Government" for making it possible. "If this sale had not been made and the reconstruction had not been carried out," he told the State Department, "the whole Yangtze Valley would have gone 'Red.'"[55]

For all the positive responses to the food relief scheme, however, the program also elicited some ambivalent and downright unfavorable reactions. Among the most poignant of these critiques surfaced in late 1932, more than a year after the floods receded. In early October, a coalition of farmers' associations from five Chinese provinces wrote the American consul general at Nanking to report the unintended and harmful consequences of the wheat sales. While stressing that they "appreciate[d] very much" the intention behind the aid, the letter writers explained that "the quantity of American wheat borrowed last year exceeded actual needs." As a result, they continued, "the excessive portion was sold at low prices" and the "prices of all kinds of grain were brought down." Now farmers throughout the region found themselves struggling to recoup their costs, "thus resulting in their bankruptcy." A year beyond the floods, they lamented, "places where there has been no famine suffer just the same as if there had been a famine." In its efforts to address the American farm crisis, it seemed, the Hoover administration had offloaded it to central and southern China. Should such conditions persist, the petitioners warned, "the farmers will not be able to earn a living and will become robbers"—or, worse still, "become Communists."[56]

The US government's experiment in selling surplus commodities as disaster relief, then, was not without controversy. Many extolled this novel form of humanitarian assistance, convinced it benefited both foreign disaster survivors and American farmers. Yet, as the critics and casualties of this program recognized, flooding foreign markets with surplus crops was not necessarily the elegant, win-win solution that proponents declared it to be. Instead, it had the potential to destabilize local economies, thus producing even greater suffering. It risked exacerbating rather than relieving the devastation that floods and other catastrophes produced. It threatened to spur crime and communist sympathies rather than pacifying them.

No matter the critiques, this new instrument of foreign disaster assistance was there to stay. In the years since the Great War, US officials had experimented with several new approaches to foreign disaster assistance, including using airplanes to deliver aid, deploying American disaster experts to other countries to superintend aid operations, and undertaking far-reaching reconstruction and development projects. Sending surplus food to affected nations represented yet another of the era's new innovations in foreign disaster assistance. In coming years, governmental sales and gifts of surplus American commodities were to become more and more commonplace, ultimately becoming a customary tool of US foreign disaster assistance. Events and experiences in China following the great floods of 1931 presaged the complicated and contested future of American food power.[57]

★   ★   ★

In November 1932, as farmers in central and southern China struggled against the burdensome consequences of American humanitarian action, voters in the United States went to the polls to elect a new president, Franklin Delano Roosevelt. During Herbert Hoover's four years in the White House, despite the financial constraints the Great Depression imposed, the US government and ARC undertook sizable disaster relief operations in the Dominican Republic, Nicaragua, and China and contributed smaller amounts of assistance to several other countries. During the mid- to late 1930s, however, US foreign disaster assistance efforts decreased in scale and frequency, a function of both the persistent economic crisis and a political culture ever more wary of foreign entanglements.

But even in this environment, the flow of foreign disaster aid never dried up entirely. The Roosevelt administration and ARC contributed disaster aid to multiple countries during the remaining years of the decade. Constrained by reduced funds and mounting geopolitical concerns, they were forced to become more targeted in their assistance than their predecessors. Yet although their humanitarian operations grew more limited in scope, members of the Roosevelt administration continued to regard disaster aid as an expedient instrument of foreign policy.

For reasons of both geography and geopolitics, the Roosevelt administration directed a disproportionate share of its foreign disaster assistance to the Caribbean Basin. When he came into office in March 1933, Roosevelt pledged to chart a new, more respectful course in his country's relations with Latin America, a guiding principle that became known as the Good Neighbor Policy. Throughout the 1930s, improving the United States' image in the Western Hemisphere remained a cornerstone of his foreign policy agenda. Responding to disasters in the nearby Caribbean Basin, members of the Roosevelt administration understood, offered one means of advancing this goal. During the mid-1930s, State Department and US Foreign Service personnel devoted particular attention to catastrophes in this region, working closely with ARC leaders to put the United States' professed neighborliness on display.

Less than six months after Roosevelt's inauguration, on September 1, 1933, members of his administration found their first major opportunity to administer disaster aid in the Caribbean after a hurricane smashed into western Cuba, leaving 100,000 people homeless.[58] The hurricane arrived at a tumultuous period in the island's politics. Just a few weeks earlier, Cubans had overthrown their unpopular president, Gerardo Machado, and established a new provisional government—a régime change the US minister, Sumner Welles, both supported and helped orchestrate.[59] When Cuba's new president, Carlos Manuel de Céspedes, requested US aid to address the catastrophic hurricane, both Welles and the State Department entreated the ARC to respond, "extremely anxious" to demonstrate US support for the new government by making "an immediate response to this appeal."[60] Acceding to this request, ARC leaders made arrangements to send $10,000, a shipment of navy medical supplies, and two "experienced disaster relief executives" to the island.[61]

Before they could do so, however, a second coup occurred—this one far less desirable in the eyes of Welles and his associates. Between September 3 and 5, a group of Cuban army sergeants staged a revolt, ousting Céspedes and subsequently installing another new government, headed by the reformer Ramón Grau San Martín. When news of these events reached Washington, the State Department turned an abrupt about-face, instructing the ARC to "hold its preparations in abeyance" until the situation became clearer.[62] Citing "the precarious situation," the head of the ARC, too, "felt that it was an unwise step" to deliver the pledged assistance.[63]

Despite harboring his own concerns about the political situation, Welles encouraged his colleagues in Washington to reconsider. In addition to alleviating Cuban suffering, the US minister argued, providing hurricane assistance carried valuable political benefits, enhancing his ability to wield diplomatic influence in Cuba. A "generous demonstration on the part of the American Red Cross," Welles told his superiors, "will be particularly useful under present conditions."[64]

Persuaded by Welles's assessment of the situation, Secretary of State Cordell Hull once again enjoined the ARC to send aid to Cuba. "Firm in his desire" to show US support for the country, the ARC's vice chair remarked, Hull "pressed very vigorously the matter."[65] Though they remained apprehensive, ARC leaders soon bowed to the State Department's pressure, sending $10,000 and an ARC disaster specialist, Maurice Reddy, to Cuba. Arriving in Havana on September 8, Reddy spent the next ten days on the island, surveying the hurricane damage and observing the Cuban Red Cross's relief efforts.[66] He came away impressed with Cuban Red Cross leaders, and left the island confident that they were "conducting very effective work."[67]

Soon after Reddy returned to the United States, however, further political shakeups undermined American confidence yet again. In late September, colonel Fulgencio Batista, a leader of the sergeants' revolt and the Cuban army's newly minted chief of staff, assumed command of the Cuban Red Cross, pushing out the leaders who had met Reddy's approval. Wary of Batista's intentions and seeing it as his duty to safeguard ARC assets, Welles—like so many US diplomats before him—took relief matters into his own hands. Assuming control of the ARC's funds, he instituted a policy of issuing aid only to those Cuban authorities whom he personally trusted. Such a procedure, he promised, should afford "ample protection against possible misuse of the American Red Cross contribution."[68] Before long, Welles expended the ARC's contribution, bringing the official US relief operation to a close.

The US government's involvement in Cuban politics, on the other hand, was another matter entirely. In a coda to the events of September 1933, Welles and other US officials quickly overcame their initial suspicions of Batista, seeing him as a far better choice to lead Cuba—and to promote US interests—than Grau.[69] In January 1934, acting with the State Department's support, Batista forced Grau to resign the presidency and subsequently began consolidating his own power in Cuba. Throughout much of the quarter-century that followed, US officials maintained their support for Batista and his increasingly authoritarian regime, embracing him as a force for political and economic order and a staunch ally against communism.[70]

Although the tumultuous political situation in Cuba complicated the Roosevelt administration's attempts to demonstrate US concern through disaster aid, subsequent catastrophes in the Caribbean Basin presented American officials with more straightforward opportunities to behave as good neighbors. While US officials were busy responding to the disaster in Cuba, in fact, a second major disaster was unfolding across the Gulf of Mexico, giving them just such an occasion.

On September 15, 1933, a hurricane made landfall near Tampico, on the northeastern coast of Mexico, producing a relatively small-scale disaster.[71] Just ten days later, a second, far more powerful hurricane hit precisely the same location, triggering a far greater catastrophe. This second storm completely destroyed Tampico, leaving several hundred people dead and making between 8,000 and 10,000 people homeless.[72] Embracing the chance to put the Good Neighbor Policy on display, the US ambassador to Mexico, Josephus Daniels, urged the Roosevelt administration and ARC to respond swiftly and generously, telling the State Department, "I do not think there ever was a time when the service of the Red Cross was more essential and would be more appreciated."[73] Both parties quickly acted on Daniels's recommendation, with Roosevelt issuing a formal message of sympathy and the ARC sending $25,000 to its sister society, the Mexican Red Cross.[74]

This humanitarian response elicited just the reaction Daniels desired. Mexican radio stations broadcast FDR's message widely, announcing it as "the new proof of America's practice of the 'Good Neighbor' doctrine."[75] Local newspapers, likewise, praised "the sympathy and effective help of our neighbor in the north."[76] Reporting on the political climate in Mexico, Daniels observed that the aid resulted in "a most friendly spirit and feeling towards our Government and people." More than this, he stressed, it offered "proof here that our regard for our neighbors is deep and genuine and expresses itself generously when opportunity is offered."[77]

Just nine months later, in June 1934, a major hurricane in Honduras gave the Roosevelt administration yet another opportunity to demonstrate it neighborliness. The disaster produced by this storm proved far less severe than either the catastrophe in Cuba or the one in Tampico; consequently, ARC leaders made only a token $1,000 contribution to the country. Eager to provide additional assistance, Secretary of State Cordell Hull requested that the War Department send a consignment of tents from the Canal Zone to Honduras.[78] In a visible display of American airpower, eleven army bombers, manned by thirteen officers and twenty-five soldiers, soon landed in Tegucigalpa with this aid.[79] Pleased with the diplomatic outcome, the US minister to Honduras reported that the arrival of US planes and assistance "seemed to have created the most favorable impression." The joint response by the army and ARC, he continued, "cannot fail to have a most happy influence upon relations between Honduras and the United States."[80]

Over the next couple years, maintaining the trend established in Cuba, Mexico, and Honduras, ARC leaders sent the majority of their foreign disaster relief allocations to countries within the Caribbean Basin. In 1935, ARC leaders sent $2,000 to Haiti, $2,500 to Cuba (again), and $1,000 to Honduras (again) when hurricanes struck those countries. And in 1936, they delivered $1,000 to Colombia and $2,500 to El Salvador following earthquakes in those nations.[81] These cash appropriations "were frequently token contributions given at the request of State Department officials," an internal ARC report later conceded, and "in some instances no doubt were partially motivated by the State Department's desire to further 'goodwill' in the country affected."[82] Whether ARC leaders liked it or not, their partners in the Roosevelt administration viewed their organization's disaster assistance as a critical element of Good Neighbor diplomacy.

Although the US government and ARC concentrated mainly on the Caribbean Basin during the mid-1930s, US officials made a few exceptions to this regional focus. In 1934, the ARC sent $5,000 to Poland, after devastating floods affected a fifth of that country, and $10,000 to India, when a major earthquake killed more than 10,000 people in Bihar.[83] They contributed another $5,000 to the Indian Red Cross the following year, after a second powerful earthquake struck Quetta, killing upward of 60,000.[84] And in 1935, the ARC sent $46,000 to China, following catastrophic flooding of the Yellow and Yangzi Rivers.[85] Although this latter figure represented a higher sum than most ARC disaster relief allocations during the decade, it still amounted to less than half of what the organization provided to China in 1931. Like that earlier contribution, moreover, the funds came nowhere close to meeting the real needs of Chinese flood survivors.

In these and other sites during the mid-1930s, the Roosevelt administration thus attempted to behave as a "good neighbor" to nations and empires far beyond the Western Hemisphere. American officials' ability to give generously in these regions was comparatively constrained, however, by the financial situation at home and by other, seemingly more pressing foreign policy priorities.

During Roosevelt's first term in office, the Great Depression, Dust Bowl, and other exigencies led US officials to reduce, yet never fully stanch, their foreign disaster relief commitments. As Roosevelt entered his second term as president in 1937, however, mounting global political instability and emerging armed conflicts posed further impediments to American catastrophic diplomacy. Between 1935 and 1939, the outbreak and intensification of the Italo-Ethiopian War, the Spanish Civil War, and the Second Sino-Japanese War, together with escalating hostilities in Europe, profoundly influenced US foreign policy. These events struck a blow to US foreign disaster assistance, too. Having already decreased in frequency and scale throughout the 1930s, official instances of foreign disaster aid became rarer still during the last few years of the decade, amid rising international tensions, fears of foreign entanglements, and the ongoing economic slump.

Yet even amid these foreign policy crises and challenges, the flow of US foreign disaster aid did not dry up entirely. During the late 1930s, the government, military, and ARC maintained the routines of relief, responding to several international catastrophes.

The most noteworthy of these humanitarian operations came in the early weeks of 1939, in an event that once again demonstrated disaster aid's value to the Good Neighbor Policy and to the Roosevelt administration's broader diplomatic goals in the Western Hemisphere. On January 24, a powerful earthquake struck south-central Chile, causing extensive damage in the cities of Chillán and Concepción. The quake left 15,000 dead and 80,000 homeless, while affecting an estimated 4.6 million people in some way.[86]

This immense disaster immediately attracted the concern of US officials in both Washington and Santiago. Although the US government and ARC had eschewed major disaster relief operations in recent years, many State Department personnel considered it diplomatically advantageous to stage a generous response to the catastrophe, for reasons connected to both Chilean and hemispheric politics. Just one month before the earthquake, Chileans had elected a Popular Front government, led by President Pedro Aguirre Cerda.[87] The victory of his left-leaning party provoked considerable anxiety among US policymakers, who anticipated that the Aguirre Cerda administration might move to nationalize US-owned copper mines and nitrate fields in Chile. At the same time, US officials worried about the potential for violence, and possibly a coup, waged by Aguirre Cerda's opponents.[88] Beyond their specific concerns with Chile, the Roosevelt administration was, in early 1939, deeply preoccupied with the goal of strengthening US relations with Latin America, endeavoring to promote hemispheric unity against the looming threat of global war.

In this context, State Department officials viewed a concerted response to the catastrophe as a way to achieve a trio of objectives: currying favor with the Aguirre Cerda administration for American economic interests, minimizing whatever tensions the disaster might fuel between rightist and leftist factions in Chile, and demonstrating to all Latin American observers the United States' commitment to the Good Neighbor Policy.[89]

Motivated by these aims, the US government, US military, and the ARC organized a rapid response to the disaster in Chillán and Concepción. The morning after the earthquake, after consulting with the State Department, ARC leaders wired $10,000 to the Chilean Red Cross and appealed to the US public for additional donations.[90] The Secretary of War, meanwhile, instructed US Army forces in the Panama Canal Zone to send supplies from their stockpiles to Chile.[91]

Highlighting the role air power had come to play in US foreign disaster relief operations over the past decade, Pan-American Airways put two of its planes at the Chilean government's disposal and assigned several others to transport ARC relief supplies from Miami to Chile, all at no charge.[92] On January 28, supplementing these commercial flights, two US Army bombers departed the Canal Zone carrying 500 army tents, gangrene and anti-tetanus serum, powdered milk, and other relief supplies. Also on board were nine US Army personnel and the ARC's field director in the Canal Zone.[93] Once again, the United States' military and humanitarian geographies in the Caribbean Basin converged, as military bases in Panama enabled officials to respond swiftly to a catastrophe as far away as South America.

Although the US government and ARC reacted quickly to the crisis, US diplomats in Santiago worried for political reasons that this initial aid was insufficient.[94] A week after the earthquake, US Ambassador to Chile Norman

Armour advised the State Department that Argentina, Peru, and Japan had all sent substantial quantities of aid to the country, threatening to overshadow the United States' contribution. The government must do more, he urged, "to avoid any conclusion that other countries were responding more generously in Chile's predicament." Officials must also act "promptly," the ambassador continued, "in order to obtain full credit" for their aid. In Armour's view, simply matching other nations' contributions was insufficient. The government needed to take bolder steps, outshining its rivals in the humanitarian field. The earthquake presented "an excellent opportunity," he advised, "for the United States to do something on a pretty big scale."[95]

Responding favorably to Armour's suggestion, US officials launched an even greater airlift, intended to respond more substantially to the emergency while maximizing the "publicity value" and "moral effect" of their aid.[96] Relying once again on the military to demonstrate the United States' commitment to Chilean relief, the State Department asked the secretary of war to send a B-15 Flying Fortress—one of the largest and most impressive aircraft in the US Army Air Corps fleet—to Chile. Envisioning the diplomatic benefits to be gained from this spectacle, State Department officials imagined that "the effect would be splendid."[97] On February 4, the bomber left Langley Field, Virginia, carrying eleven officers and enlisted men and 3,300 pounds of medical supplies, purchased with ARC funds. Forty-nine hours later, on February 6, the plane and its cargo arrived in Santiago.[98] Describing the mission, the ARC's Ernest Swift boasted that it "was probably the largest cargo of supplies ever sent on one airplane and certainly the longest distance ever flown with relief supplies to the scene of a disaster."[99]

This record-breaking relief flight made just the diplomatic splash US officials hoped. In Washington, the Chilean chargé d'affaires declared that "the United States has proved to be a good neighbor indeed," while the Chilean consul in San Francisco lauded the arrival of a plane "originally built to carry bombs for war time necessity, but carrying now medical supplies to bring mercy to the poor people of the afflicted zone."[100] In Santiago, US embassy staff reported that "the arrival of the flying fortress was prominently recorded in the press," adding that they had received expressions of gratitude from "numerous Chileans" for the relief supplies it carried.[101] Many observers outside Chile voiced their praise as well. In neighboring Peru, for instance, one newspaper cheered the "bombing planes" that went "to Chile to bomb with generosity the places visited by misfortune," while another dubbed the relief flights, more concisely, as "Panamericanism in action."[102] Eager to play the Good Neighbor—and to shore up Chilean and hemispheric support for the

United States—State Department officials welcomed the overwhelmingly positive reactions they received.

By February 6, some twenty US Army officers and troops, several US planes, and large quantities of aid supplies had converged in Chile. Yet in a marked contrast to the Dominican Republic or Nicaragua during the early 1930s, US personnel did not play a substantial role in on-the-ground relief work in Chillán and Concepción. Although the Aguirre Cerda administration welcomed the outpouring of monetary and material aid from abroad, Chilean authorities made it clear that they neither required nor desired US personnel to administer this assistance. Acceding to the Chilean government's wishes, US Army personnel remained in Chile for only a few days. Most of their time was spent conducting relief flights between Santiago and the devastated cities of Concepción and Chillán. By February 9, they had departed, returning to their regular posts in the Canal Zone and the United States.[103]

Additional ARC supplies continued to trickle in by plane and ship over the next several weeks, bringing the organization's combined contributions to just over $40,000 in cash and material supplies. More than $120,000 also flowed in through American private channels, most of it, as ARC leaders observed, from "mining companies with large interest in Chile" and "financial and manufacturing firms doing business there."[104] By late February, for all intents and purposes, the official US aid operation for Chile had concluded. Over the next several months, the House of Representatives debated extending a large government aid package to Chile, intended for both relief and rehabilitation. Yet in the end, the House failed to move forward with this aid appropriation. By the late 1950s and 1960s, as subsequent chapters will explore, rebuilding assistance like this became a regular staple of US foreign disaster aid. For most members of Congress in 1939, however, a federally funded reconstruction project in Chile appeared a bridge too far.[105]

Although the proposed congressional reconstruction assistance for Chile never materialized, the US response to the Chillán and Concepción earthquakes still represented a sizable and highly visible humanitarian undertaking. Even as the world spiraled headlong into a second global war, as this episode revealed, US foreign disaster assistance remained a flexible instrument of foreign policy, a means of fulfilling the United States' sometimes shaky pledge to behave as a Good Neighbor.

Across the mid- to late 1930s, in spite of the limitations imposed by domestic and international events, catastrophic diplomacy had endured, remaining a consistent (if somewhat less prominent) element of US foreign relations. What is more, although US international disaster aid operations constricted during Roosevelt's first two terms in office, that development ultimately proved ephemeral, a temporary deviation from the century's longer-term trends. In the years ahead, US foreign disaster assistance was poised to transform in fundamental and enduring ways, against the tumult of the Second World War and its aftermath.

9

Recasting the Pillars of US Foreign Disaster Assistance

1939–1947

As the 1930s drew to a close, an extraordinary crisis engulfed the globe, an event that was to transform the international system—and the United States' place in it—in fundamental and enduring ways. That crisis, of course, was the Second World War. Conflict had already gripped parts of Asia for much of the decade, but hostilities on that continent intensified with the outbreak of the Second Sino-Japanese War in 1937. Over the next several years, smoldering tensions among European nations, empires, and their respective allies also continued to escalate. In early September 1939, with the German invasion of Poland, the long-looming threat of another global war fully materialized. The United States did not officially enter the Second World War for another two years, joining the conflict as an Allied Power in December 1941. Yet from the late 1930s on, the war affected every facet of American politics and society—not least the system of foreign disaster assistance.

Between 1939 and 1947, from the eruption of hostilities in Europe into the immediate postwar period, the US government assumed a central role in global affairs, involved as never before in waging war, securing peace, and alleviating the human suffering the conflict produced. In the process, the United States' foreign policy apparatus experienced a metamorphosis. Across these eight years, the federal government, the armed forces, and the US security state all expanded dramatically, as did their respective overseas footprints.[1] At the same time, the government developed novel ties with the American voluntary sector, partnering with a host of new nonstate organizations to achieve their mutual objectives abroad.[2] Together, these momentous shifts carried profound implications for the conduct of US humanitarian assistance, forever altering how the three pillars of US foreign disaster aid responded to catastrophes abroad.

During the Second World War era itself, however, the ramifications of these dramatic shifts were not yet apparent or obvious. Much like the First World War era before it, only in retrospect did this period mark a clear turning point in the history of US foreign disaster assistance. Between 1939 and 1947, catastrophes caused by natural hazards took a backseat to other, more pressing diplomatic and humanitarian crises. With their energies focused principally on the winning the war and dealing with its direct human consequences, delivering foreign disaster relief represented at best a secondary priority for the US government, US military, and American Red Cross. Officials' seeming inattentiveness to catastrophes caused by natural hazards was also a result, at least in part, of fortuitous timing. The period between June 1939 and June 1946, as an ARC official subsequently observed, was "largely free from large scale natural disasters both at home and abroad."[3]

This is not to say the US government and its partners abandoned foreign disaster aid entirely. Quite the contrary. Across these eight years, while fighting war and waging peace, US officials provided aid to multiple countries affected by earthquakes, hurricanes, floods, and other natural hazards. During the remaining years of US neutrality, from 1939 to late 1941, the government, military, and ARC sent disaster aid to Brazil, Greece, Mexico, Nicaragua, Peru, Portugal, and Rumania. After the United States' entry into the conflict, American assistance flowed into Afghanistan, Argentina, Cuba, Ecuador, Peru, and Turkey. And in the two years after hostilities ceased, US disaster aid went to England, Poland, Japan, the Dominican Republic, Nicaragua, and Peru.[4]

In their planning and execution, these efforts deviated little from pre-war precursors. The three-pillared system of US foreign disaster assistance continued to function in much the same way as it had since the last world war a generation prior, in accordance with well-worn patterns and routines. American Red Cross leaders allocated roughly $200,000 toward these efforts, with most of these contributions taking the form "of token gifts of one, two, five or ten thousand dollars."[5] Striving to ensure that "friendly relations and goodwill are immediately involved," the State Department typically sent these ARC contributions to the highest-ranking US diplomat in each affected country, with instructions to deliver the aid in as visible and ceremonious a fashion as possible.[6] In many instances, US military personnel, vehicles, and aircraft assisted with the delivery and distribution of ARC-funded food and other supplies, maintaining their long-standing humanitarian collaboration.

But if the Second World War did not alter the conduct of US foreign disaster assistance in the immediate moment, its effects would in time prove deeply consequential, due to the monumental structural changes it unleashed. Involvement in this conflict revolutionized US diplomatic, military, and humanitarian relations with the wider world and remade the US foreign policy establishment. By 1947, as the Second World War era gave way to the postwar period and an emerging Cold War, the three essential pillars of the US foreign disaster aid system—the government, the military, and the American voluntary aid sector—had each undergone profound changes, becoming distinctly different versions of their prewar selves. Collectively, these developments had enormous and enduring consequences for US foreign disaster assistance, transforming how American officials responded to international disasters in the decades ahead.

❂ ❂ ❂

Among the most important of these developments was the expansion of the US military and diplomatic presence throughout the world, a function of the United States' ascendance as a global superpower. In mobilizing for the conflict, to borrow one historian's phrase, the nation became nothing less than a "warfare state."[7] As this process unfolded between 1939 and 1947, it completely—if inadvertently—remapped the United States' humanitarian geography, leaving the federal government far more capable of conducting foreign disaster assistance operations than ever before.

The meteoric growth and global spread of the US Armed Forces during the Second World War was one key facet of this process. Between 1940 and

1945, the US Army, Navy, and Marines mobilized millions of servicepeople, deploying them to Europe, the Pacific, and beyond. By May 1945, the US Armed Forces reached a peak strength of more than 12 million personnel, 7.6 million of them stationed overseas.[8] As troop levels swelled, so did military production. The number of US Navy vessels, including both warships and auxiliary vessels, grew from around 1,000 in 1940 to nearly 69,000 by war's end.[9] The size of the army and navy's air fleets expanded, too, from just 800 aircraft in 1939 to 300,000 by 1945.[10]

During these same years, the number of US overseas military installations grew astronomically, transforming the United States into what historians have called an "empire of bases" and a "pointillist empire."[11] In 1938, the government maintained very few overseas military installations, most of them in US territories. By the end of the Second World War, it controlled or enjoyed access rights to more than 2,000 bases and 30,000 smaller military installations throughout the world, a combined result of wartime base construction, military occupations, and leasing agreements with other nations.[12]

As the Second World War drew to a close, it initially appeared that this tremendous expansion of the US military would prove a temporary phenomenon. After the conflict ended, the US Armed Forces demobilized fairly quickly, contracting to 1.6 million personnel by 1947. Officials also began decommissioning many ships and aircraft, and they relinquished access rights to some overseas military installations.[13]

The US military, however, never returned to the prewar status quo. Even at its nadir, the postwar strength of the armed forces remained five times what it was before the war. Hundreds of thousands of military personnel continued serving abroad, stationed on a vast network of overseas bases and greatly enlarged naval fleets.[14] Already by 1947, moreover, the Truman administration and Congress started taking steps to reverse postwar cuts to the military. Reacting to the perceived threats of the Soviet Union and international communism, policymakers set in motion a massive Cold War remobilization of the military, which was to escalate dramatically over the next few years.[15] The warfare state was now a permanent condition.

Paralleling the military's growth, the State Department and Foreign Service expanded substantially during the Second World War era. Between 1939 and 1947, the government's diplomatic, intelligence, economic, and cultural activities overseas increased in both scope and complexity. As they did so, the foreign policy bureaucracy inflated in concert. During the war years, the Roosevelt administration established a host of emergency agencies to conduct various aspects of foreign affairs, among them the Office of Strategic

Services, the Office of War Information, and the Board of Economic Warfare. Once the conflict ended, the State Department made the work of many of these agencies permanent, transferring some of their functions to existing units while creating new bureaus and agencies to oversee the rest.

To perform the labor this mushrooming bureaucracy required, the State Department brought in an influx of new diplomatic staff. In 1940, the State Department employed just under 2,000 personnel, 840 of them in the Foreign Service. By the decade's end, the number of State Department employees exceeded 16,000, with roughly half of them stationed overseas.[16] To house them all, Congress launched a Foreign Service buildings program in 1946, financing the acquisition and construction of a wave of embassies, consulates, and other diplomatic facilities abroad.[17]

By 1947, both the State Department and the armed forces had established a formidable footprint across the world. In late July that year, reflecting the amplified importance of these institutions to postwar US foreign affairs— and the perceived need to modernize and restructure these institutions amid a dawning Cold War—Congress and the Truman administration adopted the National Security Act of 1947. This landmark legislation overhauled the military and diplomatic establishments. Specifically, it created a unified Defense Department, comprising the War and Navy Departments and a new, third service branch, the Department of the Air Force. It also established the National Security Council, designed to coordinate diplomatic and military goals, policies, and actions more effectively. Finally, it created the Central Intelligence Agency, charged with conducting strategic analyses and clandestine activities abroad.[18]

The passage and implementation of the National Security Act marked a watershed in US foreign relations, a testament to the government's new position in international affairs. Although the precise size and global reach of the foreign policy establishment fluctuated in coming years, by 1947 it had changed in significant and enduring ways, enabling the government to project diplomatic and military power in ways unimaginable less than a decade before.

These shifts also positioned the government to project its humanitarian power to an extent theretofore unprecedented. Although this outcome was never the main objective or intention of American policymakers, it is difficult to overstate the effect these momentous changes had for the government's foreign disaster aid capabilities. With an appreciably larger State Department, and with thousands of US Foreign Service personnel and CIA agents now scattered throughout the world, American officials were much better able to gather information about foreign catastrophes and to respond to these crises efficiently. With hundreds of thousands of army, navy, marine, and air force personnel now stationed overseas, and with access rights to thousands of overseas bases, airstrips, and other military installations throughout the world, US military forces were positioned to reach a far greater number of places beset by disaster, and to get to these sites much more rapidly. The military's expanded global presence also gave US forces convenient access to large stockpiles of rations, tents, and medical supplies, as well as vast, state-of-the-art fleets of ships, aircraft, and land vehicles, all of which the newly established Defense Department could use for disaster relief operations.

An incidental effect of the United States' military and political ascendancy during the Second World War era, in short, was to extend the government's humanitarian geography and to improve its command of humanitarian logistics. From the late 1940s on, using these enhanced capabilities, the state and its agents were to play a greater and greater role in foreign disaster assistance operations.

A second critical legacy of the Second World War era was a dramatic shift in the government's relationship to American foreign assistance, broadly defined. Between 1939 and 1947, Congress and the White House committed unparalleled levels of funding for a comprehensive slate of international war relief and recovery projects, adopting novel aid legislation and establishing new federal agencies to administer US assistance abroad. Although most of these projects were temporary initiatives, dismantled as the upheaval of war subsided, the US government's assumption of a more active role in foreign military and civilian assistance endured. In time, government officials would similarly expand the state's role in peacetime foreign assistance programs— including, eventually, the field of international disaster aid.

The government began taking on new foreign military and civilian aid commitments even before the United States entered the Second World War. The Lend-Lease Act, signed into law in March 1941, marked an early move in this direction. Under its auspices, the Roosevelt administration provided roughly $50 billion in military and economic assistance to more than thirty Allied nations over the course of the war. The government's entrance into wartime humanitarian relief (as opposed to military or economic aid) followed in late 1942, with the establishment of the Office of Foreign Relief and Rehabilitation Operations, a federal agency charged with providing material assistance to civilians in liberated areas.[19]

One year later, in November 1943, the US government's humanitarian portfolio expanded further still with the establishment of the United Nations Relief and Rehabilitation Administration (UNRRA). Comprising forty-four nations, UNRRA was created to provide material, economic, and medical assistance to victims of war in areas under Allied control. The new organization may have been international and multilateral, but the United States played a dominant role in it. During the organization's four years of operation, the US government committed some $2.7 billion to its operations, roughly 70 percent of UNRRA's total expenditures. Three Americans served as its directors, performing their duties from UNRRA's headquarters in New York.[20]

Into the early postwar era, the US government remained actively involved in foreign assistance projects. Between 1945 and 1947, Congress and the Truman administration provided billions of dollars in direct and indirect aid to allied and occupied countries, in the form of economic assistance, military aid, technical assistance, and humanitarian relief. During these years, the US government also contributed hundreds of millions of dollars to UNRRA's ongoing efforts, remaining its main patron until the organization ceased operations in 1947.[21]

If these actions signified the government's newly expanded commitment to postwar assistance, further evidence that American officials intended to maintain the state's involvement in peacetime (and Cold War) foreign aid came on June 5, 1947, with the announcement of the European Recovery Program, or Marshall Plan. Though it did not take effect until the following year, the proposal for this multibillion-dollar economic and technical assistance package for Europe indicated the permanent role the US government intended to assume in the foreign assistance field.[22] As 1947 drew to a close, the government's foreign aid contributions had already "reached formidable dimensions," as President Harry Truman put it.[23] All signs pointed to a more robust role for the government in American foreign assistance during the postwar era.

At the time, the policymakers involved in constructing the state's newly enlarged foreign assistance apparatus devoted minimal consideration to catastrophes caused by natural hazards. Instead, they focused primarily on the human consequences of war: providing relief to civilians and returning soldiers, assisting refugees and displaced persons, and contributing to the monumental tasks of rebuilding and reconstruction. Yet as federal foreign aid programs expanded in scope and scale in coming years, the inexorable result was to transform how the US government conducted all forms of international assistance—including, not least, its responses to foreign "natural" disasters.

A third key development to transpire during the Second World War era was the forging of new ties between the US government and an expanding American voluntary aid sector. Although the ARC retained its longtime position as the state's humanitarian auxiliary, dozens of other voluntary aid agencies entered into their own formal partnerships with the government at this time, relationships made permanent after the conflict ended. With this restructuring of American voluntary assistance between 1939 and 1947, the cast of players involved in US international humanitarian endeavors grew considerably, altering the dynamics of postwar foreign disaster aid and the associational state.

During the war and its early aftermath, even as the federal government assumed many new foreign aid commitments, officials continued to depend on the American Red Cross to provide much humanitarian aid on the nation's behalf, vigorously promoting its relief work.[24] Benefitting from its close ties with the government, the ARC administered a far-reaching program of military and civilian assistance throughout the war-torn world. During the era of US neutrality, the ARC delivered $50 million in cash and supplies to belligerent and neutral nations. After the United States entered the war, the ARC swiftly ballooned in size, as did its sphere of humanitarian activities. By 1945, 36 million Americans had joined the organization, contributing $666 million to support its domestic and foreign relief activities. Although membership and donations dropped off once hostilities ceased, the ARC maintained a robust humanitarian program throughout the early postwar years, expending more than $200 million between 1945 and 1947.[25]

Between 1939 and 1947, the ARC once again served as the US government's principal instrument for humanitarian aid, much as it had during the First World War—and indeed, much as it had since the turn of the twentieth century. Some substantial changes lay in store, however, which reduced the federal government's influence over the ARC and its future operations. More than this, they contributed to a gradual erosion of the ARC's position as the government's primary humanitarian auxiliary.

In May 1947, acting at the request of the ARC's leadership, Congress adopted a set of significant revisions to the organization's charter, which had remained largely unaltered since 1905. While preserving many elements of the original charter—including the ARC's mandates "to carry out a system of national and international relief in time of peace" and to mitigate and prevent the suffering "natural" calamities caused—the amendments overhauled the ARC's governing structure. Most notably, they replaced its eighteen-member central committee with a fifty-member board of governors. Crucially, the US president now appointed a much smaller fraction of this national leadership: 16 percent rather than the 33 percent it was previously. As a result, representatives from the Departments of State, Defense, and other federal agencies exercised a proportionally weaker voice in the ARC's governance and decision-making than before.[26]

The adoption of the charter revisions in 1947 marked an important turning point for the ARC and, by extension, for US foreign disaster assistance. Although the ARC continued to function as a unique federal instrumentality, including in the field of foreign disaster relief, in the years ahead it became somewhat more independent from the government, and from US foreign policy, than it was during the first half of the twentieth century. At the same time, while US government officials continued to cooperate closely with the ARC in international disaster response, they became less wholly reliant on the organization to finance, plan, and administer these humanitarian operations.

While the ARC's increased autonomy after 1947 partially explains this shift, two additional factors were equally responsible for it. First, as discussed above, the US government itself played an ever more active role in foreign assistance. As the state's involvement in financing and administering international disaster relief expanded, government officials assumed many of the humanitarian tasks they once relied on the ARC to perform. Second, and just as crucially, the government entered into formal partnerships with multiple *other* American voluntary associations, cooperating with these entities to fund and carry out US foreign assistance operations.[27] The entry of these new organizations into the humanitarian field further transformed the conduct of US foreign disaster aid, giving the government many additional auxiliaries in the American voluntary sector.

The roots of this expanded state-voluntary partnership lay at the beginning of the war. Following the outbreak of hostilities across Asia, Europe, Africa, and the Middle East during the late 1930s, scores of US voluntary aid organizations—established and new, faith-based and secular—began raising funds and collecting supplies for the victims of these conflicts. In early 1941, seeking to better regulate and coordinate the work of these organizations and ensure it aligned with the US government's interests and objectives, the State Department created a Committee on War Relief Agencies, a first step toward extending federal control over the nation's voluntary assistance efforts.

The level of state oversight expanded in July 1942, when Roosevelt established the President's War Relief Control Board (WRCB). All voluntary agencies that sent relief overseas were now required to register with the WRCB and receive a license from it to operate. The WRCB, in turn, served as a liaison between licensed voluntary agencies and relevant federal agencies, creating a novel channel of voluntary-state communication and cooperation. The only organization exempt from registering, significantly, was the American Red Cross, due to its special relationship with the government.[28]

With this wartime measure, the Roosevelt administration laid the foundations for a permanent, postwar alliance between the government and American voluntary organizations other than the ARC. After the conflict ended, in May 1946, President Truman terminated the WRCB by executive order, relinquishing the federal government's wartime powers to regulate private charities involved in overseas relief work.[29] Yet even as he surrendered government authority over the American voluntary sector, Truman sought to maintain some elements of this formal wartime partnership. To that end, he recommended the establishment of a new, peacetime committee, designed "to tie together the governmental and private programs in the field of foreign relief."[30]

On July 10, 1946, fulfilling the president's request, Truman's secretaries of state and agriculture established the Advisory Committee on Voluntary Foreign Aid (ACVFA). Headquartered in Washington, ACVFA served as a liaison between American voluntary agencies with overseas missions and the Departments of State, Agriculture, and other pertinent government agencies.[31] "On matters concerning the foreign policy of the Government of the United States," the committee's leaders affirmed, ACVFA was to "be guided by the Department of the State."[32]

To register with ACVFA, voluntary organizations first had to apply for membership; once accepted, these organizations were required to provide regular financial statements and reports of their international activities to the State Department. Although registering with ACVFA was not mandatory, member organizations enjoyed several distinct benefits. Specifically, the government started reimbursing ACVFA affiliates for the expense of shipping humanitarian aid supplies overseas.[33] In 1949, Congress also began granting these organizations access to surplus agricultural commodities "for the assistance of needy persons outside the United States," a program that expanded significantly in 1954.[34] The State Department, meanwhile, consulted closely with ACVFA's leadership about humanitarian issues and shared privileged information about the politics surrounding them, in ways it had traditionally done only with the ARC. In late 1949, institutionalizing this relationship, ACVFA became formally attached to the State Department.[35]

These perks were clearly attractive. By 1949, some four dozen voluntary organizations had registered with ACVFA, among them CARE, Catholic Relief Services, Church World Service, the Jewish Joint Distribution Service, Lutheran World Relief, and the American Friends Service Committee.[36] Many more were to follow in their footsteps in the decades ahead.

The creation of ACVFA did not supplant the government's unique relationship with the ARC (which remained exempt from either registering or reporting). It did, however, mark the advent of similarly close partnership between the federal government and many of the nation's other major voluntary aid agencies, significantly enlarging this third pillar of US foreign disaster assistance. "Foreign relief," as ACVFA's first chairman remarked in 1946, had "reached a new and critical stage."[37]

Initially, ACVFA's member organizations devoted their energies and resources mainly toward postwar relief and recovery efforts. As these issues became less pressing, however, many broadened or revised their institutional missions to encompass new activities, including international disaster relief. By the early 1950s, a growing number of ACVFA agencies began participating alongside the ARC, the government, and the military in the nation's foreign disaster assistance operations. From there, the role these voluntary organizations played in American catastrophic diplomacy only continued to expand.

✪　✪　✪

In the decades spanning the First and Second World Wars, the methods and practices of US foreign disaster assistance evolved in measured yet discernible ways. Between the late 1910s and the late 1930s, responding to catastrophes in other nations became a more established and routine diplomatic norm. While working within the three-pillared system of foreign aid, officials also started testing novel approaches to disaster assistance, which encompassed not only emergency relief but also more comprehensive reconstruction and development projects.

Building on these foundations, the Second World War era sowed the seeds for sweeping changes in the organization and administration of US foreign disaster assistance. Between 1939 and 1947, the entities responsible for planning and conducting the nation's foreign disaster aid operations morphed into distinctly different versions of their prewar selves. In the course of fighting the war and addressing the immense humanitarian suffering it produced, the federal government, US military, and American voluntary sector swelled in size and global reach. As they tested novel approaches to wartime and postwar relief, they also established new bureaucratic structures to carry out humanitarian efforts more effectively.

Although these seismic shifts were not all permanent, the cumulative, enduring consequences of the Second World War era were no less profound. By the late 1940s, these collective developments began to alter how the US government and its partners responded to all global humanitarian crises—including, not least, those caused by natural hazards. Over the next three decades, they wholly revolutionized American catastrophic diplomacy.

Between the late 1940s and the mid-1970s, as the chapters in part III trace, three key developments transformed the conduct of US international disaster aid. First, control over the nation's foreign disaster aid efforts became increasingly centralized under the auspices of the federal government. In Washington, new foreign assistance agencies joined the State Department in planning and administering disaster relief operations abroad, while new aid legislation gave them the resources and personnel to do so. Within other countries, US Foreign Service Officers, military personnel, and development experts—stationed throughout the world in numbers unthinkable as late as the 1930s—became active participants in these humanitarian operations. Flying from long-established sites like the Panama Canal Zone and from many new US air bases in Germany, Italy, Japan, Saudi Arabia, Libya, and other sites, US military planes regularly transported ARC supplies, military stocks, and US personnel to disaster-stricken countries.

Amid these sweeping changes, the government did not sever its long-standing ties with the American voluntary sector. It continued to collaborate closely with the ARC to respond to catastrophes abroad. From the early 1950s on, it cooperated with ACVFA member organizations in a similar manner. Even so, responsibility for funding, coordinating, and conducting US foreign disaster assistance efforts fell more and more under the state's purview. Between the late 1940s and the mid-1970s, in short, the system of US foreign disaster assistance remained intact, yet two of its pillars—the federal government and the military—became progressively more dominant in its operations.

Second, the shifting geopolitical and intellectual landscape of the postwar decades shaped both the practices of US foreign disaster aid and the government's motives for providing it. Although American officials had long wielded disaster assistance a tool of foreign policy, a pair of entangled dynamics—the global Cold War and the rising tide of anticolonial nationalism and decolonization—now influenced their political and diplomatic calculations profoundly, informing decisions about where to send disaster aid and whether it was in the United States' strategic interest to do so. At the same time, mounting concerns over the problems of modernization and international development affected American understandings about natural hazards and the function of disaster aid. In both theory and praxis, US foreign disaster relief, recovery, and prevention efforts became more tightly linked with the nation's burgeoning global development assistance initiatives.

Finally, the US government and its partners began responding to foreign disasters more frequently, more swiftly, more liberally, and more globally than ever before. As the scale and scope of these humanitarian operations increased, approaches piloted during the interwar years—military airlifts, donations and sales of surplus commodities, expert advisory missions, and far-reaching reconstruction and development projects—became a customary part of the nation's foreign disaster assistance efforts. As disaster relief operations grew more numerous and more complex, however, problems started to surface. Eventually, mounting frustrations over these issues led government officials to undertake a series of bureaucratic and legislative reforms, intended to improve the coordination of US foreign disaster assistance and to further centralize it under government auspices.

Between the late 1940s and the mid-1970s, these collective trends fundamentally changed the United States' postwar catastrophic diplomacy. None of these momentous shifts, it bears emphasizing, occurred overnight. American catastrophic diplomacy continued to evolve along a gradual path, drifting steadily if somewhat haphazardly toward these outcomes in the three decades after the Second World War.

PART THREE

Drifting toward Centralization and Coordination

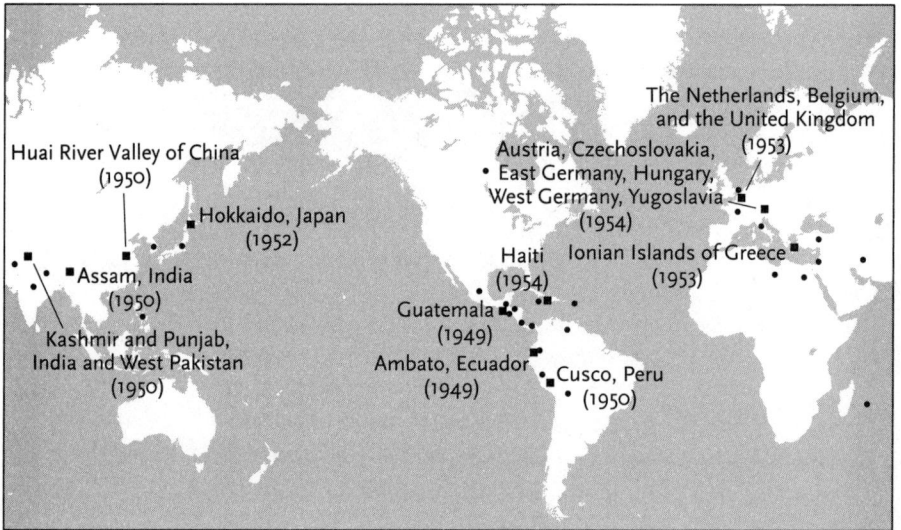

Huai River Valley of China
(1950)

Hokkaido, Japan
(1952)

Assam, India
(1950)

Kashmir and Punjab,
India and West Pakistan
(1950)

Austria, Czechoslovakia,
East Germany, Hungary,
West Germany, Yugoslavia
(1954)

The Netherlands, Belgium,
and the United Kingdom
(1953)

Haiti
(1954)

Ionian Islands of Greece
(1953)

Guatemala
(1949)

Ambato, Ecuador
(1949)

Cusco, Peru
(1950)

**Recipients of official US foreign disaster assistance
referenced in chapter 10, 1948–54**

10

New Mechanisms, New Motivations

1948–1954

The launch of the Marshall Plan in 1948 is commonly cast as the US government's first real foray into foreign assistance.[1] Yet as this book has shown, there were myriad precedents for this landmark program. Already by the late 1940s, the government had a long history of involvement in international disaster relief and in foreign aid more broadly. Across the first half of the twentieth century, government and military officials, together with the American Red Cross, frequently contributed humanitarian, economic, and technical assistance to other nations. Decades before the Marshall Plan, foreign assistance had become an established instrument of US foreign policy.

That said, it is also true that 1948 marked a decisive turning point. The next quarter-century witnessed a revolution in US foreign aid—and, with it, US international disaster assistance.

Signs of change first appeared between the late 1940s and the mid-1950s, the focus of this chapter. Between 1948 and 1954, spurred by the hardening of the global Cold War and a first major wave of decolonization, US government officials established a growing roster of federal foreign aid programs and agencies, charged with administering postwar relief, economic and technical assistance, and food aid abroad. For the most part, significantly, these bureaucratic and legislative developments did not pertain explicitly to natural hazards or international disaster assistance, both of which remained something of an afterthought for foreign policy planners at the time.

Nevertheless, the wholesale restructuring of the foreign assistance architecture had many indirect consequences for American catastrophic diplomacy, as new governmental aid agencies, their resources, and their personnel began assuming a more prominent role in US international disaster response. If the established methods and practices of US foreign disaster assistance did not transform immediately during these years, a steady drift toward state centralization and coordination over foreign disaster had clearly begun in hindsight. Between 1948 and 1954, amid a postwar revolution in US foreign aid, American catastrophic diplomacy was embarking down its own, more gradual evolutionary path.

What is more, although bureaucratic change occurred rather slowly, the frequency and reach of US foreign disaster assistance operations grew swiftly and dramatically during these years. Gravely concerned with the apparent threats posed by communism and anticolonial nationalism and eager to win friends and garner influence across a rapidly changing world, US officials embraced disaster aid as a tool for achieving multiple diplomatic and strategic objectives. Between 1948 and 1954, driven by these novel political motivations and relying on both new and established mechanisms of aid, the government and its partners responded to dozens of catastrophes globally. They aided survivors of devastating storms in Northern Europe and Haiti; destructive earthquakes in Greece, India, Japan, and Peru; catastrophic floods in South Asia, Central America, and Central and Eastern Europe; and others too numerous to mention. Together, these episodes reveal the changing character of US foreign disaster assistance at midcentury.

✪   ✪   ✪

The late 1940s and early 1950s were a turbulent time for the architects of US foreign policy. Cold War divisions solidified in Europe, as American and Soviet leaders sparred over the future of Germany, vied for influence in the

Western and Eastern blocs, and threatened one another with nuclear annihilation. With the Communist Revolution's success in China in 1949 and the outbreak of the Korean War the following June, Asia became the second main theater in a now global Cold War. In conjunction with these events, bloody struggles for national liberation in French Indochina, together with concerns about the political fates of several newly independent nations—among them the Philippines, India, Pakistan, and Indonesia—made Asia a locus of US foreign policy concern. In the Western Hemisphere, US officials grew increasingly apprehensive about the threats that communism, populism, and economic nationalism appeared to pose. They therefore kept a wary eye on Latin American countries and their internal politics, determined to forestall the hemispheric spread of these "radical" ideas.

In response to these emerging concerns and threats, Congress and the Truman administration undertook a remarkable expansion of the government's peacetime foreign aid programs. Building on foundations established during the Second World War era, they launched multiple new aid initiatives, embracing economic and technical assistance as critical instruments of postwar foreign policy. These included, most notably, the European Recovery Program or Marshall Plan, administered from 1948 to 1951 and designed to spur economic recovery in Western Europe; the Point Four Program, launched in 1950 to provide technical assistance to "underdeveloped areas" of the world; and the creation, in 1951, of the Mutual Security Agency, responsible for all US foreign economic assistance other than Point Four technical aid.[2] By the time Truman left office in 1953, foreign assistance was firmly ensconced in the federal bureaucracy.[3]

Significantly, none of these governmental aid initiatives directly addressed natural hazards or the disasters they precipitated. The legislation establishing these new federal aid programs contained no explicit language about international catastrophes or disaster relief. Within the government's new foreign aid agencies, moreover, there were neither specific departments nor personnel devoted to these particular humanitarian problems.

This neglect of foreign disasters at midcentury is in some ways surprising, for during these same years, policymakers devoted considerable attention to catastrophes *within* the United States and its territories. In 1950, Congress passed the Federal Disaster Relief Act, a landmark piece of emergency management legislation. The law authorized the president to issue disaster declarations and provide supporting assistance to state and local governments grappling with major catastrophes. It also empowered federal agencies to participate in these efforts.[4] With the passage of this legislation, Congress greatly expanded the federal government's role in domestic disaster management, endeavoring to make it a more orderly and systematized process. Foreign disaster aid, by contrast, remained glaringly absent from both federal law and bureaucracy.

Policymakers eventually took steps to rectify this oversight, but in the short term, they continued responding to international catastrophes on an ad hoc basis, contributing assistance through more traditional channels. Throughout the late 1940s and early 1950s, the State Department retained the primary responsibility for coordinating the nation's official responses to foreign disasters, working with the US Foreign Service and the Defense Department to deliver aid and administer humanitarian operations. The American Red Cross, meanwhile, remained the state's principal auxiliary in international disaster relief. As it had for decades, the ARC served as the primary source of funds and supplies for the nation's official relief efforts.

Although the bureaucratic and legal mechanisms of US foreign disaster assistance changed only minimally during the late 1940s and early 1950s, the frequency of disaster relief efforts increased precipitously during these years. During Truman's second presidential term, US disaster aid reached survivors of floods in Bolivia, Canada, France, Guatemala, India, Italy, Mexico, Palestine, Panama, and Pakistan; of tropical cyclones in Jamaica, Japan, the Philippines, and Réunion; and of earthquakes and volcanic eruptions in Colombia, Costa Rica, Ecuador, El Salvador, Nevis and Saint Kitts, Peru, and Turkey.[5] The postwar United States, as this roster suggests, was fast expanding its humanitarian reach.

Motivating US officials to assist many of these places, and guiding their actions and decisions on the ground, were the era's nascent and most pressing foreign policy concerns. Cold War fears and anxieties profoundly influenced US responses to catastrophes in multiple nations across the world. In others, the politics of decolonization and a desire to cultivate postcolonial ties and allegiances shaped the government's humanitarian calculations. Elsewhere, emergent security concerns and shifting political alliances led officials to respond to catastrophes in divergent and disproportionate ways. A survey of US responses to major catastrophes in Latin America, South Asia, and East Asia brings these political dynamics more clearly into focus while illustrating the midcentury system of US foreign disaster aid in action.

Between August 1949 and May 1950, a trio of disasters gripped Central and South America. The first and largest of these catastrophes began August 5, 1949, when a strong earthquake occurred in Ecuador, its epicenter near the city of Ambato. The quake, together with the aftershocks and landslides it generated, killed between 5,000 and 6,000 people and left an estimated 100,000 homeless.[6] Then, in mid-October, a period of unusually heavy rainfall brought widespread flooding to Guatemala, triggering a major disaster in that country. Leaving 1,000 dead and up to 70,000 homeless, the floods caused significant damage to the nation's coffee and banana crops, the backbone of Guatemala's export economy.[7] The third of these hemispheric calamities struck seven months later, on May 21, 1950, when a powerful earthquake occurred in the Peruvian city of Cusco. Though relatively few people died, the material damages were tremendous, with nearly 20,000 losing their homes.[8]

The official US responses to these catastrophes followed a fairly similar trajectory. In all three cases, the government's first step was to dispatch survey missions from the Panama Canal Zone (now headquarters of the US Caribbean Command) to the scene. Composed of army and air force personnel and ARC staff assigned to the military, these teams were charged with delivering emergency supplies, investigating conditions, cooperating with local authorities, and assessing the need for additional US aid.[9]

Based on the preliminary reports of these survey missions, the US government and ARC went on to provide additional aid to each country, the second stage of their joint response. In Ecuador, this phase of the relief operation took the form of an extensive airlift. From the Canal Zone, the US Air Force ultimately made seventy-four relief flights to Ecuador, delivering $135,000 worth of tents, blankets, medical supplies, and sanitation equipment, all donated by the ARC.[10] In neighboring Peru, four C-47s, two C-54s, and a B-17 flew in thousands of dollars' worth of aid supplies right after the Cusco earthquake, with additional relief flights to follow.[11] To Guatemala, the ARC shipped 150 tons of corn to extend emergency food supplies. The US military also loaned equipment and heavy machinery to the Guatemalan government, to use in carrying out needed repairs.[12]

Supplementing this material aid, the government and ARC deployed American advisers to each country to assist local authorities in planning relief and recovery efforts. After the Ambato earthquake, the ARC assigned two of its "most experienced disaster relief experts" to Ecuador, where they remained for several weeks.[13] A few months later, President Truman dispatched a delegation of twenty prominent US medical professionals to Guatemala to assist with flood recovery efforts. Led by the deputy surgeon general of the air force and Truman's personal physician, their team consulted with Guatemalan officials on matters relating to public health and disaster recovery.[14] In both Ecuador and Peru, meanwhile, health and sanitation missions of the

Institute of Inter-American Affairs (IIAA)—a US government development agency that was the hemispheric precursor to the Point Four Program—were already present when the earthquakes destroyed Ambato and Cusco. In each country, IIAA personnel traveled to the disaster zone, where they administered typhoid vaccinations to disaster survivors, dusted DDT to prevent typhus, set up temporary chlorination systems to provide safe drinking water, and performed other public health–related tasks.[15]

Just as they always had, diplomatic motives and strategic calculations lay behind the US government's decisions to extend this disaster assistance to Ecuador, Guatemala, and Peru. For foreign policymakers at midcentury, much like their predecessors, a primary objective of disaster assistance was to burnish the United States' image in the Western Hemisphere. Following the Ambato earthquake, for instance, Secretary of State Dean Acheson reasoned that contributing aid would "be a concrete indication of our good will and would have a beneficial effect on our foreign relations, not only with Ecuador," he added, "but with the American republics as whole."[16] Former president Herbert Hoover felt much the same way. The earthquake, he told Truman, presented "an opportunity to advance greatly the 'good neighbor' policies."[17] Through disaster aid, these men and their associates imagined, the government and ARC could bolster support for the United States throughout the entire region.

Complicating this more conventional diplomatic reasoning, however, were newer political concerns: the escalating Cold War and the perceived threat communism posed to Latin America. Desperate to maintain allies and forestall the spread of Soviet influence in the Western Hemisphere, many US government officials viewed disaster assistance as a useful tool for achieving those objectives. Deciding how best to achieve these twin goals, however, sometimes led them in contradictory directions. For US diplomats in Peru, simmering Cold War anxieties underscored the urgency of US disaster aid. Advising the government and ARC to send aid to Cusco, a "well-known communist stronghold," the US ambassador in Quito considered the disaster "a good opportunity for us to get in some good licks for our side by proving to the people of Cuzco—Communists and all—that we do not intend to let them down in their moment of need."[18]

In Guatemala, by contrast, Cold War concerns led US officials to make very different recommendations. There, worries about communist infiltration sowed grave doubts about the necessity and wisdom of providing disaster aid at all. Like his counterparts in Peru, the US ambassador to Guatemala, Richard Patterson, initially requested ARC aid for flood survivors, calling it "an effective good will gesture" that "would appear desirable politically." Yet he also expressed apprehension, informing the State Department that the catastrophe had "been deliberately exaggerated for political reasons."[19] Once American aid arrived, Patterson and his associates escalated their criticisms. Rehashing age-old tropes and stereotypes, and echoing their predecessors in many prior US disaster relief operations, they accused Guatemalan authorities of distributing American supplies in "an unscrupulous and corrupt manner" and selling them "for the personal gain of Government officials."[20]

Underlying these critiques were much deeper anxieties, widely shared among US officials, about Guatemala's apparent leftward turn. "Communist penetration has made startling progress in Guatemala," Patterson warned the State Department, adding that it "may become the focus of Soviet activities in the Western Hemisphere." Blaming "the leftist administration" of President Juan José Arévalo for the "build-up of an ordinary flood into a national disaster," Patterson reversed his initial stance and advised against sending additional disaster aid to Guatemala.[21] Persuaded by his warnings, the State Department and ARC accepted his recommendation, bringing the relief effort to a close.

Collectively, the US responses to the disasters in Ecuador, Guatemala, and Peru provide a good illustration of the mechanisms of US foreign disaster aid and the motives animating it during the late 1940s and early 1950s. Guided by Cold War concerns and a desire to boost US regional influence, the Truman administration used multiple tools at its disposal—military airlifts, gifts of ARC supplies, and expert advisory missions—to respond to catastrophes throughout Latin America. Such patterns and dynamics, though, were hardly confined to the Western Hemisphere. They resurfaced across many US disaster relief operations throughout the midcentury world.

Much as they did in Central and South America, a complex array of diplomatic and humanitarian motives guided the Truman administration's responses to disasters in South Asia. A few months after the earthquake in Cusco, a cascade of devastating catastrophes struck India and Pakistan. While concerns about the Cold War's expansion into Asia shaped the government's responses to these crises, in this instance the politics of decolonization proved equally consequential. Just three years earlier, both India and Pakistan had achieved independence from Great Britain, becoming two of the world's newest nation-states. In this postcolonial moment, US officials sought to win the allegiance of both countries and cultivate them as anticommunist allies. In 1950, they looked to disaster aid to achieve those objectives.

The string of calamities in South Asia began on August 15, when a powerful earthquake occurred in the northeast Indian state of Assam. Calling the earthquake "the greatest within the history of India," contemporaries estimated that 1 million people lost their lives and another 5 million became homeless.[22] Compounding this already grave humanitarian crisis, heavy rains brought widespread flooding to the region later that month, affecting not only Assam but also the nearby Indian states of West Bengal, Bihar, and Uttar Pradesh, as well as much of Pakistan's East Bengal province.[23] Then, in mid-September, catastrophic floods ravaged Kashmir and Punjab, transnational regions forming the northern border of India and West Pakistan. Declaring it the "greatest disaster from natural causes . . . in known history," US diplomats reported that the floods left thousands of villages and farms uninhabitable, while washing away communication lines, railways, and vast fields of staple crops.[24]

Watching as these disasters uprooted millions of lives throughout India and Pakistan, US diplomats in both countries urged the State Department to offer aid. They had multiple reasons for their appeals. On a humanitarian level, many recognized the enormous challenges confronting the new nation-states, difficulties these new crises only exacerbated. It was hard to overstate "the overwhelming weight of the burden . . . [caused] by these disasters," the consul general in Bombay noted, for populations "already struggling with such huge human, social, and economic problems."[25]

At such a precarious and pivotal moment, US officials also spied a propitious diplomatic opportunity. By aiding survivors of these concurrent crises, they reasoned, the Truman administration stood to enhance the United States' image in postcolonial South Asia, gaining the esteem of populations "friendly to [the] US and whose friendship is of value."[26] As in Latin America, concerns about the threat of communist penetration made the quest to nurture allegiances in the region appear all the more urgent. "For political reasons I believe it would be extremely helpful if we could show humanitarian interest," the US ambassador to India told the State Department. Delivering relief supplies, he explained, "would help us in off-setting intensive Communist agitation against us."[27]

Eager to win allies and outshine political rivals in this newly postcolonial region, and "anxious to be of as much assistance as possible," State Department officials planned to deliver prompt and generous disaster aid to both India and Pakistan, just as they had recently done for Ecuador, Guatemala, and Peru.[28] Much to their chagrin, however, they encountered multiple delays and setbacks, which greatly frustrated their intentions.

After the mid-August earthquake in Assam, State Department officials initially asked the Defense Department to contribute supplies from its stockpiles in Asia and airlift them to India. Citing "heavy commitments in Korea" due to the conflict that just erupted there, the Defense Department agreed only to provide a few aircraft for the relief operation, loaning them from US air bases in Japan.[29] Rebuffed, State Department officials turned to the ARC, whose leaders agreed to contribute blankets and medicines to the cause.[30] Yet, due to miscommunications with the Indian Red Cross, several weeks passed before Indian authorities formally accepted the US aid offer. Eventually, in late October, a MATS C-54 Skymaster departed Massachusetts carrying the six tons of ARC supplies. By the time it finally arrived in New Delhi, two and a half months had passed since the disaster commenced.[31]

Delays similarly plagued the US response to the mid-September floods in Kashmir and Punjab. From both New Delhi and Karachi, US embassy officials urged the State Department to send relief as quickly as possible. It was not until November, however, that the first official American aid reached Pakistan. It took the form of a $35,000 ARC allocation, used to purchase blankets and medicines.[32] In December, at the State Department's request, the ARC agreed to send additional relief supplies to India and Pakistan, transported to the region once again on MATS planes.[33]

But this time, the ARC was not the only American voluntary organization US government officials turned to. They also partnered with the National Catholic Welfare Council (NCWC), one of the American voluntary organizations registered with the new Advisory Committee on Voluntary Foreign Aid. Undertaking a "joint cooperative effort" with the Defense Department, the NCWC sent four tons of clothing blankets, medicines, and food to India and Pakistan, carried on air force planes and distributed to flood survivors in Kashmir and Punjab.[34]

By the end of 1950, the State Department, together with its partners in the military and voluntary sector, finally succeeded in delivering disaster aid to the Indian subcontinent. Yet, hampered by delays and relatively limited in scope—particularly in proportion to the immense scale of these catastrophes—these relief operations made less of a diplomatic and humanitarian impact than they might have. Officials had hoped to use aid to convince Indian and Pakistani peoples "of [the] genuine feeling of friendship and sympathy entertained by [the] people of America."[35] In practice, however, their attempts to demonstrate American concern for these postcolonial nations proved something of a flop.

Undeterred by these logistical setbacks, US officials turned their humanitarian attention to East Asia, reacting to major catastrophes in China and Japan. Once again, the shifting winds of Cold War politics indelibly affected American responses to disasters in these nations. In this case, mounting fears over the political situation in China, coincident with a rapidly evolving military and diplomatic relationship with Japan, caused those responses to be wildly disproportionate. US officials pursued vastly different approaches to assisting these neighboring nations, the former an emerging enemy, the latter a nascent ally.

For the first several decades of the twentieth century, China ranked as one of the world's leading recipients of US foreign disaster aid. American assistance to China had declined since the late 1920s, however, and as the twentieth century reached its midpoint, it essentially ground to a halt due to the Chinese Communist Revolution. As early as 1947 and 1948, before the outcome of China's civil war was assured, ARC leaders began declining requests for flood and famine assistance from the United States' anticommunist ally, the Nationalist government, citing the "considerable difficulty" of distributing aid in war-torn parts of the country.[36] By July 1949, even as particularly destructive flooding in central and south China left 20 million people homeless, officials grew even firmer in their resolve not to send disaster relief to China. Although officials in the State Department and ARC tentatively discussed sending relief supplies to the few areas still under Nationalist control, they ultimately decided against it, citing the challenges of transporting aid around "Communist dominated areas."[37]

After Mao Zedong proclaimed victory in the civil war a few months later, in October 1949, future US disaster aid to the new People's Republic of China appeared increasingly untenable. Even so, when major floods displaced 10 million people in the Huai River valley a few months later, at least some US government officials contemplated making a sizable offer of disaster aid to the country, hoping to garner support for the United States and its anticommunist allies in China. In the Foreign Economic Assistance Act of 1950, acting on this impulse, Congress earmarked $8 million "for relief on humanitarian grounds" to "any place in China suffering from the effects of natural calamity."[38]

Ultimately though, this liberal aid appropriation made little headway. Concluding "that no channel existed" for them to "undertake a relief program on behalf of the Chinese people" and recognizing the act was little more than "a political gesture," officials in the State Department, ARC, and ACVFA all declined to administer the funds Congress allocated.[39] In the end, it failed to matter anyway. Announcing that it would "permit only friendly nations to send relief for its people," the Chinese government ultimately rendered the entire question of US flood aid moot.[40]

If the geopolitical situation in China spelled the end of US disaster relief in that country for the time being, the opposite result occurred in nearby Japan. Under US military occupation since 1945, Japan had, by the early 1950s, undergone a profound political, economic, and social transformation. During that time, US occupying forces carried out multiple disaster relief operations, conducting these efforts in coordination with both ARC personnel and local Japanese authorities.[41] While US officials regarded these humanitarian operations as one part of the wider project of reconstructing Japan, they also viewed their aid as a means to curry favor among Japanese civilians and government officials, hoping to bolster the United States' fledgling alliance with its former geopolitical adversary.

A final opportunity for US occupying forces to advance these diplomatic goals came in early March 1952, when a powerful earthquake and tsunami struck the northern Japanese island of Hokkaido, affecting an estimated 9,000 people. Soon after it hit, US Army planes flew into the devastated region with emergency relief supplies. Shortly thereafter, the US First Cavalry dispatched a "Mercy Train" to Hokkaido, carrying army medical teams and 8,500 pounds of additional material aid, much of it supplied by ARC chapters associated with the US military occupation. American Red Cross personnel also worked closely with their counterparts in the Japanese Red Cross, cooperating with them to conduct a coordinated response to the crisis.[42]

Less than two months later, in late April 1952, the American occupation of Japan formally concluded. Disaster relief operations like the one in Hokkaido, however, did not end with it. To the contrary, they remained a regular feature of US-Japanese relations, and of Asian-American relations more broadly. In coming years, military bases in Japan—which remained under US control as per the terms of the Treaty of Peace—became key staging grounds for US foreign disaster relief operations, not only in Japan but for many other countries throughout Asia. In contrast, decades were to pass before official US foreign disaster assistance once again reached China. The disparate US responses to catastrophes in these two nations underscore the extent to which Cold War geopolitics, alliances, and enmities determined the flow of US humanitarian aid.

Taken together, the US relief efforts in Central and South America, South Asia, and East Asia show the workings of US foreign disaster aid at midcentury.

More than this, these episodes reveal the political and diplomatic imperatives driving US disaster assistance at this time. During the late 1940s and early 1950s, members of the Truman administration regularly employed foreign disaster assistance to advance foreign policy goals, not only in these regions but in countries all over the globe. In a world rapidly transforming under the forces of decolonization, anticolonial nationalism, and the global Cold War, US officials viewed disaster response as a way to win friends, garner influence, and maintain political stability in other nations—in theory, if not always in practice. For their successors in the Eisenhower administration, such assumptions not only persisted, but became more firmly entrenched, a pattern visible as early as Eisenhower's first two years in the White House.

✪   ✪   ✪

Dwight D. Eisenhower had been president of the United States for less than two weeks when, in early February 1953, his administration confronted its first crisis in Europe—triggered not by nuclear threats or saber rattling but by heavy rains and wind. Over the night of January 31, a powerful storm occurred in the North Sea, producing massive storm surges in the Netherlands, Belgium, and the United Kingdom. By morning, seawaters had inundated hundreds of thousands of acres of land along the coasts of each nation. Although the floods caused major disasters in all three countries, they hit the Netherlands especially hard, killing more than 1,800 people and causing extensive damage to farms, homes, and other property.[43]

News of the North Sea floods quickly reached Washington, where US officials organized a vigorous response. Within two days, the ARC contributed $100,000 to its sister societies in Europe.[44] Other voluntary organizations registered with ACVFA, among them Church World Service and CARE, provided aid as well, soon sending more than $300,000 to Europe.[45] Eisenhower and his advisers, meanwhile, began discussing plans to supplement this aid with governmental assistance, optimistic about the "political benefits . . . and the amount of goodwill which would be created for us in Europe."[46] On February 6, the president appointed a cabinet-level committee, chaired by Secretary of State John Foster Dulles and including the director of mutual security and the secretaries of defense and agriculture, to plan and coordinate the government's response to the catastrophe.[47]

Taking advantage of the US military's extensive presence in Western Europe, the members of this interagency task force agreed on military action as the most expedient way to aid flood survivors. To the Netherlands, they sent more than 2,000 US Army NATO troops to assist the Dutch government in its response. Using military helicopters and DUKW boats, army personnel transported relief supplies to isolated areas and ferried both humans and livestock to dryer ground. Others helped repair the many miles of seawalls, dikes, and roads damaged by the storm. In the United Kingdom, army troops performed similar tasks, while air force pilots flew in sandbags, medical equipment, and other supplies from US military stocks in Europe.[48] Diplomats called for comparable measures for Belgium, reasoning that they could "use this disaster to demonstrate our concern with Europe is not only military, but humanitarian." Belgian authorities politely declined the offer, however, explaining they did not require outside help.[49] Even so, US officials judged the overall response a resounding diplomatic success, certain that it "strengthened the friendship of the affected countries toward the United States."[50]

Back in the United States, meanwhile, members of Congress introduced several pieces of relief legislation, attempting to respond to the disaster in more experimental ways. One bill proposed permitting immigration from the Netherlands above the nation's annual quota, allowing an additional 25,000 Dutch flood victims to resettle in the United States. Another authorized the donation of $10 million worth of surplus commodities, held by Commodity Credit Corporation, to peoples affected by the disaster. Ultimately, this legislation made little headway, as Dutch authorities advised that neither proposal would be particularly useful. Still, as the mere drafting of these bills suggests, US policymakers were thinking more creatively about how to use the government's powers and resources to respond to foreign catastrophes.[51]

Eisenhower and his advisers could not have known it at the time, but international disasters like the North Sea floods were to become frequent foreign policy flashpoints of his presidential administration. Within the first two years of his presidency, the US government and its partners responded to catastrophes in multiple countries across the world. Between early 1953 and late 1954, major catastrophes in Greece and Haiti, and across Central and Eastern Europe, all prompted extensive US humanitarian operations. During that same time frame, US officials sent smaller amounts of disaster assistance to many other nations, from Central America to South and East Asia to the Middle East and North Africa. Although the ARC and many ACVFA affiliates participated in most of these humanitarian efforts, a growing share of US assistance now flowed directly through governmental and military channels.

Facilitating the US government's more active role in foreign disaster assistance during these years were two key federal initiatives. The first was the establishment, in 1953, of a new foreign aid agency, the Foreign Operations Administration. The second was the passage of landmark foreign aid legislation, Public Law 480, or the Agricultural Trade and Development Act of 1954. Both gave the government new channels for contributing disaster aid to other nations. Although natural hazards and disaster assistance remained peripheral priorities for the policymakers who devised these programs, these initiatives altered (if inadvertently) the government's approach to international catastrophes by broadening its repertoire of humanitarian options. They also signaled the state's steadily expanding involvement in foreign disaster assistance, a trend that only accelerated in the years ahead.

In June 1953, Eisenhower submitted a series of reorganization plans to Congress. Among them was a proposal to coordinate and consolidate the existing hodgepodge of economic, technical, and military assistance programs, transferring their functions to a new, centralized aid agency: the Foreign Operations Administration (FOA). Legislators readily endorsed the president's plan and took steps to implement it. Among its many functions, the FOA was to help administer the government's foreign disaster assistance operations, assuming a more active, direct role in this humanitarian field than its predecessor agencies.[52]

On August 1, 1953, the FOA came into existence—just in time to contend with a devasting disaster in the eastern Mediterranean. Over several days in mid-August 1953, dozens of earthquakes rocked the Ionian Islands, the archipelago stretching across the western coast of Greece. By the time the most powerful of these tremors occurred, on August 12, multiple towns and villages were entirely razed, leaving 100,000 people homeless.[53] The US government joined many other nations in responding to this sudden emergency—initially, at least, through the channels of its newly minted foreign assistance agency. The very first US official to reach the Ionian Islands was an FOA field representative who was already stationed in the region. Other personnel of the FOA mission to Greece soon arrived on the scene as well, where they assisted in extinguishing fires, treating injuries, and evacuating refugees by motorboat.[54]

Foreign Operations Administration staff were among the first American responders to the Ionian earthquakes, yet their contributions were quickly dwarfed by the US military's. On August 13, the flagship of the US Navy's Sixth Fleet anchored off the Ionian Islands. Other vessels of the fleet, including an aircraft carrier and multiple transport ships, followed in quick succession. Next came helicopters, cargo planes, and other aircraft, flying in from US bases in Germany and Libya. All told, some 10,000 US sailors, pilots, and Marines converged on the devastated region. Most remained in the Ionians for only a few days, but in that time, they performed many humanitarian tasks, including conducting search-and-rescue operations, establishing emergency communications and medical facilities, and treating the wounded. Before departing, they delivered $500,000 worth of food, tents, and other material assistance to Greek relief agencies, later supplementing this initial gift with an additional $2.5 million worth of military supplies.[55] Substantial contributions from partner voluntary organizations, including $415,000 from the ARC and more than $1 million from various ACVFA-registered agencies, amplified these already considerable US governmental commitments.[56]

Hoping for more than emergency relief, Greek authorities asked the FOA mission to fund a long-term rehabilitation and reconstruction program for the Ionian Islands.[57] This request, however, gave rise to a heated debate among US officials over the proper purposes and limits of foreign disaster aid. For strategic reasons tied to Cold War anxieties, some officials considered the proposed reconstruction aid a critical investment. "We cannot turn our backs on the problem now," one diplomat argued, "because the prime US objective in Greece is to maintain, and if possible, to increase the effectiveness of the Greek armed forces against potential Communist aggression."[58] Others proved much less sympathetic to Greek appeals. Grousing that American aid "has been taken virtually for granted," another US diplomat maintained that the government had already contributed more than its fair share, complaining, "The bulk of such assistance is naturally expected to come from the United States."[59] In the end, concluding that they must counteract this "disturbing attitude on the part of the Greek Government," FOA leaders rejected the request for additional aid. A "major reconstruction effort," they insisted, "must come from the Greeks themselves."[60]

With this decision, the US aid effort in the Ionian Islands came to a rather unremarkable conclusion. Even as US officials proved increasingly willing to consider longer-term reconstruction projects, such appropriations were by no means automatic, as this episode revealed. Despite a strong desire to counteract communist influence through their aid, US officials declined to finance a large-scale rehabilitation program in Greece, deeming it more important "to impress upon them the necessity of self-help in meeting the problem."[61]

Questions about the US government's proper role in disaster relief and reconstruction, however, were far from resolved. Similar debates resurfaced in the aftermath of many future catastrophes, as government officials continued to weigh the political costs and benefits of assuming more extensive humanitarian and development commitments.

Less than a year after the Ionian earthquakes prompted the FOA's first venture into disaster assistance, policymakers added another powerful tool to the US government's foreign aid arsenal. On July 10, 1954, Eisenhower signed Public Law 480, the Agricultural Trade and Development Act.[62] A historic piece of foreign assistance legislation, PL-480 authorized the president to dispose of hundreds of millions of dollars' worth of surplus US agricultural commodities annually as international food assistance. Precedents for this legislation dated to at least the early 1930s, when the Hoover administration sold surplus US wheat to the Chinese government for use as flood relief. In the intervening years, and increasingly since the Second World War, the US government sold or donated surplus commodities for humanitarian purposes on multiple occasions. With the passage of PL-480, the practice of disposing of American agricultural surpluses as humanitarian aid became official policy, enshrined in federal law.

Although most of the "aid" that PL-480 authorized was to be sold to foreign governments on credit, the legislation also explicitly authorized some gifts of commodities for emergency disaster relief. Title II of the legislation, specifically, empowered the president to donate $300 million worth of food per year for victims of "famine and other urgent relief requirements" at no cost to recipient governments. Title III of the law, moreover, authorized the government to transfer surplus commodities to the ARC and ACVFA-registered organizations "for use in the assistance of needy persons outside the United States," expanding a smaller program in place since 1949.[63]

Federal officials did not have to wait long to employ the provisions of PL-480 in a disaster scenario. In early July 1954, as this legislation was nearing passage, unusually heavy rainfall began drenching Central and Eastern Europe. By midmonth, the incessant rains caused severe flooding along the Danube River, the Elbe River, and their tributaries, producing a transnational catastrophe across Austria, Czechoslovakia, East and West Germany, Hungary, Romania, and Yugoslavia. Although the Danube and Elbe floods claimed few human lives, they affected an estimated 1 million people on both sides of the Iron Curtain.[64]

American officials responded swiftly to the crisis, initially through conventional channels. In West Germany and Austria, still under US military occupation, 5,000 members of the US Armed Forces performed such tasks as rescuing civilians, airlifting sandbags and other supplies, and setting up field kitchens.[65] From the United States, the ARC and several ACVFA affiliates contributed money and supplies.[66] As the US military and American voluntary sector carried out these more traditional humanitarian activities, however, members of the Eisenhower administration were simultaneously laying the groundwork for a major and experimental food assistance program—a response, they hoped, that might include Soviet Bloc countries as well.

Planning for this relief effort began in earnest on July 16, when Secretary of State John Foster Dulles proposed making an "offer to [the] devastated countries, both free world and communist, of agricultural commodities," under the Title II provisions of PL-480. Extending such an offer, Dulles and other US diplomats agreed, would create an "enormous propaganda advantage" for the United States by demonstrating the nation's "great humanitarian concern toward the flood victims of Europe, both East and West."[67] Most importantly, they concluded, it would show "the peoples of Eastern Europe . . . we are willing to help them, even though their Russian masters may be incapable of doing so."[68] Acting promptly on Dulles's proposal, officials at the State Department and FOA, together with ARC leaders, spent the next two weeks secretly hammering out the details of the relief program.[69]

On July 29, Eisenhower announced the relief offer to the world.[70] Under its terms, the US government pledged to donate millions of dollars' worth of agricultural commodities to the various flooded countries and to cover the costs of shipping them to Europe. But once the food reached European shores, US officials vowed not to participate in its distribution, sensing that a noticeable American presence in this operation would lead many Europeans to perceive it as merely a political ploy. Instead, they delegated that responsibility to the Geneva-based League of Red Cross Societies, the arm of the International Red Cross Movement dedicated to disaster assistance and other peacetime humanitarian activities.

Dulles and other US diplomats viewed the League's involvement in the distribution of commodities as crucial, enabling them to demonstrate that US aid was "based primarily on humanitarian rather than propaganda considerations."[71] Equally keen to ensure relief recipients understood the "US origin of [the] food," however, Congress and the State Department mandated that the phrase "A Gift from the People of the United States" appear on every food package in the appropriate local language.[72] Policymakers may have insisted on the apolitical character of their aid, in other words, but they wanted aid recipients to know who buttered their bread.

Even with the League's participation as a neutral intermediary, US officials were initially uncertain whether Soviet Bloc nations would accept Eisenhower's offer. Since the announcement of the Marshall Plan in 1947, Soviet leaders had repeatedly rejected US offers of aid for the USSR and its satellites. Many US diplomats and State Department personnel assumed that this time would be no different.[73] It therefore came as something of a shock when, on August 5, East Germany's prime minister confirmed his government's willingness to accept American food aid.[74] Over the next few weeks, the Hungarian, Czechoslovakian, Yugoslavian, Austrian, and West German governments all agreed as well.[75] Only Romania, which experienced relatively little damage from the floods, declined. For the first time in seven years, the US government had received authorization to deliver humanitarian aid to the Soviet Bloc.

In late August, with the consent of participating governments secured, State Department and FOA personnel started taking steps to acquire and transport food to Europe. Due to assorted bureaucratic and logistical hurdles, it was not until October 30—almost four months after the floods—that the first shipment of food finally left the United States.[76] Despite the delays, a major US food aid program had commenced. Over the next few months, a total of twenty-eight ships traveled from the United States to the European port cities of Wismar, Stettin, Rijeka, and Bremen.[77] They carried 67,000 tons of American surplus commodities, including corn, wheat, rye, rice, cottonseed oil, and butter.[78] Once US officials deposited the aid at the docks, per their agreements, representatives of the League of Red Cross Societies and local volunteers loaded the food into train cars for distribution throughout the flooded regions. By the time these distributions were completed, in mid-March 1955, several thousand train cars had carried $8 million worth of American foodstuffs to some 1 million Europeans affected by the floods.[79]

Looking beyond its humanitarian achievements, US officials judged this disaster relief operation a major diplomatic coup, a strategic win for the United States in Cold War Europe. "The program," John Foster Dulles crowed, "accomplished its humanitarian purpose to the satisfaction of all concerned—including certain governments that had previously been unreceptive to our gestures of good will."[80] Undersecretary of State Herbert Hoover Jr. likewise dubbed the scheme "a singular success," declaring it "gratifying that American relief goods, marked 'Gifts of the American people' have reached the people behind the Iron Curtain."[81]

In mid-1954, the Danube and Elbe floods presented the Eisenhower administration with its first opportunity to deploy PL-480 commodities as emergency disaster relief—and, just as critically, as a tool of Cold War statecraft. It would not be the last. From that point forward, both the US government and its partners in the voluntary sector regularly donated and sold surplus commodities to nations struck by catastrophe. In Cold War hot spots and across the decolonizing world, US foreign disaster assistance and American food power grew increasingly intertwined.[82]

In October 1954, as State Department and FOA personnel were arranging shipments of food to Central and Eastern Europe, a catastrophe much closer to home simultaneously captured their attentions. On the night of October 11–12, a powerful hurricane named Hazel struck Haiti, its eye passing over the country's southwestern Tiburon Peninsula. This devastating storm completely destroyed the city of Jérémie and many surrounding towns and villages, leaving 100,0000 people homeless. Across Haiti, thousands more were killed or injured, both by the hurricane and in successive landslides.[83] The hurricane presented US government officials with yet another opportunity to deploy FOA advisers, PL-480 food, and other disaster assistance—this time in a very different geopolitical context.

In Hurricane Hazel's wake, the US government, military, and ARC carried out a sizable and multifaceted relief operation. Within Haiti, US officials delegated several technical experts from the Point Four mission to survey the disaster area and determine what types of assistance might be needed.[84] Much as they had in Greece a year earlier, and reflecting the US military's extensive footprint in the Caribbean Basin, the US Armed Forces played a prominent role in this emergency response. On October 13, the aircraft carrier *Saipan* arrived in the waters near Jérémie. Using its helicopters, air force and marine pilots surveyed the disaster area, evacuated refugees, and airdropped rations and medical supplies into the disaster zone.[85] The following day, air force and navy planes arrived from Puerto Rico carrying hundreds of thousands of pounds of emergency rations, drawn from army stockpiles.[86] Planes from the Canal Zone came as well, bringing a Caribbean Command disaster team and the ARC's director of operations to Haiti to advise and assist relief work.[87]

On October 15, boosting these initial contributions considerably, FOA leaders in Washington announced an allotment of $2 million worth of surplus commodities for Haiti—13.5 million tons of food—under the terms of PL-480.[88] By month's end, the FOA contributed an additional $1 million worth of seeds and agricultural tools to the country, to help with the recovery of farmland. The total cost of US government relief supplies now exceeded $3 million.[89]

At first, US officials felt pleased with the diplomatic results of their relief efforts. From Port-au-Prince, the US ambassador called the *Saipan*'s deployment a "master stroke [of] US-Haitian relations."[90] Other US officials commended the FOA, praising the aid agency for its "marvelous job of allocating and getting these emergency supplies to Haiti."[91]

As distributions of assistance continued over the next several weeks, however, trouble started brewing. Members of the FOA team began reporting "considerable difficulty in getting supplies into the field."[92] By and large, they blamed the Haitian Red Cross and Haitian Army officials for these challenges. Peppering their criticisms with longstanding racialized stereotypes, they cited "rampant corruption and inefficiency," disparaging Haitian officials as "lethargic and unmindful of the dire distress of their own people."[93] Because of these issues, FOA personnel warned, American aid was "not being administered and distributed properly."[94] Adding to these complaints, they reported that the wife of Haitian president Paul Magloire had requisitioned much of the American food and then "rebagged [it] in unmarked bags for redistribution" by her own charitable foundation, erasing all signs of its American provenance.[95]

Members of the FOA mission intended to funnel their protests, confidentially, to Washington. Their complaints, however, also reached President Magloire, sparking a heated diplomatic row. Magloire became "irate," US chargé d'affaires Milton Barall reported, chastising FOA personnel for their "disrespectful and non-cooperative behavior" and, especially, for impugning his wife's reputation.[96] Summoning Barall to his office, Magloire demanded that the FOA mission "be brought back into line" and be made to "show the proper respect for the sovereign Government of Haiti."[97] Reminding the chargé d'affaires that Haiti was "no occupied country, in which Haitian officials could be treated in such a cavalier manner," he declared that "Haiti's sovereignty could not be bought with $2 million worth of food."[98]

Reflecting on Magloire's remarks, Barall conceded, "It is easy to understand his anger," particularly in light of the "painful memories" of the US military occupation of 1915–34. He and his associates also regretted the "lack of tact" on the part of FOA representatives, admitting that their actions had "unavoidably complicated" the US mission to Haiti and "detracted from the overall effectiveness of the very charitable and worthwhile emergency program."[99] American personnel continued to distribute food in Haiti through December and into January, but as Barall recognized, the political damage was already done.[100] Whatever humanitarian good it might have achieved in

Haiti, the US government's response to Hurricane Hazel failed miserably as an act of diplomacy.

By the time US relief efforts in Haiti drew to a close in early 1955, just two years into Eisenhower's presidency, catastrophes had punctuated his administration's relations with other countries time and again. While the North Sea floods, Ionian earthquakes, Danube and Elbe floods, and Hurricane Hazel triggered some of the most extensive US foreign disaster aid operations in these years, they were hardly the only ones. In 1953 and 1954 alone, US officials also contributed humanitarian assistance—in the form of PL-480 commodities, FOA and military support, ARC funds, and other material aid—to disaster survivors as far afield as Algeria, Costa Rica, Egypt, Guatemala, Honduras, India, Iran, Italy, Japan, Mexico, Nepal, Pakistan, the Philippines, South Korea, and Turkey.[101]

With concerns about the global Cold War and decolonization foremost on their minds, US government officials treated these disaster assistance operations as auspicious political opportunities, a chance to demonstrate their allegiances and buttress their alliances with countries affected by natural hazards. In some cases, those efforts met with considerable diplomatic success. In others, they resulted in frustrating political setbacks, as bureaucratic obstacles, lengthy delays, and tensions with aid recipients undermined the psychological and material impact of their assistance.

Despite the mixed humanitarian and diplomatic results these relief operations achieved, it was clear that the Eisenhower administration intended to make a robust commitment to foreign disaster aid. Still, the precise place of natural hazards and international disaster relief in the expanding US foreign assistance apparatus remained rather vague and undefined. Throughout Eisenhower's remaining time in office, government officials began exploring ways to address these particular humanitarian problems more explicitly and effectively.

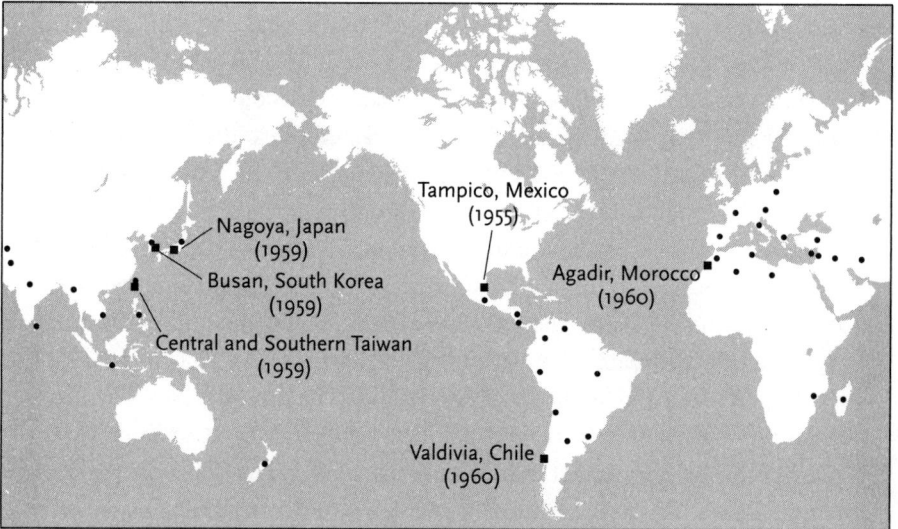

**Recipients of official US foreign disaster assistance
referenced in chapter 11, 1955–60**

11

Stumbling toward Standardization

1955–1960

Having increased in frequency since the Second World War ended, US international disaster relief efforts continued to proliferate throughout the remainder of Eisenhower's presidency. Between 1955 and 1960, the State Department and a new foreign assistance agency, the International Cooperation Administration (ICA), orchestrated responses to catastrophes around the world, collaborating with US diplomatic and development missions, the US Armed Forces, the American Red Cross, and ACVFA-registered organizations to carry out these humanitarian operations. Among the largest of these foreign disaster assistance efforts were those following tropical storms in Mexico, Japan, and South Korea, floods in Taiwan, and earthquakes in Morocco and Chile. But US disaster aid also reached dozens of other nations, as the federal government continued expanding its humanitarian reach.

Across the late 1950s—as the onward march of decolonization further transformed international politics, and as concerns over communism and socialism, neutralism and nonalignment together steered US foreign policy planning—foreign disaster aid remained a critical facet of the Eisenhower administration's relations with the world. As interest in the questions of modernization and nation-building grew during these years, US responses to global catastrophes also became more tightly linked to American international development assistance projects and agendas. By the time Eisenhower left office in early 1961, the connections between disasters and development were becoming increasingly difficult to ignore.

During these same years, the process of centralizing and coordinating foreign disaster assistance at the federal level plodded steadily along. Although US policymakers had mostly disregarded natural hazards in their foreign aid planning up to this point, in the late 1950s they started addressing these specific humanitarian problems in more direct, concrete ways. Members of the Eisenhower administration attempted to standardize the procedures guiding foreign disaster assistance operations, while the State Department and Congress established new funding sources for those efforts. The ICA, meanwhile, played a more active role in international disaster relief, recovery, and reconstruction than any of its predecessor foreign aid agencies.

With these reforms, US officials endeavored to improve the humanitarian effectiveness and, as crucially, the political impact of foreign disaster aid. In the process, the government's direct involvement in international disaster aid expanded considerably. Though further changes lay in store, catastrophic diplomacy was becoming, in fits and starts, a more formal instrument of US foreign policy.

❂   ❂   ❂

In mid-1955, the Eisenhower administration implemented a sweeping overhaul of the government's foreign aid bureaucracy. In early May of that year, Eisenhower dissolved the Foreign Operations Administration by executive order, transferring its economic and technical assistance programs to a new aid agency: the International Cooperation Administration.[1] This reorganization had two significant effects. First, whereas the FOA had existed as an independent federal agency, the ICA operated as a semiautonomous unit within the State Department, answerable to the secretary of state but led by its own director. Second, while many policymakers regarded the FOA a temporary organization, the Eisenhower administration intended the ICA to be "a permanent government establishment." The goal of the reorganization, as Eisenhower saw it, was to align the US government's foreign policy objectives more closely with its foreign aid program, ensuring that it remained a "continuing and integral part of the fabric of our international relations."[2]

Commencing operations on July 1, 1955, the ICA assumed responsibility for coordinating and conducting all nonmilitary US foreign assistance activities, including "emergency programs for relief or rehabilitation as directed by the President."[3] Although the bulk of its activities and budget were devoted to economic, technical, and development aid, the ICA also became the government's designated agency for foreign disaster assistance.[4]

Throughout the remainder of Eisenhower's presidency, the ICA took charge of many of the tasks involved in international disaster relief, including most importantly the administration of PL-480 food aid and relevant Mutual Security funds. Like the FOA before it, however, the ICA did not respond to catastrophes on its own. Its parent agency, the State Department, retained the primary responsibility for determining "which foreign disasters, because of their relationship to American foreign policy, warrant US Government disaster relief."[5] The military, other federal agencies, and the government's partners in the voluntary sector also cooperated with the ICA and State Department to conduct disaster aid operations abroad. Even as its precise shape evolved, the three-pillared system of US foreign disaster assistance remained firmly intact.

Testing the limits of this humanitarian system, the Eisenhower administration poured considerable resources into international disaster relief. In the five years after the ICA's creation, US foreign disaster aid operations mushroomed, as American officials responded to dozens of catastrophes in all corners of the world. By the time Eisenhower left office, US humanitarian assistance reached survivors of earthquakes in Afghanistan, Burma, Lebanon, Morocco, Turkey, and, on multiple occasions, Chile, Greece, Iran, and Peru. It went to victims of tropical cyclones in Ceylon, Japan (four times), Mauritius, Mexico, New Zealand, Pakistan, the Philippines (three times), South Korea, and South Vietnam. It assisted sufferers of a tremendous blizzard and cold wave affecting Spain, France, Italy, and other parts of Europe. American aid reached still more countries and colonies experiencing flood disasters, among them Algeria, Argentina, Austria, Brazil, Colombia, Cyprus, France, Iran, Iraq, Japan, Libya, the Malagasy Republic (Madagascar), Mozambique, Nicaragua, the Philippines, Spain, Taiwan, Tunisia, Uruguay, and Venezuela. In addition, Costa Rica, Greece, India, Indonesia, Pakistan, Poland, and South Korea each received US flood aid multiple times during these years.

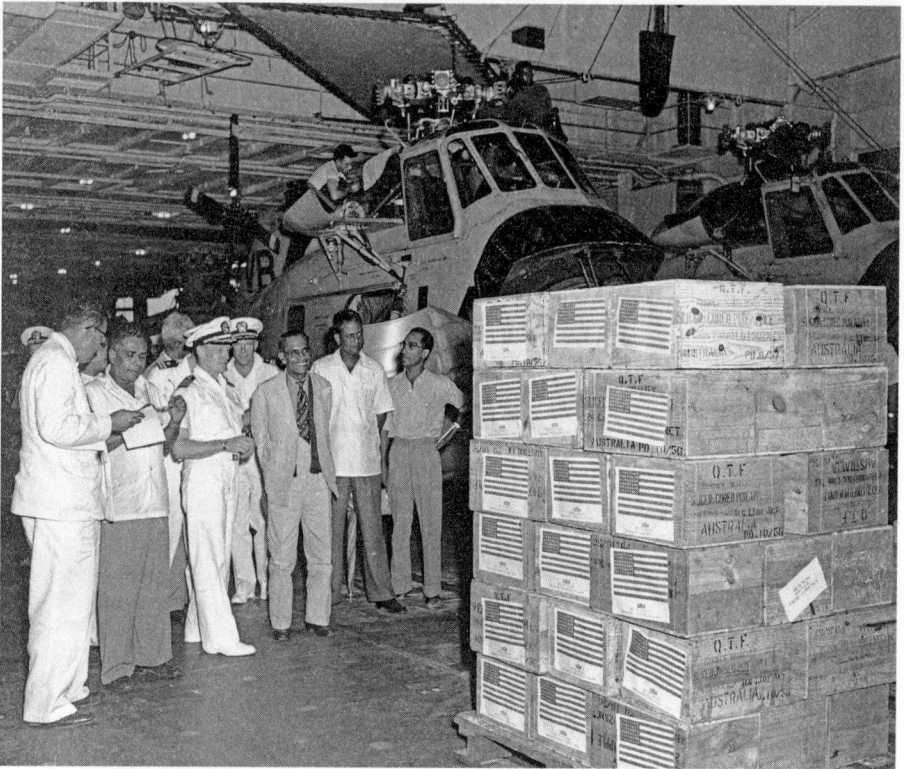

Personnel of Commander Carrier Division 15, showing the
prime minister of Ceylon the supplies that the US Navy
was delivering to flood victims in his country in early 1958.
Image courtesy of Naval History and Heritage Command.

Between 1955 and 1961, as this inventory suggests, the reach of US foreign disaster assistance grew by leaps and bounds.[6]

The scale of these official aid efforts varied considerably. They ranged from rather nominal responses, entailing limited contributions of money or aid supplies, to large-scale humanitarian operations, valued in the millions or tens of millions of dollars. The types of aid the United States contributed to each country differed in every case as well and included cash grants and soft loans, surplus commodities and other material aid, logistical support, and technical assistance. Although foreign disaster assistance continued to be restricted mainly to emergency and short-term relief activities, the US government and its partners also subsidized a growing number of long-term reconstruction projects during these years, many of which overlapped with existing US development assistance programs.

Yet if their specific details differed, collectively the US responses to these catastrophes exhibited some discernible patterns. Perhaps most notably, a far greater proportion of American foreign disaster aid came from US government and military sources than in the past. Reflecting the ongoing trend toward rising federal involvement, the ICA's Washington office allocated hundreds of thousands of tons of wheat, corn, rice, powdered milk, and other agricultural commodities to countries affected by calamity, distributed under the Title II provisions of PL-480. The ICA also drew frequently from the contingency funds allocated to it by the Mutual Security Act of 1954 and its succedents—roughly $150 million annually—using this money to finance a diverse array of disaster relief activities.[7] On the ground in other countries, personnel of the US Foreign Service, ICA missions, and US Operations Missions assisted with food distributions, provided medical care and technical advice, loaned out vehicles and heavy machinery, and performed other humanitarian activities after disasters struck.

While the State Department and ICA maintained the primary responsibility for coordinating and overseeing disaster relief operations, the Defense Department regularly lent its immense logistical capabilities to these efforts. Between 1955 and early 1961, US Air Force and MATS planes arrived at the scene of many catastrophes carrying all sorts of humanitarian cargo: PL-480 commodities, supplies donated by the American voluntary sector, and US military stores, including tents, cots, blankets, and medical equipment. While the air force and its planes were now omnipresent in US foreign disaster aid operations, the navy and marine corps also sent ships, DUKW boats, aircraft carriers, and helicopters to many disaster-stricken countries, using these vehicles to deliver supplies and conduct search-and-rescue operations. Although the army participated in these efforts less often than the other service branches, its troops took part in several disaster relief efforts during the late 1950s, typically when they were stationed on military bases nearby.

Even as a growing share of US foreign disaster assistance came through state channels, government officials continued cooperating closely with partners in the voluntary sector to plan and administer these efforts. As it had for more than half a century, the American Red Cross remained "an instrument of enormous value" and "an important and unique government asset," serving as a significant source of funds, supplies, and expertise for many disaster relief efforts.[8] The ARC, however, now operated in an increasingly crowded field. During the late 1950s, many organizations registered with the State Department's Advisory Council on Voluntary Foreign Aid—now numbering more than fifty—expanded their foreign disaster relief programs considerably, with

Catholic Relief Services, Church World Service, and CARE all playing an especially active role in this sphere. During these years, they and other ACVFA affiliates distributed considerable quantities of food, clothing, medical supplies, and other aid to disaster survivors abroad.[9]

Although they relied on the American public to support their efforts financially, these voluntary organizations were, like the ARC, deeply reliant on benefits they accrued from the federal government. Under the Title III provisions of PL-480, ACVFA affiliates enjoyed access to millions of dollars' worth of surplus agricultural commodities annually, which they routinely distributed as disaster relief. To transport food and other material aid overseas, voluntary organizations either rented space on commercial ships—financing the costs with the generous ocean freight subsidies the Mutual Security Act afforded them—or loaded their supplies on air force and MATS cargo planes, made available to them by the Defense Department.[10] The government's growing involvement in foreign disaster assistance, in other words, had significant consequences for the American voluntary sector as well, with voluntary organizations becoming progressively more dependent on state resources to fulfill their humanitarian missions.

Across the late 1950s, the US government, US military, and their partners in the voluntary sector thus worked in close collaboration to conduct disaster relief operations all over the globe. While each of the three pillars continued to play their assigned part in these humanitarian efforts, the government had assumed a far more dominant role in planning, financing, and administering them. A trend toward increased centralization and coordination of foreign disaster aid under the state's auspices was, by the late 1950s, undeniable.

As the frequency and complexity of these humanitarian efforts mounted, however, a growing chorus of US officials began voicing frustrations. Some complained they lacked clear legal authority for their actions. As one US Army officer observed in 1955, "The Department of Defense is accused of breaking a lot of laws, but never do we break so many as in our disaster relief operations."[11] Others criticized the lack of formal policies and protocols to guide their work. Writing to Secretary of State John Foster Dulles in 1956, the ICA's director griped that disaster relief operations under his watch were slower and costlier than necessary, confessing that "the aid given is somewhat hit or miss." Recommending that "disaster aid is an area which needs organizational study," he proposed that "some machinery should be set up to clear all emergency requests, and that certain criteria should be established."[12]

Over the next several years, responding to such criticisms, the Eisenhower administration took concerted steps to standardize the procedures governing US responses to catastrophes abroad. First, in September 1956, Eisenhower's Operations Coordinating Board—the body responsible for synchronizing security activities across various federal agencies—produced an official manual for "Foreign Disaster Relief Operations." It spelled out, for the first time, "inter-agency guidelines in connection with emergencies occurring overseas in which US Government assistance may be of importance." Defining "emergencies" to include "all types of natural disasters such as flood, earthquakes, hurricanes, fires, volcanic eruption, and pestilence," these guidelines clarified the specific responsibilities of the State Department, ICA, Foreign Service and military personnel, the ARC, and ACVFA-registered organizations in international disaster assistance efforts.[13] Complementing these federal regulations, the Defense Department's combatant commands issued their own orders and directives, designed to guide military disaster assistance operations in their respective regions.[14]

Additional reforms followed in 1958. That June, declaring the need for "supplemental procedures to those now existing to meet emergency needs immediately after the occurrence of a disaster," Secretary of State John Foster Dulles issued a set of official instructions for all overseas diplomatic missions, which further clarified their duties in the event of "a disaster . . . resulting from phenomena of nature such as flood, hurricane, typhoon, earthquake, fire, volcanic eruption or pestilence."[15] Affirming customary yet often implicit practices, Dulles charged all chiefs of mission with the responsibility for determining whether outside assistance was necessary, what specific needs existed, and whether it was "in the interest of the United States to assist in meeting such initial emergency needs."[16]

Within these instructions, Dulles also introduced a new policy: he authorized ambassadors and other mission chiefs to commit up to $10,000 to any disaster relief operation, distributing this aid at their discretion and without prior approval from Washington. These funds were to be drawn from a new Foreign Disaster Emergency Relief Account, administered by the ICA and replenished each fiscal year.[17] Practically, the intent of this measure was to allow diplomats to respond more swiftly to sudden, unexpected calamities as soon as they occurred.[18] Yet there was also an explicit strategic rationale behind the new policy. As a subsequent State Department directive explained, an ambassador or mission chief should use this money "only when, in his judgment, speed is of such importance in obtaining the right psychological impression that no time should be lost in obtaining Washington clearances."[19]

Together, these new guidelines and procedures were meant to resolve the issues impeding the humanitarian effectiveness of US foreign disaster assistance. Yet, as the remarks of Dulles and his colleagues made clear, the reasoning behind these changes was as much—if not more—about improving its political effectiveness. With these structural reforms, the ICA and State Department sought to ensure that US aid made a good "psychological impression" on disaster survivors and their governments, that it burnished the United States' image in other nations, and, above all, that it served the interests of US foreign policy.

★ ★ ★

These developments in US foreign disaster aid policy and programming took place against the backdrop of a complex and rapidly changing geopolitical context, defined by the concurrent throes of decolonization and the global Cold War. In Western Europe, where the success of postwar recovery was ushering in a period of substantial economic growth, US officials enjoyed relatively cordial and stable relations. Following Josef Stalin's death in 1953 and Nikita Khrushchev's subsequent ascent to power, Cold War tensions throughout the European continent relaxed somewhat, giving rise to cautious optimism about the possibility of peaceful coexistence with the Soviet Union. As events like the 1956 Hungarian Revolution or the U-2 incident of 1960 made clear, however, US-Soviet frictions were far from resolved.

Elsewhere in the world, the global Cold War continued to escalate and evolve, its contours shaped by the added dynamics of decolonization and anticolonial nationalism. In Latin America, US policymakers' concerns over communism, socialism, and anti-Americanism smoldered throughout the 1950s, only to be inflamed by the Cuban Revolution in 1959. In Asia, US relations with the People's Republic of China remained in a dismal state. Determined to contain the spread of communism and revolution, US officials worked to bolster ties with such regional allies as Japan, the Philippines, South Korea, South Vietnam, and Taiwan while vying for influence in India, Indonesia, and other neutral or nonaligned nations. In the Middle East, hoping to defuse Nasserism, pan-Arabism, and Soviet influence, the Eisenhower administration concentrated on propping up regional allies such as Israel, Iran, and Turkey, a policy formalized in the Eisenhower Doctrine in 1957. And in Africa, violent wars and bloody struggles for national liberation raged in Algeria, Kenya, and other sites. Despite fierce European resistance to decolonization, between 1955 and 1960, some two dozen African nations won independence from colonial rule, seventeen of them in 1960 alone.

Amid the shifting winds of the global Cold War and decolonization, and with mounting interest in the problems of modernization and international development, US officials redoubled their involvement in foreign disaster aid, seeking to promote multiple strategic objectives. In Cold War hot spots, they relied on disaster relief to achieve "a clear and decisive propaganda victory over the Soviets," as the US ambassador to Iran proclaimed in 1956 after severe flooding affected that country.[20] In newly independent nations, it served as a "gesture of friendship and solidarity," as the US consul told citizens of the Malagasy Republic, following a catastrophic tropical cyclone in 1959.[21] And in many other parts of the world, US officials used disaster aid to maintain alliances, win friends, and spread American influence.

While these trends were reflected across the dozens of global catastrophes eliciting American humanitarian action during these years, an in-depth look at some of the era's most extensive US foreign disaster relief operations—in Mexico, East Asia, Morocco, and Chile—brings these patterns into clearer focus.

In September 1955, the newly established ICA experienced the first major test of its disaster relief capabilities when a pair of tropical storms struck northeastern Mexico. On September 19, Hurricane Hilda made landfall in Tampico, one of Mexico's largest port cities. Hilda's winds and rains caused considerable damage and flooding in Tampico and its environs, leaving 18,000 people homeless.[22] Shortly after the storm hit, the Mexican government appealed for outside assistance. American officials swiftly responded. By September 21, three air force helicopters had arrived in the area and begun conducting search-and-rescue operations.[23] Air force planes, meanwhile, carried food, clothing, and medical supplies to Tampico, delivering forty-four tons of aid within a few days' time.[24]

As this relief effort unfolded, a second and even more powerful storm was making its way across the Gulf of Mexico. On September 29, Hurricane Janet also slammed into Tampico, transforming what was already a precarious situation into a tremendous catastrophe. Seventy-five percent of the city now stood underwater, and roughly 100,000 people throughout the region were homeless. The situation in Mexico constituted an immense humanitarian crisis, one of the worst disasters the country had faced in many years.[25]

In the wake of Hurricanes Hilda and Janet, State Department leaders spied a chance to improve US relations with Mexico, which had withered following the US-backed coup in Guatemala, the forced repatriation of Mexican immigrants under Operation Wetback, and other recent controversies. Secretary of State Dulles, for one, considered it an "excellent opportunity" to show Mexico the "sympathetic desire of our government" to help.[26] Acknowledging that "Mexicans had always been rather sensitive about our sending ships into their ports"—a thinly veiled reference to the 1914 Tampico incident and subsequent US invasion and occupation of Veracruz—State and Defense Department personnel also reasoned that a major relief effort might "break up the Mexican prejudice against [US] naval vessels" on its coasts.[27]

Given the high political and diplomatic stakes, American officials resolved to leave nothing to chance, and responded to the disaster with a centrally planned and coordinated humanitarian operation. From Washington, the State Department and ICA took the lead in this effort, cooperating closely with the US military and ARC to carry it out. Signaling the diplomatic importance of this undertaking, they dubbed it Operación Amistad (Operation Friendship).[28] As Janet neared Mexico's Gulf Coast, the Defense Department ordered two aircraft carriers and several other ships to the region. Arriving on October 2, they brought thirty helicopters, multiple DUKW boats, and many additional tons of food and other material aid. These helicopters and amphibious craft remained in the region for two weeks, transporting supplies around Tampico and ferrying thousands of flood survivors to dryer ground.[29]

On October 7, supplementing this aid, the ICA announced a donation of $175,000 worth of PL-480 commodities, to be carried into Mexican ports on US Navy ships. To distribute this food, US Army personnel and ARC staff, deployed to Tampico from military bases in Texas, established and administered field kitchens throughout the city. Over the next few weeks, they served hot meals to 350,000 people and provided 1.4 million glasses of milk to children.[30] By the time Operación Amistad officially closed, on October 30, the US government and ARC together furnished $570,000 worth of aid to Mexico.[31]

Government officials judged the relief operation a smashing diplomatic success. Operación Amistad had "done much to offset a degree of anti-American[ism] which exists in this consular district," observed the US consul in Veracruz, "and which has been kept alive by frequent public references to 'Invasions' by the United States in 1847 and 1914."[32] The US ambassador to Mexico agreed, affirming that the aid went "far to overcoming the bitterness that has existed since the unfortunate incidents [in Tampico] and in Veracruz in 1914."[33] Relaying these reports to Eisenhower, John Foster Dulles brimmed with enthusiasm. "At no period in the history of United States–Mexican relations has the United States been held in such high regard by both the Mexican Government and people," the secretary of state told Eisenhower. The outpouring of US disaster aid to Mexico, he extolled, "may well have an important bearing on the future course of our relations."[34]

Whether it did, of course, remained to be seen. Regardless, as their accolades revealed, Dulles and his associates placed considerable faith in disaster aid's power to promote US foreign policy interests in Mexico. They did the same, time and again, as catastrophes struck other parts of the globe.

This included East Asia, a leading recipient of US foreign disaster aid during these years. Four years after Hurricanes Hilda and Janet, three of the US government's key allies in this region—Taiwan, South Korea, and Japan— experienced a string of extreme weather disasters, all within two months. These successive catastrophes occurred in a region of critical strategic importance to the Eisenhower administration, which housed sizable US diplomatic and development missions and many US military installations. Taking advantage of this extensive American footprint, government officials responded swiftly and generously to the crises in these three nations, desirous to advance the United States' diplomatic designs in Cold War Asia.

The troubles began in early August 1959, when a period of torrential rains brought catastrophic flooding to the Republic of China (Taiwan), leaving 134,000 people homeless.[35] Eager to assist this important Cold War ally, US officials responded quickly to the catastrophe. The ICA released $200,000 to Taiwan's Nationalist government to finance disaster relief work, while the Defense Department sent military medical teams, trucks and heavy equipment, and an aircraft carrier with twenty-one helicopters to assist flood survivors.[36] At the urging of the US embassy and the ICA mission in Taipei, the ICA then committed an additional $9 million worth of PL-480 commodities to the country.[37] In response to an explicit request by the State Department, the ARC and several ACVFA agencies provided tens of thousands of dollars' worth of aid to Taiwanese flood survivors, supplementing the already significant outpouring of governmental assistance.[38]

As US officials were dealing with the flood disaster in Taiwan, two powerful tropical cyclones formed in the Pacific Ocean and began making their way west. On September 16, the first of these storms, Typhoon Sarah, slammed into South Korea. Sarah's winds, rain, and storm surge wreaked enormous destruction in the Busan area, leaving 800,000 people homeless and killing nearly 2,000 people.[39] Ten days later, Typhoon Vera hit the Nagoya area of central Japan. This time, 5,000 people lost their lives and another 1.5 million their homes. Reporting the crisis to the State Department, Ambassador Douglas MacArthur II—the nephew of the general who commanded the postwar occupation of Japan—called it the "most extensive natural disaster since [the] 1923 Great Tokyo earthquake."[40] Much as they had in Taiwan, US officials responded vigorously to both catastrophes, staging prolonged and far-reaching foreign aid operations. Once again, government and military agencies played the principal role in these efforts, fulfilling the responsibilities assigned to them in newly established federal guidelines.

In Japan, making use of the ICA disaster emergency funds available to all ambassadors since the previous year, MacArthur immediately obligated the $10,000 at his disposal. He then requested additional aid from Washington. Acknowledging that Vera was "a disaster of major proportions in an area of great concern to the United States," the ICA and State Department determined "that it is [in] our interest to respond promptly and effectively." They committed another $100,000 to the cause.[41] These funds underwrote a sizable military relief operation, carried out jointly by personnel of the US Forces Japan, Seventh Fleet, and Fifth Air Force and the Japanese servicemembers of the country's Self-Defense Forces.[42]

Though the US military completed its relief mission by mid-October, additional American assistance continued to reach Japan over the next several weeks. The ARC delivered $65,000 worth of cash, school supplies, and other aid to Vera's survivors, while ACVFA affiliates operating in Japan provided their own material aid.[43] Governmental agencies gave still more. Having already committed $100,000 in disaster aid, the ICA increased its financial contribution by almost twofold. Additionally, the agency allocated 4 million pounds of PL-480 wheat flour to feed Japanese cyclone survivors.[44]

After Typhoon Sarah, American aid flowed into Korea through many of these same channels, though the scale of these efforts was initially less extensive than in Japan. From Seoul, Ambassador Walter Dowling turned over his $10,000 in emergency funds to local Korean charities. The ARC, meanwhile, wired $20,000 to its Korean sister society, while ACVFA affiliates operating in Busan distributed food and clothing to cyclone survivors.[45]

For many US officials, these actions failed to go far enough. As the weeks went by, US diplomats and development personnel in South Korea started appealing for supplemental aid from Washington, warning that the initial response was inadequate. In late October, making their requests concrete, members of the US Operations Mission in Korea presented a comprehensive, multimillion-dollar plan for relief and rehabilitation assistance to their superiors in Washington.[46] Without a significant infusion of additional aid, they cautioned, both "unnecessary hardship" and diplomatic fallout were sure to result.[47] Although it took several months for ICA leaders to act on this request, eventually they agreed to do so. In February 1960, the ICA allocated $10 million to South Korea, earmarked for multiple typhoon relief, recovery, and associated development projects.[48]

In mid-1959, the Eisenhower administration and its partners had carried out extensive and costly disaster assistance operations in three East Asian nations, all key allies of the United States. Just a few hundred miles to the west, the US reaction to another contemporaneous catastrophe could hardly have looked more different. In mid-1959, severe river floods also displaced several million in the People's Republic of China, one of the United States' chief geopolitical rivals.[49] Between June and September that year, as they orchestrated responses to calamities in Taiwan, Japan, and South Korea, State Department personnel simultaneously debated whether to respond to this catastrophe in the world's most populous communist country.

Hoping to demonstrate that the Eisenhower administration was "prepared to help the Chinese people even though we do not recognize their Government," the State Department confidentially discussed several possible ways to deliver American aid to flood survivors.[50] One proposal involved making a highly public offer of assistance to China's leader, Mao Zedong, while inviting the Soviet Union's premier, Nikita Khrushchev, to act as the United States' intermediary in the exchange. Realizing Chinese and Soviet leaders would surely see this scheme for what it was—"a psychological and propaganda exercise"—the State Department abandoned this plan, concluding, "There would be very little chance of the Communist government accepting such an offer."[51] A second proposal entailed making "air drops of food over Communist controlled territory" without the Chinese government's authorization. Ultimately, they scrapped this idea, too, reasoning that both the "Peiping regime" and "a substantial body of free world opinion" would judge this humanitarian invasion a "hostile and aggressive act."[52]

In the end, as they had multiple times since the 1949 revolution, US officials decided against any attempts to aid Chinese flood victims. The decision *not* to offer aid to certain countries, as this case reveals, was as much a political act as the choice to provide it to others.

In 1959, Cold War politics, alliances, and enmities thus steered the Eisenhower administration's responses to concurrent catastrophes in Taiwan, South Korea, Japan, and China. These dynamics, however, were hardly unique to East Asia. They informed the course of numerous other US foreign disaster aid operations, all over the world, during the late 1950s and early 1960s.

On February 29, 1960, less than six months after Typhoons Sarah and Vera ripped through Busan and Nagoya, a powerful earthquake occurred in Agadir, Morocco. The ensuing catastrophe claimed at least 12,000 lives, if not more, and resulted in widespread damage and destruction.[53] Having won its independence from France less than four years prior, the Moroccan government was grappling with several other humanitarian crises when the earthquake struck. These included the lingering effects of a mass poisoning crisis, caused by contaminated cooking oil in 1959, and the challenge of aiding 100,000 Algerians in Morocco, refugees from Algeria's ongoing war for independence from France.[54]

Suddenly, the Moroccan government stared in the face of yet another immense catastrophe. As they had in Mexico and East Asia, US government officials responded quickly to this crisis in North Africa. The morning after the earthquake, judging the situation a "most serious national disaster," Ambassador Charlie Yost turned over the $10,000 emergency funds at his disposal to US military forces in Morocco.[55] Yost then contacted his superiors in Washington, urging the State Department to commit another $500,000 toward the relief effort. The State Department and ICA not only complied with this request but, at the ambassador's urging, subsequently tripled that amount, authorizing Yost to expend up to $1.5 million for Moroccan relief.[56] Although the ARC and several ACVFA affiliates provided additional monetary and material aid, this large infusion of governmental funds represented the vast majority of American assistance to Morocco.[57]

Even as US officials made these substantial humanitarian commitments, behind the scenes they were hesitant to provide too much aid to Agadir. Their reticence was based primarily on diplomatic concerns, grounded in the complex politics of decolonization. More specifically, they sensed that the United States must not overshadow the French government's response to a crisis in its former colony. A generous French relief effort in Agadir, State Department leaders calculated, would likely "be of great political and psychological benefit to France in its relations with Morocco and the Moslem world generally."[58] Although an equivalent or greater American response might win the US government friends in North Africa and elsewhere in the postcolonial world, it also risked antagonizing France, a critical European ally.

And indeed, a similar situation had arisen in the region just two years earlier. In early 1958, devastating floods swept through Algeria, then four years into its bloody liberation war. At the time, both the State Department and ICA considered offering PL-480 food for Algerian flood survivors yet feared that French officials would perceive this action as an affront to their authority. "There is considerable sensitivity on the part of French public with regard to aid to Moslems," an ARC leader involved with this decision observed, particularly "with regard to Algerian refugees, who are considered by the French to be rebels."[59] Ultimately, prioritizing good relations with the French government over aiding French colonial subjects, the State Department and ICA decided against providing governmental assistance to Algeria, though they did eventually cooperate with several ACVFA affiliates to deliver privately donated relief supplies.[60]

In 1960, the dynamics were somewhat different, for Morocco had been an independent nation since 1956. Reasoning that both the "well-being [of] Morocco" *and* "French relations with Morocco" were "of major importance to our own interests," State Department officials concluded that they should provide governmental aid in this instance. However, they also determined they must neither impede nor outshine the French government's efforts.[61] A more limited humanitarian response was necessary, Yost agreed, to "convince [the] French we are not trying to cut them out."[62]

In the end, the $1.5 million the ICA pledged underwrote a large—though relatively brief—emergency relief operation in Agadir, conducted by the US military in coordination with Moroccan authorities and French military forces. In 1960, Morocco was home to two major US military installations, Port Lyautey Naval Base and Nouasseur Air Base, legacies of basing agreements made during French colonial rule. From these sites, as well as from the US European Command's Paris headquarters and the US Navy's Sixth Fleet in the Mediterranean, hundreds of US Navy, Marine Corps, and Air Force personnel converged on Agadir to assist with the emergency.[63]

Remaining in Morocco for roughly two weeks, these service personnel located and disposed of bodies, rescued survivors, distributed food, and cleared away rubble and debris. As they worked, US Air Force and MATS planes carried in more than 2 million pounds of food, medical supplies, construction equipment, and other material aid from Europe and the United States. By mid-March, the military concluded its relief operations and withdrew from Agadir. In contrast to South Korea the year before, US officials eschewed longer-term aid commitments, leaving Moroccan and French authorities to sort through the messier details of recovery and reconstruction.[64]

American soldiers standing in the ruins of a house in Agadir, Morocco, following the 1960 earthquake.

Image courtesy of Sueddeutsche Zeitung Photo via Alamy Stock Photo.

In Agadir, the complex politics of decolonization steered the course of US disaster aid decisively, with US officials striving to strike a delicate balance two competing interests: preserving good relations with European allies and demonstrating solidarity with postcolonial countries. Just as they had in Morocco, these tensions shaped the Eisenhower administration's responses to disasters in many newly independent nations and throughout the so-called Third World.

✪   ✪   ✪

Between 1955 and 1960, catastrophes in Central America, East Asia, and North Africa provoked sizable US humanitarian responses, each ranking among the largest US foreign disaster aid efforts during the ICA's tenure. Vastly overshadowing these aid operations, however, was the US response to a tremendous catastrophe in Chile. Spanning the final eight months of Eisenhower's presidency—and indeed, for years beyond it—the American relief and reconstruction efforts that followed this crisis were in many ways unparalleled. Eager to influence the political situation in Chile, the US government once again assumed the central responsibility for planning and executing this humanitarian operation. Perhaps more than any other contemporary catastrophe, this episode showcased the varied mechanisms of US foreign disaster aid, and the Eisenhower administration's motivations for providing it, as the 1960s dawned.

On May 21 and 22, 1960, a series of earthquakes occurred across a vast stretch of southern Chile. The strongest of these quakes struck the city of Valdivia, whose name became the metonym for a far broader national crisis.[65] The seismic activity, together with the tsunamis and landslides it spawned, affected 72,000 square miles of territory, roughly one-third of the entire country. Although recent estimates place the death toll at roughly 2,000, estimates at the time claimed that more than 5,000 people lost their lives. Some 450,000 Chileans became homeless, while roughly 800,000 were affected in some way.[66] Ambassador Walter Howe spoke for many contemporaries in calling it "the worst physical disaster in Chilean history."[67]

The Valdivia earthquake occurred at a time of significant handwringing in Washington over the direction of Latin American politics. The success of the Cuban Revolution in 1959 stoked US policymakers' fears about the threat communism posed to the entire Western Hemisphere, anxieties that had only intensified by May 1960. In Chile itself, the Eisenhower administration enjoyed relatively cordial relations with the conservative president

Jorge Alessandri, considering him a reliable anticommunist ally. However, Alessandri's narrow victory over Socialist candidate Salvador Allende in the elections of 1958—and Allende's enduring popularity ever since—made US officials apprehensive about Chile's potential to slide leftward.[68] These concerns, together with a desire to demonstrate US sympathy for Chile and Latin America, underpinned one of the largest and costliest foreign disaster relief operations the federal government had ever administered.

From the start, the US government's response to the catastrophe in southern Chile was a major undertaking. Commencing on the second (and worse) day of the earthquakes, when Ambassador Howe used the $10,000 at his disposal for local relief efforts, the humanitarian operation really got underway on May 24 with the launch of a massive US military airlift, named Operation Amigos.[69] From the Caribbean Command's headquarters in the Canal Zone, seventy aircraft flew 900 tons of supplies into Chile, a cargo that included "all available" army tents and blankets, ten helicopters, and two fully staffed army field hospitals, erected in the cities of Valdivia and Puerto Montt. Accompanying this material aid were an eleven-man disaster survey team and 700 US service personnel, many of whom remained in Chile for several weeks.[70]

By early June, the State Department had expended $3,760,000 from the ICA's Contingency Funds to finance Operation Amigos—and this was just one piece of the US government's emergency response.[71] The ICA released large stockpiles of PL-480 flour, rice, and corn meal from its warehouses in Chile and shipped an additional two million pounds of rice and other PL-480 foodstuffs from the United States.[72] The Export-Import Bank, the US government's export credit agency, also authorized a $10 million line of credit to Chile, earmarked for eventual reconstruction projects.[73]

Seeking to supplement the state's already extensive relief commitments, the Eisenhower administration not only enlisted American voluntary organizations to provide additional aid, but also directly intervened in the voluntary sector's operations to achieve this goal. On May 27, Eisenhower deputized the ARC's president, General Alfred Gruenther, to lead a nationwide relief campaign, appointing him to coordinate the efforts of all American voluntary organizations involved in Chilean relief.[74] Convening multiple times at the White House and the ARC's headquarters during the next few weeks, Gruenther met with heads of multiple ACVFA affiliates and representatives of the ICA, State Department, and Defense Department to devise a collective, coordinated response to the situation in Chile.[75]

Consonant with its leadership role, the ARC played a principal part in the voluntary relief work that followed. In addition to allocating more than

$500,000 in cash and medical supplies to Chile in 1960 (and more in the years ahead), ARC leaders sent seven disaster experts to the country to assist their sister society, the Chilean Red Cross.[76] Many ACVFA affiliates also contributed assistance, chief among them CARE, Catholic Relief Services, Church World Service, and the Seventh-day Adventists. By mid-June, these and other American voluntary organizations raised $4.5 million for Chilean relief and sent large quantities of PL-480 food, clothing, household furnishings, tools and construction materials, and other supplies to Chile. By month's end, the value of this undertaking topped $15 million.[77]

The tremendous US response to the Valdivia earthquake, on the part of both the US government and the American voluntary sector, generated considerable praise for the United States in Chile. As the Chilean politician and future president Eduardo Frei opined, "Many years of visits and propaganda were not able to accomplish what was accomplished in a few days."[78] American diplomats in Chile widely agreed with this assessment. Judging the relief effort "a smashing public relations success," Howe and his embassy colleagues reported that the American response was "unanimously acclaimed by all sectors (except the Communists) of the Government, press, and public."[79]

Even more importantly, in the Eisenhower administration's eyes, US humanitarian actions in Chile appeared to have boosted the nation's image elsewhere in Latin America, too. As one US embassy official in Panama remarked, echoing his colleagues elsewhere in the Western Hemisphere, "it will not be quite as easy in the future for the America-haters here (or elsewhere) to claim that the United States has no sympathy for Latin American problems."[80] To be sure, some outspoken critics remained; yet even they failed to dampen US diplomats' enthusiasm. "Perhaps the best proof of our effectiveness," one State Department official observed with a measure of pride, "has been the fact that Fidel Castro's agents have denounced our assistance as an attempt to 'increase U.S. domination of Chile.'"[81]

Pleased with the goodwill their response achieved in the short-term, US officials in Chile and Washington began considering an ambitious reconstruction and development program for Chile, at a scale that far surpassed earlier efforts in Italy, Japan, South Korea, and elsewhere. Though the government most often limited its humanitarian involvement to the emergency relief phase, the Eisenhower administration saw multiple good reasons to remain involved in Chile's longer-term recovery. In addition to burnishing the United States' image, they reasoned, reconstruction assistance stood to buttress Chile's economic development and political stability, which appeared imperiled by the tremendous costs and challenges of rebuilding.

The Alessandri government's "success or failure" to achieve recovery "while maintaining fiscal responsibility and encouraging private enterprise, will influence [the] future of Chile and neighboring countries for many years," Howe advised the State Department. "Aside from [the] needs of Chile," the ambassador counseled, "I believe strongly that it is in [the] US political and economic interest to be prepared to help on a substantial scale."[82]

State Department and ICA leaders generally agreed with Howe's assessment. Observing, "Our prompt disaster aid has had a most favorable impact, not only in Chile but worldwide," the assistant secretary of state for inter-American affairs recommended that "a prompt and realistic follow-through now is essential to assist the moderate and pro-US Alessandri Government in its reconstruction efforts."[83] While conceding it was "neither practicable nor desirable" for the United States to finance all of Chile's reconstruction needs, State Department leaders believed that "it should nevertheless be possible for us to make a contribution large enough to win us considerable good will."[84]

And so, when representatives of Alessandri's administration approached the State Department with an official request for reconstruction aid, members of the Eisenhower administration were eager to oblige.[85] On June 23, the State Department announced an additional $20 million grant, drawn from the ICA's Contingency Fund, "to assist Chile in its reconstruction efforts."[86] Supplementing the ICA's contributions to Chilean recovery, the US Navy secretary provided twelve ships to the Chilean government, loaning these vessels for five years at no cost.[87] The Eisenhower administration also sent $477,000 to the Organization of American States, a contribution to its special fund for Chilean reconstruction.[88]

Even as US officials made these substantial and fairly unorthodox contributions toward Chilean reconstruction, the Alessandri administration urged the US government to do more, tactically appealing to American officials' anxieties about the tenuous political and economic situation in Chile. Meeting once again with State Department leaders in late June, Chilean government officials made a strategic plea for supplemental aid. "Leftists, headed by extremist Senator Allende, were already demanding that the country pass over its debt payments and raise the income taxes on business, meaning especially the U.S.-owned copper companies," warned Chile's ambassador to the United States, Walter Müller. Though he assured US officials that President Alessandri "has refused to consider such steps," Müller cautioned, "it will be difficult for him to hold the line unless additional visible financial support is forthcoming."[89] Jorge Schneider, a special representative of the Chilean government, echoed the ambassador's points. Explaining that "this was the first decent Government Chile had had in many years," he stressed, "it was very important that it receive timely help from the U.S." Reconstruction aid, he added, was even "more important psychologically than economically."[90]

State Department officials found these arguments persuasive. But even as they wished to respond favorably to the Chilean government's appeals, they recognized that their usual sources of assistance—the ICA's Contingency Fund and sales of PL-480 food—were insufficient for this task and possibly illegal to use for these purposes.[91] So instead, State Department officials turned to Congress for help. Specifically, they urged legislators to authorize $100 million in special assistance for Chilean reconstruction.[92]

One hundred million dollars represented an extraordinary sum. The decision to request such a large congressional aid package for postdisaster rebuilding, moreover, marked a notable departure from precedent. Although the US government and its partners had funded foreign rebuilding projects before, they typically did so only when excess money remained after emergency relief efforts ended, as had occurred in Italy in 1909 or Japan in 1923. When proposals for comparable large-scale reconstruction packages did arise—as they had in the Ionian Islands in 1953 or in Chile following the last major earthquake, in 1939—US officials historically rejected them, viewing US governmental involvement in the reconstruction phase a bridge too far.

In contrast to their predecessors, many members of Congress in 1960 proved receptive to the proposal. While sympathetic to the humanitarian and political imperatives of assisting Chilean recovery, they also demonstrated a greater willingness to allocate US government funds in this way. And so, acting on the State Department's request, the House and Senate began drafting a joint resolution for Chilean aid.

Although it took some time to wind its way through Congress, the proposed aid package eventually became a reality on September 8, 1960, with the passage of Public Law 86-735. Notably, this legislation was not limited to Chilean relief. Signifying the US government's growing embrace of international development as a tool of foreign policy, the law authorized a comprehensive program of development assistance for Latin America as a whole, earmarking $500 million for projects throughout the Western Hemisphere. Additionally, illustrating the growing links between humanitarian relief and development aid, the act authorized $100 million in special assistance, "for use . . . in the reconstruction and rehabilitation of Chile on such terms and conditions as the President may specify."[93] Three and half months after the

Valdivia earthquake, the Eisenhower administration made good on its pledge to assist Chile's long-term recovery.

Or so it seemed. Although Congress authorized this assistance for Chilean reconstruction, lawmakers had yet to disburse the funds. Before this could occur, the US and Chilean governments first had to agree to the terms and conditions of this aid package. Much to the State Department's dismay, negotiations on this agreement abruptly stalled. While there were multiple reasons for this impasse, one was decidedly paramount: a belief, rooted in age-old ideological tropes, cultural assumptions, and racial stereotypes, that Chileans were not doing enough to help themselves. By late 1960, many American policymakers were growing disillusioned with what they considered the slow progress of Chilean recovery. By and large, they blamed this outcome on the Chilean government and its people. Citing an "apparent lack of leadership" in the country's reconstruction effort, and criticizing Chileans for their unwillingness "to make the sacrifices required under the disaster situation," some members of Congress sought to make the appropriation "dependent on positive action by Chile to assist itself." Others went further still, threatening to reduce the amount of aid they had promised or even rescind it entirely.[94]

Together, this combination of delays, insults, and ultimatums risked negating the diplomatic gains US disaster aid had achieved. By late November 1960, two and a half months after the passage of PL 86-735 and a full seven months since the catastrophe commenced, Howe warned of "widespread bitterness and sense of deception" among Chileans should Congress fail to deliver the aid it had pledged. Further delays or reduced levels of assistance, he cautioned the State Department, "will seriously prejudice [the] goodwill thus far attained" while endangering Chilean political and economic stability.[95]

When Congress still had not acted a month later, Howe and his colleagues in Chile redoubled their appeals. This time, they tried a different line of argumentation, emphasizing the diplomatic benefits of reconstruction aid and its function as anticommunist propaganda. "The long range results of this program will lead to a substantial strengthening of the general U.S. psychological position in Chile," embassy officials stressed. More than this, it promised to show "Chileans how effectively the free world can help them reconstruct the South."[96] Given US policymakers' preoccupation with Chile's potential leftward turn, this last point was perhaps the most compelling argument of all.

Despite Howe's best efforts, congressional inaction persisted, ultimately outlasting both the Eisenhower presidency and Howe's tenure as ambassador.

CHAPTER 11

The funds eventually arrived in Chile, but not until the following year, under a new US presidential administration.

Between the ICA's creation in 1955 and the time Eisenhower left office, in January 1961, the US government and military had responded to dozens of catastrophes globally. Although the ARC and ACVFA affiliates cooperated in many of these operations, American foreign disaster aid had begun to flow increasingly through state channels.

Taking stock of these changes, at least some US officials lamented the state's creeping involvement into American foreign disaster aid. "The U.S. Government should not get into the habit of dashing to the rescue every time there is an earthquake, fire or flood," ICA director John B. Hollister insisted in 1957. "We are not an international Red Cross," he argued, and "we can't go running around doing this sort of thing all over the world."[97] Most officials, however, appeared to embrace—or at least, resign themselves to—the federal government's expanding humanitarian role. Whatever their opinions on the matter, few would have disagreed with ARC leader George Elsey's observation, in 1959, that "the U.S. government is becoming increasingly interested in the entire problem of disaster-caused needs in other countries."[98] Within the next few years, the trends Elsey and Hollister identified became all but impossible to ignore.

Recipients of official US foreign disaster assistance referenced in chapter 12, 1961–63

12

Disaster Assistance
in the Decade of
Development

1961–1963

George F. Kennan (the younger) is typically remembered for many things—his influence on US Cold War policy, his views on the Soviet Union, his containment strategy—but not for his role as a volunteer blood donor. Yet on July 26, 1963, the penultimate day of his two-year stint as US ambassador to Yugoslavia, Kennan led forty members of the US embassy in Belgrade to a Yugoslavian Red Cross blood drive.[1] Earlier that morning, a powerful earthquake occurred several hundred miles away in Skopje, Yugoslavia's third largest city. The result was an incredible disaster, which left Skopje in ruins. In the wake of this catastrophe, Kennan's blood donation marked the symbolic start of a major US disaster relief operation in Yugoslavia, one that unfolded during the last few months of John F. Kennedy's life and presidency—and for several years beyond it.[2]

In the fifteen years before this catastrophe, across the late 1940s and 1950s, US policymakers had steadily expanded the state's role in foreign disaster assistance, as well as its financial commitments in this humanitarian sphere. These trends continued apace throughout Kennedy's presidency, from January 1961 to November 1963. During these three years, US officials further centralized foreign disaster aid under the federal government's auspices, including most notably the new US Agency for International Development (USAID). Together with their partners in the American voluntary sector, the government and military provided disaster aid to dozens of countries, staging particularly noteworthy responses to the Skopje earthquake, another catastrophic earthquake in Iran, and a hurricane that struck both Cuba and Haiti. In these and many other nations, US humanitarian actions continued to be guided by Cold War political concerns, by the dynamics of decolonization and anticolonial nationalism, and—more than ever before—by ascendant theories and practices of international development.

As US foreign disaster aid efforts multiplied during the early 1960s, so did the administrative headaches and bureaucratic confusion associated with them. The individuals and agencies responsible for administering these humanitarian operations complained frequently of delays, unnecessary expenditures, poor planning, and redundancies. Although US officials had taken tentative steps to address these problems during the previous decade, a growing consensus emerged about the need for more substantive reforms. Officials called for improved coordination, more consistent funding, and the creation of a designated bureau or office, charged with managing the nation's official responses to international catastrophes. By the time of Kennedy's assassination in late 1963, these critics had laid the groundwork for a complete overhaul of the US foreign disaster aid system, a process that was to culminate under the succeeding administration.

★　★　★

When he became president of the United States in early 1961, John F. Kennedy inherited a host of complex international challenges, rooted in the interplay between the global Cold War and decolonization. During his brief presidency, Kennedy and his administration dealt with multiple foreign policy crises arising from these issues: confrontations with the Soviet Union over the status of Berlin and the mutual threat of nuclear weapons; escalating tensions with Cuba over the Bay of Pigs invasion and subsequent missile crisis; an expanding war in Vietnam and the spread of communism elsewhere in Indochina; and the aftermath of a bloody US-backed coup in the Congo.

Alongside these foreign policy flashpoints, the Kennedy administration devoted considerable attention to another pressing concern: improving US relations with the so-called Third World. Determined to supplant Soviet, Chinese, and Cuban influence in the Middle East, Africa, Asia, and Latin America, the Kennedy administration made diplomatic overtures to nonaligned nations and nationalist movements throughout these regions, striving to win friends and allies for the United States. The government also funneled large amounts of military aid to several key battleground countries, aiming to support moderate, anticommunist, and avowedly "modernizing" leaders in these nations. And to a far greater extent than either the Eisenhower or Truman administrations before them, Kennedy and his advisers embraced a third strategy for winning Third World loyalties: economic aid and development assistance.[3]

Kennedy and his advisers agreed that development assistance must form a cornerstone of US foreign policy. Yet in order to achieve this goal, they believed, they first needed to fundamentally reform the nation's approach to foreign aid. As Kennedy's close adviser, the economist Walt Rostow, argued shortly after the election of 1960, the government needed to achieve a "breakthrough on the question of aid to the under-developed nations."[4] Once he was sworn into office, Kennedy and his administration set to work on this matter. Just a few days after his inauguration, Kennedy issued an executive order establishing the Food for Peace Program, responsible not only for administering PL-480 food assistance but also for reinvigorating and expanding this Eisenhower-era aid program.[5] By mid-March 1961, Kennedy launched two new foreign aid initiatives, the Peace Corps and the Alliance for Progress, both designed to provide economic and technical assistance to developing nations.

Although these actions marked a decisive start, Kennedy and his administration had even bolder plans in mind. On March 22, 1961, the president issued a Special Message to Congress, calling for a complete restructuring of US foreign assistance. Criticizing the nation's existing foreign aid programs as "bureaucratically fragmented, awkward and slow," Kennedy called for a "fresh approach," one better suited to "the needs of the underdeveloped world." Making this proposal more concrete, the president urged Congress to establish a new foreign aid agency, with a more logical organizational structure and a more professional staff. Above all, he implored, the new agency must place development aid at the center of its priorities. Citing the United States' moral, economic, and political obligations to "the underdeveloped nations," he urged legislators to act on these proposals, declaring that "the 1960s can be—must be—the crucial 'Decade of Development.'"[6]

The dramatic changes Kennedy proposed promised to transform the administration of all foreign assistance programs, including international disaster relief and recovery aid. But first, they had to be implemented. Although Congress responded favorably to Kennedy's entreaties, it took nearly six months for the Senate and House to pass the sweeping aid legislation he recommended. Not until November 1961, moreover, did the new aid agency finally come into being.

In the intervening period, US foreign disaster assistance continued to function largely as it had during the past six years: carried out under the auspices of the State Department and ICA and with the collaboration of the US military, ARC, and voluntary organizations registered with ACVFA. American officials found no shortage of catastrophes to which to respond. During the first ten months of 1961 alone, the government and its partners provided aid to survivors of earthquakes in Ethiopia, Indonesia, and Iran; of tropical storms in British Honduras, East Pakistan, and Japan; and of floods in India, Indonesia, Libya, South Korea, South Vietnam, and Turkey.[7] Through the channels of the new Food for Peace Program, the US government sent PL-480 commodities to several of these countries. Officials also used the ICA's contingency funds, allocated each year by the Mutual Security Act, to finance many of these relief efforts. In other instances, official aid took the form of $10,000 grants from US ambassadors (drawn from the ICA's Foreign Disaster Emergency Relief Account), cash donations from the ARC, or deliveries of food and material supplies from ACVFA affiliates.

Throughout 1961, while responding to these and other catastrophes, US officials also continued planning and administering aid projects in southern Chile, part of the ongoing reconstruction efforts for the 1960 Valdivia earthquake. As described earlier, Congress authorized $100 million in special assistance for Chilean recovery in September 1960, only to have legislators drag their feet for months before appropriating those funds. They finally did so, at Kennedy's express urging, in late May 1961.[8] The US government's involvement in Chilean reconstruction, however, did not end with this action. Before those funds could be disbursed, US and Chilean officials first had to reach an agreement on how to spend them.

Representatives from the two governments had very different visions for this aid and its purposes. Chilean officials, for their part, requested "considerable latitude in determining the types of projects eligible for this assistance."

The Alessandri government, they explained, desired to blend disaster reconstruction projects into a broader, ten-year economic development program for the devastated region. Hoping to facilitate "orderly planning for the reconstruction and development of the area," Chilean officials urged their US counterparts to accept a "liberal definition of 'reconstruction' and 'rehabilitation.'"

Representatives from the State Department and ICA had other ideas. More concerned with the diplomatic symbolism of their aid, they prioritized rebuilding projects with a "maximum impact in terms of easy identification of US assistance."[9] Moreover, despite their professed commitments to Chilean development and inter-American cooperation—exemplified in the new Alliance for Progress—US officials desired to maintain a substantial degree of control over any American-funded reconstruction activities. For the Kennedy's administration, American public diplomacy and oversight trumped Chilean-led development planning.

Despite their divergent goals and interests, US and Chilean officials eventually achieved a consensus on how to use the congressional appropriation, formalizing this agreement in early August 1961. Under its terms, the US government extended a $100 million line of credit to Chile, repayable over forty years without interest. This loan, US and Chilean officials concurred, should finance the reconstruction of housing, schools, hospitals, and other public works projects in the disaster zone. All building projects were supposed to be "durable" and "easily identifiable" with the United States and to "have a favorable impact on people living in devastated area." Projects with a potentially negative connotation, such as jails and police stations, were expressly excluded. To ensure these projects met with US officials' expectations, finally, the agreement stipulated that the US government would reimburse Chile for approved expenditures rather than remitting the funds in advance, a measure that allowed Americans continued supervision over the programming and administration of this aid.[10] Under these terms, US-sponsored reconstruction work in southern Chile finally began during the second half of 1961, more than a year after the disaster commenced.

The US government's longtime humanitarian auxiliary, the American Red Cross, also remained deeply involved in Chilean recovery efforts. In the early aftermath of the Valdivia earthquake, the ARC committed more than $1.5 million toward Chilean relief and sent seven staff members to Chile to lend their expertise. Although the ARC normally restricted its disaster aid to emergency forms of relief, in this case the organization followed the US government's lead and expanded its humanitarian work in Chile. In December

1960, ARC leaders pledged to invest an additional $500,000 toward Chilean reconstruction.[11] Throughout 1961, these funds subsidized many rebuilding projects, including new homes, Chilean Red Cross centers, dormitories, gymnasiums, cultural centers, and orphanages. The ARC also financed several special aid programs for Chilean youth, providing school supplies, scholarships, and Christmas gifts to children affected by the earthquakes.[12] Several ARC staff members remained in Chile through late 1961 as well, supervising the design and construction of building projects and ensuring that the funds were spent as ARC leaders intended.[13]

By the end of 1961, both the federal government and ARC had launched a wide array of reconstruction projects in southern Chile. Although US officials had initially prioritized their own diplomatic interests over the Chilean government's development plans, many of the projects they sponsored nevertheless blurred the lines between disaster recovery and international development—an outcome American officials supported, so long as the specific form of those projects met their approval. From this point forward, US officials reined in their direct involvement in Chile. The rebuilding activities they financed, however, continued well into 1963, mainly under Chilean supervision.[14] By the time the US government and ARC closed the books on their respective responses to the Valdivia earthquake, it ranked as one of the most extensive foreign disaster assistance operations they had ever undertaken.

As 1961 drew to a close, with reconstruction projects now well underway in Chile, the US foreign aid community's attention shifted to events in Washington. Although President Kennedy first called for the wholesale restructuring of US foreign assistance in early 1961, the transformation he proposed finally came to fruition on September 5, with congressional passage of the Act for International Development of 1961.[15] With this landmark legislation (better known as the Foreign Assistance Act of 1961), Congress completely overhauled the existing US foreign aid system. Among the many changes policymakers implemented, two were particularly consequential. First, underscoring the government's professed commitment to international development, the act drew a clear distinction between military and nonmilitary forms of assistance. Second, the legislation abolished the International Cooperation Administration while authorizing the creation of a new agency, dedicated specifically to nonmilitary aid, to replace it.

Two months later, on November 3, 1961, Kennedy established this new entity—the US Agency for International Development—by executive order.[16] Charged with administering the US government's foreign economic, technical, development, and humanitarian aid programs, USAID brought many disparate foreign assistance initiatives together under one roof. Though it was housed within the State Department, USAID was technically an independent agency and in theory more autonomous and more insulated from diplomatic pressures than its predecessors.

The passage of the Foreign Assistance Act of 1961 and subsequent creation of USAID transformed US foreign aid in many appreciable ways. And yet, once again, these developments had few immediate consequences for the conduct of foreign disaster assistance. As with previous foreign aid reforms in the late 1940s and 1950s, these watershed initiatives affected disaster assistance only indirectly, not intentionally. Focused primarily on long-term development projects and long-range development planning, the architects of these new programs continued to neglect the specific problems of natural hazards and the catastrophes they triggered. Opting for continuity in this area over any substantive reforms, policymakers simply retained the assorted disaster-related provisions that were implemented in piecemeal fashion during the Truman and Eisenhower presidencies.[17]

More specifically, with the Foreign Assistance Act of 1961, Congress reauthorized the presidential "Contingency Fund," first introduced by the Mutual Security Act of 1954, for use in unspecified events determined "to be important to the national interest."[18] It remained the principal source of funding for the government's foreign disaster aid operations in the years ahead. The Foreign Assistance Act also retained the practice of reimbursing the ARC and ACVFA-registered organizations for the costs of transporting disaster relief supplies overseas, a precedent dating to the late 1940s.[19] Additionally, USAID, the ARC, and ACVFA affiliates maintained the authority to access and distribute surplus commodities as foreign disaster relief, as stipulated by Titles II and III of PL-480 since 1954.[20] The State Department, finally, continued to grant chiefs of mission the authority to commit $10,000 in emergency aid for any disaster situation, a practice initiated by John Foster Dulles in 1958.[21]

This is not to say the reforms of 1961 left American catastrophic diplomacy totally unchanged. In a major bureaucratic shift, USAID replaced the ICA as the government's designated agency for foreign disaster aid, with the agency's administrator assuming responsibility "for funding and operational coordination of emergency disaster relief operations."[22] The new emphasis on development planning and programming, fundamental to USAID's mission, also influenced foreign disaster assistance in perceptible ways. As

USAID established new missions in many countries throughout the world, its personnel regularly became involved in both short-term and long-range disaster aid efforts, further blurring the already hazy lines between relief, reconstruction, and development.

Still, the Foreign Assistance Act of 1961 made no explicit mention of natural hazards. Nor did it identify disaster assistance as a distinct category of aid. In turn, USAID had no dedicated bureaus or personnel (aside from its administrator) assigned specifically to foreign disaster assistance. The effect, as far as US foreign policy was concerned, was that catastrophes remained contingencies, treated in federal law and policy as unforeseeable events. The Kennedy administration, accordingly, continued responding to foreign disasters on a case-by-case basis, through the channels of a humanitarian system that had taken form piecemeal over the course of the twentieth century. As a result, problems of planning and coordination, which plagued the ICA during its tenure, continued to cause headaches for USAID, the State Department, and their partners in the military and voluntary sector.

Over the next two years, from November 1961 to November 1963, frustrations over these issues steadily mounted. As officials searched for ways to resolve these problems, they laid the foundations for a more fundamental transformation of US foreign disaster assistance in the decade ahead.

It was only two weeks after USAID's establishment that a major catastrophe in eastern Africa precipitated the agency's first large-scale international disaster relief operation. In mid-November 1961, floods along the Juba and Shabelle Rivers triggered a serious disaster in the Somali Republic, affecting an estimated 250,000 people.[23] At the request of both Somali authorities and the US embassy in Mogadishu, the US government joined several other countries and international organizations in responding to the crisis. USAID initially authorized $50,000 for the emergency response and subsequently allocated tens of thousands of dollars more. These funds underwrote a sizable military airlift, carried out by US Air Force and Army planes from US bases in Europe. By the end of November, these aircraft had ferried in large quantities of PL-480 grain and oil, medical supplies, and two helicopters, used to distribute aid throughout the flooded region.[24]

Despite the efforts of Somali authorities, the United States, and other countries, the situation continued to worsen. By mid-December, incessant rains considerably aggravated the floods, which now affected an estimated 600,000 people. Food supplies grew scarcer, and rates of malaria, dysentery, and infectious disease were on the rise.[25] Eager to provide some additional American aid to this newly independent African nation, the State Department asked the ARC to send a disaster specialist to the Somali Republic to assist the ongoing relief effort. "Pleased at this opportunity to cooperate with the Department of State and AID," ARC leaders dispatched the organization's deputy director of disaster services, Robert Pierpont, to the scene.[26] He arrived in Mogadishu on December 18 and remained there until the end of the year, offering his counsel to the Somali government while also assessing the US response.

What Pierpont observed in the Somali Republic did not fill him with confidence. Cataloging myriad problems with the US relief effort, he came away convinced that something must be done to improve the coordination and execution of US foreign disaster assistance. Upon his return to Washington, Pierpont wrote a lengthy report to USAID's administrator, offering both critiques and recommendations. "There should be established in the State Department . . . a unit authorized to deal immediately and positively with disaster situations of world-wide import in other countries," Pierpont advised. "This unit should bring into concert the diplomatic, AID and other elements of State, [and] the logistics support of Defense," he continued, and should "avail itself of the disaster capabilities of the American National Red Cross." Explaining to the administrator that "there will unquestionably be future natural disasters in the world which will be international in scope," Pierpont also warned that "disaster can provide a breeding ground for discontent, unrest and political disease if the essential human needs in disaster go unmet." To prepare for these challenges, he urged, USAID must "eliminate the piece meal handling" of foreign disaster aid.[27]

Pierpont was hardly the first to recommend such reforms. Since the mid-1950s, US officials had experimented with various strategies to improve the coordination and administration of foreign disaster aid. Shortly after USAID came into being in 1961, ARC and ACVFA leaders met in Washington to discuss this matter, reaching many of the same conclusions Pierpont did in Mogadishu. "One of the major problems," they agreed, "was the absence of a coordinated and rapid machinery through which relief could be provided to victims of disasters in other countries." To resolve these difficulties, they proposed the "setting up of some effective machinery within A.I.D. for rapid action in the event of disaster, as well as the creation of some coordinating body within A.I.D. to centralize relief efforts by other government agencies."[28]

The issues Pierpont observed in the Somali Republic, in other words, were only the most recent example of a more chronic problem; the recommendations he made to USAID's administrator, in turn, formed part of a much broader conversation over how to improve the existing US foreign disaster aid system. Over the next few months, dialogue on this issue continued. Leaders of the ARC and USAID met multiple times in early 1962 to discuss "cooperation in times of international disaster" and "the need for better coordination among government and voluntary agencies in international disaster relief."[29] By April, ARC leaders were pleased to learn that the State Department was planning "a study of the present regulations governing the actions of the entire government, of State, AID, and Defense, in disaster situations." They noted approvingly that USAID leaders were "taking these discussions seriously."[30] Dialogue about improving US foreign disaster aid continued throughout the remainder of Kennedy's presidency. Substantive reforms, however, had to wait a bit longer.

While these conversations transpired in Washington, natural hazards continued to occur throughout the world. Between USAID's creation in November 1961 and Kennedy's assassination in November 1963, US officials responded to many of the disasters these hazards precipitated. In addition to the floods in the Somali Republic, the State Department, USAID, and their partners aided survivors of flooding in Dahomey, Greece, India, Kenya, Morocco, Pakistan, Peru, South Vietnam, Spain, Tanganyika, Tunisia, and West Germany. During this same two-year span, US aid reached victims of earthquakes in Iran, Turkey, and Yugoslavia; a landslide in Rwanda; tropical cyclones in Cuba, East Pakistan, Haiti, Japan, South Korea, and Thailand; and volcanic eruptions in Costa Rica and Indonesia.[31] To these and other countries, US officials contributed tens of millions of dollars' worth of disaster aid, in the form of PL-480 commodities, USAID contingency funds, money and material supplies from the ARC and various ACVFA affiliates, and the logistical support of the US military.

Of all the disaster relief and reconstruction efforts American officials undertook during these two years, among the most diplomatically significant and administratively complex were the responses to a powerful earthquake in Iran in September 1962, another major earthquake that struck Yugoslavia the following July, and a hurricane that devastated both Haiti and Cuba just a few months later, in October 1963. A closer look at these humanitarian operations illustrates the evolving mechanisms of US foreign disaster assistance, the ongoing problems with its coordination and execution, and the complex politics influencing disaster response during the Kennedy presidency.

☆ ☆ ☆

On September 1, 1962, an earthquake occurred in the area of Buin Zahra, in Iran's northwest Qazvin Province. This seismic event resulted in more than 12,000 fatalities, leveled scores of villages, and left some 100,000 people homeless.[32] Commencing on the eve of Iran's White Revolution—a wide-ranging series of reforms, launched by Shah Mohammad Reza Pahlavi in early 1963, with the professed aim of "modernizing" Iran—the US humanitarian response to the Buin Zahra earthquake became enmeshed with this controversial program and its precursors, once again obscuring the boundaries between disaster relief and development assistance.[33]

The Buin Zahra earthquake came at a pivotal moment in US-Iranian relations. Ever since Kennedy came into office, his State Department had eyed the political situation in Iran nervously, concerned that widespread opposition to the shah threatened to incite revolution and communist takeover. Though they conceded that many Iranian grievances against the shah had merit, US officials saw few good alternatives to his regime.[34] By September 1962, hoping to stabilize this precarious situation—and thereby promote US interests in Iran and the Middle East—the Kennedy administration adopted a policy of "sticking with the Shah" while encouraging him to commit to greater democratization and development.[35]

In this context, US officials saw a generous response to the earthquake as essential, enabling the Kennedy administration to demonstrate its concern for both the shah and the Iranian people. Through disaster aid, they hoped to curry favor for the United States in this crucial Cold War battleground state. Such a "symbolic gesture," Vice President Lyndon Johnson mused, "might afford [a] basic dramatic contrast of US humanity at [a] time Communists are shooting escapees at [the] Berlin Wall."[36]

With these aims foremost in their minds, US government officials launched an "immediate and formidable" relief operation, as Secretary of State Dean Rusk described it, in northwest Iran.[37] The morning after the earthquake, Ambassador Julius Holmes turned over the $10,000 emergency fund at his disposal to Iran's Red Lion and Sun Society.[38] Meanwhile, three sanitation advisers from the USAID mission to Iran traveled to the disaster area, where they helped Iranian authorities locate survivors, excavate bodies from the ruins, and bury the dead.[39] The USAID mission subsequently established a larger Task Force on Earthquake Rehabilitation, composed of technical specialists in such fields as housing, agriculture, water, and public health, to assist Iranian authorities with related facets of the disaster response.[40]

From Washington, USAID authorized the distribution of all existing Food for Peace stocks in Iran—17 million pounds of flour, milk, beans, and oil—and made plans to ship additional PL-480 commodities from the United States.[41] Supplementing this governmental aid, the ARC sent more than $200,000 in cash and medical supplies to the scene. Multiple ACVFA affiliates contributed money or material aid as well.[42] Chief among them was CARE, which sent $100,000 in cash donations and 500,000 food packages, provided under the terms of PL-480.[43]

Perhaps the most visible element of this US emergency response was a large-scale military airlift, dubbed Operation Ida, executed by the US European Command. From US bases in Germany, twenty-eight US Air Force planes arrived in Iran shortly after the earthquake. Among other supplies, they carried in 1,000 tents, 10,000 blankets, two helicopters, and a field hospital staffed with 200 US Army medical personnel.[44] Declaring Operation Ida a "manifest success," Ambassador Holmes called the airlift "an impressive demonstration of the capability and readiness of US armed forces to mobilize and dispatch vitally needed resources to distant and foreign stricken areas." Just as importantly, in his estimation, the US military's muscular response served as "a manifest reminder to Iran of the worth of its powerful and friendly ally."[45]

These emergency relief efforts marked a first step toward showcasing the Kennedy administration's support for Iran, but the US response did not end there. As it had in Chile following the 1960 Valdivia earthquake, the US government became involved in a major long-term rebuilding project in Iran—and, in the process, in the shah's ambitious and contested schemes for rural development.

In mid-September 1962, the shah announced plans to incorporate disaster rebuilding projects into his broader programs of land reform and modernization. The "earthquake stricken area," he declared, was "to be developed as a model area for the country."[46] Iranian government officials, in turn, appealed to multiple foreign governments to finance and participate in this initiative, including the United States. In Tehran, Holmes and his embassy colleagues greeted these proposals enthusiastically, pleased to see the shah's commitments to Iranian development. Noting, "The earthquake disaster . . . has dramatized the need for country wide rural reforms," Holmes urged the

State Department and USAID to commit additional aid to Iran and "support the Shah's announced aim of model agricultural development in the quake area." American participation in this project, he stressed, would have "both immediate political advantage and longer term development implications."[47]

In Washington, State Department and USAID leaders agreed with Holmes's assessment and moved to act on his recommendations. Over the next few weeks, US and Iranian government officials negotiated a formal agreement on American reconstruction aid. Under its terms, USAID agreed to donate 44,000 tons of PL-480 wheat to Iran, to be sold on local markets. The proceeds were to finance the construction of hundreds of earthquake-resistant, single-family houses, arranged in several villages throughout the devastated area.[48] A portion of the wheat would also go to Iranian laborers hired to build these homes, their compensation under a food-for-work scheme.[49] Additionally, the USAID mission agreed to loan several American engineers to the project, to advise on the planning and construction of the new villages.

In both Iran and Washington, USAID personnel viewed their continued involvement and oversight in these rebuilding efforts as critical. It would not only "help to give the entire undertaking an American label," they reasoned, but also enable them to supervise the use of US government funds.[50] Yet significantly, even as they insisted on a certain degree of control, members of the USAID mission emphasized the importance of Iranian involvement and buy-in. Admitting that previous reconstruction efforts "have been failures," USAID personnel underscored the need for Iranian acceptance of the new home and village designs, stressing that this was "crucial to the success of the entire operation." Influenced by theories of community development then coming into vogue, the USAID mission in Iran made it a policy "that to the maximum extent possible villagers be involved in the planning of the housing and in participating in some measure in building it."[51] By actively including Iranian citizens in both the design and construction of their new homes, USAID mission personnel endeavored to circumvent criticism while cultivating a more genuine, lasting appreciation for the United States among Iranian earthquake survivors.

By the end of 1962, US officials and their Iranian counterparts completed most of their planning, allowing Iranian laborers to begin clearing land and constructing the new homes. It took roughly a year, however, before the US-sponsored villages were completed. During that time, the shah officially launched the White Revolution following a deeply contested national referendum, held in January 1963. Moving forward with his bold plans to modernize Iran, the shah expanded existing land redistribution projects while embarking on an ambitious and controversial slate of social, political, and economic reform initiatives.[52] As per his intentions, the new American villages in northwest Iran became one showpiece of the White Revolution's broader program of rural modernization and development, designed to serve as a demonstration site for the entire country.

On November 15, 1963, the shah and Ambassador Holmes held a ceremony in Qazvin Province, where they dedicated four newly completed American villages and presented deeds for more than 1,000 homes to their new owners. In addition to earthquake-resistant houses, the new villages boasted many "modern" amenities, including graveled roads, toilets, piped water, electrical systems, and four school buildings. In the ruins of the Buin Zahra earthquake, model villages now advertised the promises of the White Revolution, to proponents and critics alike.[53]

By the time these construction efforts concluded, the US government had expended more than $5 million on relief and rebuilding assistance and remained involved in Iranian recovery efforts for fifteen months. Yet by this point, in the latter half of 1963, most US officials had turned their eyes away from the situation in Iran. Their attentions were now fixed on a series of immense catastrophes in Yugoslavia, Haiti, and Cuba, which struck between July and October that year. Together, the US aid efforts that followed these three crises further underscored the messy politics of disaster relief in the Decade of Development.

✪   ✪   ✪

On July 26, 1963, as Iranian laborers worked to rebuild their homes and lives in Buin Zahra, an earthquake 1,800 miles away in Yugoslavia leveled the city of Skopje, the main economic, political, cultural, and administrative center of Macedonia. The disaster left more than 1,000 people dead, injured several thousand more, and made another 200,000 homeless. The quake destroyed or badly damaged 80 percent of Skopje's buildings, including the city's university and its hospitals, libraries, and industrial plants. Later that day, as George F. Kennan and his embassy colleagues in Belgrade rolled up their sleeves to donate blood for injured survivors, their superiors in Washington began discussing how else the US government should respond to this tremendous cataclysm.[54]

The situation in Skopje immediately became a matter of profound concern for the Kennedy administration, entangled with broader issues in

US-Yugoslav relations. Since 1948, Yugoslavia, led by President Josip Broz Tito, had navigated a path of independent socialism, distancing itself from the Soviet Union and pursuing relatively cordial relations with the United States. During that time, Yugoslavia became a major recipient of American economic and military aid, receiving well over $2 billion in assistance by 1961. Many US policymakers hoped that with this aid, they could bolster Yugoslavia's independence from Moscow.

By the early 1960s, however, this relationship began to sour. Congress became critical of the closer ties Tito appeared to be forging with the Soviet Union, as well as his leadership within the Non-Aligned Movement. In 1962, demonstrating their dissatisfaction, US legislators voted to cut off aid to Yugoslavia and revoked the country's most favored nation status. As a result, US-Yugoslav relations rapidly deteriorated, eventually prompting a frustrated Kennan to tender his resignation as ambassador.[55] Though Kennan's departure from Belgrade was planned well in advance, its timing just so happened to coincide with the earthquake in Skopje and the onset of one of the most immense catastrophes ever to befall Yugoslavia.

In this context, US diplomats and State Department leaders agreed that a swift and liberal response to the catastrophe would be "highly desirable," as Chargé d'Affaires Eric Kocher put it, giving them an opportunity to improve the US government's image in Belgrade.[56] Within a few hours of the quake, embassy officials extended an offer of aid to Yugoslavia's foreign minister, who promptly accepted. Shortly thereafter, US emergency assistance began arriving. At the State Department's request, Defense Secretary Robert Mc-Namara ordered troops at Ramstein Air Force Base in West Germany to airlift a US Army field hospital to Skopje, complete with 220 army medical personnel.[57] It arrived the next day, transported to Yugoslavia by twenty-seven C-130 aircraft, and was soon up and running in the nearby city of Kumanavo.[58]

Between July 28 and August 16, US military aircraft made 100 additional relief flights to and within Yugoslavia. They brought large quantities of medical supplies, food, and clothing, a portable kitchen, thirty vehicles, and 250 twelve-man tents, transported from US military bases in Germany, Italy, and the United States. The government also made available nearly $1 million worth of surplus commodities to disaster victims, under the Title II provisions of PL-480. This governmental aid was supplemented by an initial $100,000 in cash and material supplies, donated by the American Red Cross and multiple ACVFA affiliates.[59]

The American relief effort in Yugoslavia was both substantial and highly visible, and by mid-August, as the emergency subsided, US officials in both

A US Air Force transport plane delivering a truck and other assistance to Skopje, Yugoslavia, following the 1963 earthquake.
Image courtesy of Keystone Press via Alamy Stock Photo.

Belgrade and Washington felt very satisfied with the public relations victory they achieved. The arrival of US aid made a "tremendous impact," Kocher crowed, reporting that "our initial generous and timely response will be long remembered by the Government and people of Yugoslavia."[60] Just as importantly from a diplomatic perspective, Kocher observed that Soviet assistance had been "inept, inadequate, [and] tardy," making the US response appear all the more favorable by comparison.[61] Conveying these messages to other members of the Kennedy administration, Under Secretary of State George Ball cheered, "We have gotten splendid echoes from Yugoslavia and I am sure this was a ten strike for the United States."[62]

But if this relief operation generated goodwill for the United States in the short-term, complications soon arose over the matter of longer-term aid. A few days after the earthquake, Yugoslavian government officials started extending feelers to the United States and other countries, as well as the International Monetary Fund and World Bank, asking for assistance in Skopje's eventual reconstruction.[63] At first, it appeared that the US government intended to respond favorably to this request. On August 11, Agriculture

Secretary Orville Freeman paid an official visit to Tito in Yugoslavia, where he pledged an astonishing $50 million in loans and grants, derived from local sales of PL-480 wheat and other foodstuffs, to help rebuild Skopje.[64]

State Department, USAID, and embassy personnel all expressed considerable enthusiasm for Freeman's offer. In addition to recognizing Skopje's real need for rebuilding assistance, they believed that the US government stood to accrue additional diplomatic goodwill by providing it. Some US officials also spied lucrative economic opportunities, imagining that American companies could reap a tidy profit by selling building materials and construction equipment to Yugoslavia. As one US embassy official put it, summarizing these motivations, "We think there is every reason for us to extend this form of help."[65]

Unfortunately, there was a problem. In 1962, Congress had amended the Foreign Assistance Act to prohibit USAID from providing anything other than short-term humanitarian relief to any communist country, unless the president determined that "such assistance is vital to the security of the United States."[66] Over the next six weeks, State Department legal advisers deliberated whether the pledged reconstruction assistance ran counter to this congressional mandate—and, if it did, whether they could legitimately request a presidential waiver on national security grounds.[67] In the meantime, the United States' widely publicized $50 million aid offer was put on hold indefinitely.

Growing increasingly concerned that the assistance Freeman promised might prove "difficult to justify," Ball and other State Department officials began "considering ways to respond by means other than AID funds."[68] Collaborating with loyal partners in the voluntary sector represented one such channel. In late August, in cooperation with the State Department and US embassy, the ARC and several ACVFA affiliates devised plans to provide warm meals to the children of Skopje for up to nine months, acquiring this food under the provisions of PL-480.[69] Kocher expressed full support for this "political gesture," believing it could provide "at least some semblance, small though it may be, of US participation [in] Skopje [during] this period."[70] Still, as the chargé d'affaires and his colleagues understood, a child-feeding program administered by American voluntary organizations was hardly equivalent to a multimillion-dollar reconstruction commitment from the federal government.

By mid-September, complicating matters further, the US embassy in Belgrade reported that 170,000 Skopje residents were still without housing. Moreover, embassy officials reported, a dozen other countries had already taken up temporary housing or reconstruction projects in the area—among them Romania, East Germany, and the Soviet Union.[71] The United States, Kocher warned, was becoming "increasingly conspicuous by its absence."[72] After a brief moment of glory, the Kennedy administration now appeared on the brink of another public relations crisis in Yugoslavia.

Several weeks later, in early October 1963, US officials were still contemplating how to avoid a diplomatic catastrophe in Yugoslavia. As they considered how to assist Skopje's recovery, another major disaster was unfolding halfway around the world—this one caused by a hurricane named Flora. While continuing to weigh their options in Eastern Europe, US officials simultaneously directed their energies toward this newly emerging crisis in the Caribbean.

Hurricane Flora was (and still is) one of the deadliest Atlantic hurricanes in history. This powerful storm affected several Caribbean nations, including the Dominican Republic, Jamaica, and Trinidad and Tobago, but it produced the greatest disasters, by far, in Haiti and Cuba.[73] On October 2 and 3, Flora smashed into Haiti's southern peninsula, causing significant destruction. The hurricane then moved west over Cuba, where it stalled for several days, causing extensive damage in the eastern provinces of Oriente and Camagüey. In Haiti, Flora claimed 5,000 lives and left another 100,000 homeless. In Cuba, the death toll was smaller, at roughly 1,750 people, but 300,000 lost their homes.[74]

As they had with Yugoslavia, members of the Kennedy administration saw compelling reasons to provide disaster relief to both Haiti and Cuba. Complicating any aid offer, however, were the United States' preexisting relations with both countries, led by François Duvalier and Fidel Castro, respectively. Divergent US foreign policy interests vis-à-vis Haiti and Cuba ultimately compelled the Kennedy administration to pursue different approaches to aiding the two nations, approaches that also contrasted sharply with the government's ongoing responses to the catastrophes in Yugoslavia and Iran.

In Haiti, where the hurricane hit first, relations with the United States had grown decidedly frosty in recent years.[75] In mid-1962, the Kennedy administration—troubled by Duvalier's authoritarian rule and by allegations he had misappropriated US assistance—suspended virtually all US aid to Haiti. The only exceptions were a malaria eradication project and a small PL-480 food aid program, administered by ACVFA-registered organizations. Just a few months before Flora struck, relations deteriorated further still, when

Duvalier demanded the withdrawal of the US Navy and Air missions in Haiti and the recall of the US ambassador, Raymond Thurston.

By October 1963, the Kennedy administration had grown deeply hostile toward the Duvalier regime, with at least one official classifying it as the "the worst dictatorship in this hemisphere." Much as they had with the shah in Iran, however, they also feared "the threat of an eventual Communist take-over," led by Duvalier's opponents.[76] Members of Kennedy's State Department, National Security Council, and the CIA therefore felt it necessary to proceed gingerly in Haiti, so as not to stoke political tensions that might foment revolution and further unrest.[77]

Given the complexities of the situation, US officials disagreed on how to respond when, a few days after Flora struck, the Duvalier government formally requested disaster aid. Under-Secretary of State George Ball, for one, was in favor of fulfilling this request. "Assistance would be provided in [the] interest Haitian people," he argued, "who, after all, [are] not responsible for [the] shortcomings [of the] Duvalier regime."[78] The US chargé d'affaires in Port-au-Prince, by contrast, took a more cautious stance. The Kennedy administration must tread carefully, he advised, avoiding any actions that might be construed as "demonstrating US 'support' for [the] Duvalier regime."[79] Eventually, officials concluded that given the disaster's scale, they had no choice but to "respond [in] some measure to the [Haitian Government's] request, no matter how distasteful politically."[80] From the beginning, however, they agreed "minimal US official participation [was] desirable."[81]

Guided by these intentions, US government officials carried out a limited relief operation in Haiti, restricting their efforts solely to the emergency phase. The Defense Department first sent two aircraft carriers to Haiti, from Guantánamo Bay and Norfolk, Virginia, which brought helicopters to conduct aerial surveys of the devastated region.[82] Then USAID allocated $250,000 to the humanitarian effort. These funds covered the expense of sending several US Navy cargo planes to Haiti, which delivered $750,000 worth of PL-480 commodities and a large consignment of medicines and other relief supplies, donated by the American voluntary sector.[83] Officials regarded this airlift as a valuable publicity stunt. The presence of "U.S. military aircraft and military personnel," as Secretary of State Rusk saw it, "will make it clear Haitian people where [the] assistance [is] coming from."[84]

Having demonstrated this political point, however, US government officials did not feel compelled to do much more. To distribute the assistance these planes carried, the State Department relied not on US troops or US

Foreign Service personnel but rather on the three ACVFA affiliates still operating in Haiti: Church World Service, Catholic Relief Services, and CARE.[85] At the State Department's request, the ARC also dispatched two disaster specialists to Haiti. They spent several weeks in the country, advising the Duvalier government and Haitian Red Cross in their response to the catastrophe while, at the same time, acting as informants to the US embassy and State Department.[86]

By late October, with relief efforts now in the hands of the American voluntary sector, the government's direct humanitarian involvement in Haiti concluded. For the next several weeks, the Duvalier government pressed the US government for additional aid, looking ahead to the rebuilding stage.[87] American officials, though, steadfastly refused. Agreeing that they had "no intention of being drawn into long-range development plans of [the] Duvalier regime," US diplomats and State Department leaders concluded that it was "difficult see any advantage to U.S. interests in providing . . . assistance in the 'reconstruction phase.'"[88]

Unfolding along a very different trajectory from its aid efforts in Haiti was the US government's response to the concurrent disaster in Cuba. By early October 1963—nearly five years after the Cuban Revolution, two and a half years since the Bay of Pigs invasion, and twelve months since the Cuban Missile Crisis—US-Cuban relations were in a dismal place.[89] In Hurricane Flora's wake, American intelligence analysts monitored the situation closely, predicting that the "devastation will probably present the Castro regime with its most serious challenge to date."[90]

Keen to outshine Castro and destabilize his government, many US officials deemed it politically expedient for the Kennedy administration to extend an offer of aid to Cuba. The disaster, as the commander-in-chief of the Atlantic Command reasoned, presented an "opportunity for [the] U.S. to exhibit humanitarianism in [a] Cold War situation." Speaking for many of his colleagues in Washington, he argued that a generous assistance offer stood to "give the US psychological advantage" in Cuba while "enhancing our position in [the] minds of [the] Cuban populace."[91] Given the dire state of US-Cuban relations, however, officials also understood that any governmental aid offer would undoubtedly encounter enormous legal, political, and public relations obstacles, both at home and in Havana.[92]

Thus, when American Red Cross leaders requested the government's permission to extend aid to Cuba, both the State Department and National

Security Council were quick to approve, seeing this as a means to showcase American concern without going through more complicated governmental channels.[93] On October 7, with the State Department's blessings, ARC leaders sent a cable to their Cuban counterparts, offering to send relief supplies and experienced disaster specialists from the United States to Cuba.[94]

Two days later, the Cuban Red Cross responded with a brusque refusal, declaring: "We reject the hypocritical offer of assistance, at this time of disaster caused by nature, from those who constantly mean to aggravate, with their blockades and aggression, the misery and ruin of the Cuban nation."[95] Making the rejection public a few days later, the Cuban press lambasted the ARC's aid offer, calling it "nothing more than a cheap way to confuse international public opinion about the policy of the United States toward Cuba."[96] Echoing these remarks in concurrent radio broadcasts, Castro himself disparaged the offer as an act of "hypocrisy and a fake gesture of generosity," deriding the Kennedy administration for attempting "to dress up the beast of imperialism and make it look like a sweet kitten."[97]

Cuba's very public rejection of ARC assistance left US government officials fuming. With little sense of irony, State Department leaders denounced Castro for politicizing the humanitarian crisis Flora triggered and for turning the tragedy into an opportunity for "whipping up anti-US sentiment."[98] They grew even more incensed when Cuban officials went on to accept assistance from various international organizations and some two dozen other countries, among them the Soviet Union, China, and several of the United States' NATO allies.[99] But although Cuban authorities refused the ARC's assistance, US officials had not abandoned their desire to get aid into Cuba. Undeterred, they searched for other channels for delivering American disaster assistance.[100]

In mid-October, they found another potential solution, when the American Friends Service Committee (AFSC), an ACVFA-registered organization, requested the US government's permission to attempt another aid offer to Cuba. Hoping that this Quaker voluntary association might appear less threatening to Cuban authorities than the "semi-official" American Red Cross, government officials gave their confidential assent to this proposal.[101] Allowing the American Friends to deliver disaster assistance, they estimated, would enable them to "score some points with the Cuban people," putting the United States—as the National Security Council's Latin American specialist so eloquently phrased it—in "the position of the 'goodie' while he [Castro] is the 'baddie.'"[102]

Over the next few weeks, a series of complicated back-channel negotiations ensued, with the American Friends acting as intermediary between the Cuban and US governments. Finally on October 28, following "intensive negotiation," all sides reached an agreement.[103] Under its terms, the Cuban government consented to permit a small group of American Quakers to come to Cuba and distribute relief supplies, on the condition that any aid they delivered must be acquired from private sources, not from the US government or ARC. The Departments of State, Commerce, and Treasury, in turn, agreed to temporarily relax export controls and to grant the AFSC all necessary visas and licenses.[104] Aside from these actions, US government officials affirmed that they were "not participating in this private effort."[105]

On November 1, nearly a month after Flora struck Cuba, a four-person team of American Quakers arrived in Havana on a chartered Pan-Am DC-6, armed with $92,000 worth of meat, cornmeal, beans, oil, and medical equipment. Remaining on the island through November 19, they cooperated with local authorities to distribute these supplies, the sole authorized US aid for Cuban survivors of this devastating hurricane.[106]

In October 1963, US policymakers thus pursued very different paths to aiding Haiti and Cuba. Their responses to the parallel disasters precipitated by Hurricane Flora reflected divergent strategic objectives, foreign policy interests, and diplomatic relations with both countries. But as the federal government planned and orchestrated its relief efforts in the Caribbean, the question of how to provide long-term assistance to Yugoslavia remained unresolved. On October 5, as Flora's winds and rains lashed southeast Cuba, Chargé d'Affaires Eric Kocher wrote the State Department from Belgrade, once again imploring his superiors to act. "Despite [the] lapse in time," he advised, "it is still not too late for US to recapture [the] considerable impact it made" with its initial emergency response.[107]

A move in that direction, albeit a partial one, finally came via executive action. On October 17, Yugoslavian president Josip Tito paid a historic state visit to the White House, a meeting, diplomats in both countries hoped, that would mark a step toward improving US-Yugoslav relations. Declaring it the "desire of [the] United States to maintain and develop friendly relations between our countries," President Kennedy sought to make that commitment concrete by extending an offer of temporary housing assistance for Skopje.[108] More precisely, Kennedy pledged to donate 250 prefabricated houses from US military stocks in Europe, capable of housing 2,000 earthquake survivors. He also proposed to send a team of US Army personnel to supervise construction of these buildings and a loan of all necessary construction equipment.[109] Kennedy's offer of short-term housing fell far short of the $50 million in reconstruction aid Secretary of Agriculture Freeman pledged two months earlier. Nevertheless, Tito quickly accepted it. The following day, Kennedy directed the Defense Department to launch this humanitarian operation, to be financed by USAID.[110]

Diplomatic officials in both Belgrade and Washington were pleased with this breakthrough, however limited. Yet to their chagrin, the humanitarian mission that subsequently unfolded, codenamed Operation Home Run, proved far less successful than its optimistic moniker suggested.[111] After receiving Kennedy's orders, the commander of the US Army in Europe assigned twelve US officers and 135 enlisted men from the Seventh Engineer Brigade to Skopje, with orders to supervise the work of 400 Yugoslav laborers hired to erect the temporary homes. Their stated objective was to build all the promised structures before winter weather set in.[112] Due to a series of delays, however, the US team and supplies did not even begin arriving until mid-November. By mid-December, they had completed just 14 percent of the planned construction.[113] The project was not finished until late February 1964, as winter in Yugoslavia was drawing to a close and well after most other donor nations completed their own temporary housing projects.[114]

While providing temporary housing represented a contribution toward Skopje's recovery, moreover, the $50 million in grants and loans that Orville Freeman pledged for Skopje's reconstruction had yet to materialize.[115] It was not until late 1965 and early 1966—two and a half years after the earthquake—that Congress finally appropriated this money, after a new amendment to the Foreign Assistance Act invalidated some of its earlier restrictions on disaster assistance to communist countries.[116]

American reconstruction assistance thus eventually came to Skopje.[117] Its severely belated arrival, however, meant that it failed to make the diplomatic splash or permanent impression for which US officials initially hoped. "The fact remains," as one embassy official lamented in late 1966, that "no visible and imposing mark of American generosity exists today in Skopje."[118] From both a humanitarian and a political perspective, US disaster aid failed to assist survivors of the Skopje earthquake when it would have helped most.

In their attempts to provide reconstruction assistance to Skopje, US officials encountered numerous bureaucratic obstacles, legal complications, and embarrassing delays, causing considerable frustration for the State Department and USAID personnel tasked with making these efforts a success.

These issues, though, were hardly unique to Eastern Europe. The difficulties plaguing the US aid operation in Yugoslavia mirrored the government's experiences in multiple disaster-stricken countries during the early 1960s. The problems that arose in Skopje, moreover, only confirmed what many American officials had insisted for years: the critical need for improved planning, coordination, and execution of US foreign disaster assistance, at every stage of the process.

As 1963 came to an end, the time was ripe for change. Throughout Kennedy's presidency, challenges affecting many relief and rebuilding operations, together with ongoing discussions and growing pressure for reform, paved the way for a wholly new approach to the administration of US foreign disaster assistance. Years in the making, this transformation was finally set in motion in early 1964.

13

Mr. Catastrophe Goes to Washington

1964–1976

In December 1975, for the first time in its history, the US Congress affirmed the federal government's authority to provide international disaster assistance.[1] For readers of this book, this date may sound surprising. Since the dawn of the twentieth century, the US government and military regularly allocated cash, food, military support, and other disaster aid to other nations, both on their own and in collaboration with preferred partners in the American voluntary sector. Such actions became even more commonplace in the roughly two decades following the Second World War, between 1945 and the early 1960s. In these same years, the federal government's direct involvement in foreign disaster aid expanded considerably, as the state and its agents assumed a proportionally greater role in responding to international calamities.

The roots of American catastrophic diplomacy ran deep, as this book has shown, with disaster aid occupying an important place in US foreign relations throughout the early to mid-twentieth century. And yet, after all these decades, after participating in hundreds of relief and recovery operations around the world, and despite ongoing attempts and centralization and coordination, US officials *still* responded to foreign catastrophes in a strikingly ad hoc, impromptu way. Before the mid-1960s, the federal government had no agencies, bureaus, or departments devoted exclusively to foreign disaster assistance. Before the mid-1970s, moreover, no federal laws explicitly affirmed the US government's powers to conduct international disaster aid operations.

In the years between 1964 and 1976, US policymakers brought an end to this long-standing and ambiguous state of affairs. Through a series of reorganizations and legislative reforms, they carved out a formal place for foreign disaster assistance within the US federal government's bureaucracy and legal architecture, institutionalizing its role as an instrument of US foreign policy. Completed by the mid-1970s, these developments shaped the US government's responses to global catastrophes throughout the remainder of the twentieth century and beyond. Despite the shifting winds of domestic politics, geopolitics, and American and global humanitarianism, the changes set in motion between the mid-1960s and mid-1970s continue to structure US foreign disaster assistance up to the present day.

✪ ✪ ✪

During the first few years of the 1960s, under the auspices of the new US Agency for International Development, the US government and its partners in the voluntary sector conducted dozens of disaster assistance operations globally. Yet, as the previous chapter recounted, critiques of the haphazard nature of these humanitarian responses were steadily mounting. Both in Washington and overseas, the officials responsible for these efforts complained regularly of delays, excessive costs, and general disorder.

The principal reason for these problems, critics concurred, was a lack of effective and consistent coordination and an absence of clear policies and protocols. "There was no accumulation of experience, no continuity of expertise," USAID personnel later recalled. "Each disaster had a new cast of players making decisions, launching rescue and relief missions, dispatching food and personnel to disaster sites." Even though the US foreign disaster aid system had become more standardized and centralized by the early 1960s, responses to international catastrophes remained improvised and decided on a case-by-case basis. Officials widely agreed that the existing approach was no longer sustainable. "There was a consensus," USAID staff recollected, "that remedial action was desperately needed."[2]

Long advocated, a substantive reform of the system of US foreign disaster assistance finally came to fruition in early 1964. On January 22, USAID's administrator established a new position within the agency: the foreign disaster relief coordinator.[3] He then appointed Stephen R. Tripp to the job. Tripp had been a federal employee since 1931, serving for many years with the Interior Department. In 1956, he joined the International Cooperation Administration and continued working for USAID after its establishment, in 1961. Now, in 1964, he became the US government's first official point person for international disaster aid, a position he held until 1971. Together with a small staff—initially numbering just three—Tripp was responsible for mobilizing and systematizing US governmental, military, and voluntary responses to global catastrophes. Moreover, he was charged with resolving the myriad issues that had plagued disaster relief operations in recent years.[4]

Tripp quickly set to work on these tasks. One of his first assignments was to create an official USAID manual on "Foreign Disaster Emergency Relief," which he completed in October 1964.[5] Though this was not the first attempt to create such guidelines—officials had drafted several such documents during the Eisenhower administration—Tripp's manual was far more detailed and comprehensive than these earlier iterations.[6] It established formal procedures for the administration of foreign disaster aid, designed to ensure that it functioned as intended: "as a humanitarian service consistent with United States foreign policy goals."[7] Containing a host of policies, regulations, and guidelines, the manual spelled out the specific responsibilities and authorities of the foreign disaster relief coordinator, USAID, and other diplomatic, military, and voluntary actors during times of international catastrophe.

For those involved in the field, most of this information was neither new nor surprising. As the ARC's director of disaster services observed, Tripp's intent was "to formalize in a written document . . . those procedures and practices which have already been in effect." Even so, his efforts to bring clarity and consistency to US international disaster aid met with a warm reception. "The publication of these documents plus the institution of a disaster coordinator's office," the same ARC official ventured, "should reduce the confusion, delay and seemingly haphazard action that has sometimes existed in the past."[8]

But Tripp did more than compile existing policies. Within his first few months on the job, he introduced several new measures designed to improve the administration of US foreign disaster aid. For instance, he started sending alerts and regular updates to the government agencies and voluntary organizations involved in responding to catastrophes. In these disaster memos, he shared up-to-date information about the humanitarian situation, listed immediate and ongoing relief needs, and cataloged the aid the United States, other countries, and international organizations contributed.[9] Tripp also began researching earlier aid efforts and preparing historical case studies, attempting to create a fuller, more detailed record of past US humanitarian operations. Finally, at Tripp's urging, USAID's administrator agreed to increase the emergency funds the agency made available to US diplomatic missions, from $10,000 to $25,000 per disaster event. This enabled US Foreign Service personnel to respond more efficiently to catastrophes as soon as they occurred.[10]

These reforms were only the beginning. Over the next four years, Tripp collaborated closely with other government officials and agencies to further improve the mechanisms of US international disaster response. To that end—and reflecting an intensifying interest in this issue within government circles—the State Department Policy Planning Council, USAID, and the organizers of the White House Conference on International Cooperation all commissioned comprehensive studies of catastrophes and disaster relief. Carried out between 1965 and 1968, these studies surveyed the state of US foreign disaster assistance and made recommendations for reforms. Among their proposals was the need for a larger and "adequately equipped" foreign disaster relief coordinator's office, signaling an appreciation of both the coordinator's recent contributions and the office's future potential.[11]

During these same years, Tripp and other USAID personnel worked with the Defense Department to enhance the military's disaster response capabilities. Following Tripp's lead, the Defense Department issued its own directives on "Foreign Disaster Relief Operations" in late 1964, establishing more formal policies for employing military resources in these situations.[12] Additional developments followed in 1967, when USAID and the US Southern Command established, by joint arrangement, a permanent stockpile of emergency relief supplies and a standing Disaster Assistance and Survey Team in the Canal Zone. These measures were designed to enable US forces to respond more quickly and effectively to disasters occurring throughout Latin America.[13]

Tripp also worked closely with the government's partners in the voluntary sector, cognizant of the crucial collaborative role they played in international disaster response. Leaders of the ARC and ACVFA organizations met regularly with Tripp and his staff in Washington. Generally pleased with this working relationship, they observed that "he has established an extremely close relationship with American voluntary agencies in general, and with the American Red Cross in particular."[14] Tripp and his office were not entirely without critics. Leaders of several voluntary organizations voiced concerns about the government's creeping role in foreign disaster aid, apprehensive about its increasing politicization and worried that their own relief activities risked being overshadowed.[15] For the most part, however, Tripp enjoyed the American voluntary sector's enthusiastic support. As a staff member for Catholic Relief Services told his ACVFA associates, "There are few people in Washington, D.C. that have been able to do so much with so little in so short a time as did Mr. Tripp."[16] Echoing this praise, the ARC's director of international services affirmed, "He has brought order out of chaos in international relief."[17]

These commenters found plenty of opportunities to observe Tripp in action, for between 1964 and 1968, he and his staff coordinated the United States' official responses to dozens of disasters globally. Among the most extensive were those carried out after floods in South Vietnam in 1964 and Italy in 1966, a cyclone in East Pakistan in 1965, a hurricane in Mexico in 1967, and earthquakes in Turkey and Iran in 1968.[18] These large-scale humanitarian operations, however, represented only a small fraction of the broader program of US foreign disaster aid during the Johnson presidency. Within his first four years on the job, in fact, Tripp orchestrated US responses to more than 200 catastrophes in seventy-eight countries.[19]

From a small, four-room office a few blocks from the White House, Tripp and his staff liaised with the State Department and USAID, the Defense Department, the ARC, and ACVFA affiliates to coordinate the distribution of all sorts of disaster assistance. Their aid included Food for Peace commodities, USAID contingency funds, military airlifts and search-and-rescue operations, loans of tents, blankets, and other military stores, and donations of money and material aid from the American voluntary sector. Describing his work as "an avalanche of phone calls, a flood of cables, and rapid-fire changes in plans and schedules," Tripp understood his primary mission in simple terms: "to get emergency help to disaster victims in hours instead of days or weeks."[20]

In January 1968, *Time* magazine printed a short piece, introducing its millions of subscribers to a figure named "Mr. Catastrophe." Later that year, in October, *Reader's Digest* did the same, this time with the more specific moniker "America's 'Mr. Catastrophe.'" The subject of these articles was none other than Stephen Tripp, who had by now acquired the nickname from friends and colleagues. In their biographical sketches of USAID's foreign disaster relief coordinator, the two magazines presented "the brain behind the helping hand our government extends to other nations in time of disaster." Tripp's mission, *Reader's Digest* told readers, was "the relief of human suffering in all countries, whatever their political coloration."[21] Describing the foreign policy aspects of this work more candidly, *Time* explained that "the major diplomatic impact—not to mention the humanitarian aspects—of his coordinating function depends on speed in time of crisis."[22] With this line, *Time* underscored an important truth: in 1968, just as it had throughout the twentieth century, US foreign disaster aid remained inseparable from American diplomatic considerations and strategic objectives.

By the end of 1968, America's Mr. Catastrophe had certainly earned his title. In his first five years on the job, Tripp ushered in multiple reforms of the US foreign disaster aid system, endeavoring to enhance its humanitarian and political effectiveness. During that time, he and his staff—which increased from three to five—oversaw American responses to scores of catastrophes around the world. For these efforts, Tripp received many accolades, from his associates in both the government and the voluntary sector. Still, as Tripp and his colleagues recognized, additional challenges remained, particularly in the legislative realm. Seeking to address these persistent issues, officials continued working to improve US foreign disaster aid for the better part of the next decade.

By the time Richard Nixon became president of the United States, in January 1969, the foreign disaster relief coordinator position at USAID had existed for five years. In that time, the creation and subsequent development of this office had gone a long way toward institutionalizing international disaster aid within the federal bureaucracy.

Yet as the 1970s dawned, USAID's leaders held some lingering concerns. Most notably, the government's authority to provide foreign disaster assistance remained maddeningly ambiguous, having no formal basis in federal law. Although the establishment of the foreign disaster relief coordinator's office did much to streamline the administration of US foreign disaster aid, its small staff and budget limited the potential scope of its activities and operations. Across the Nixon and Gerald Ford presidencies, between 1969 and 1976, USAID's leadership devoted considerable energy toward resolving these issues, striving to enhance USAID's disaster relief capabilities and to establish clear legislative authority for US foreign disaster aid.

These ongoing efforts to reform American catastrophic diplomacy occurred amid significant transition and turmoil in US foreign affairs. While navigating the end of the Vietnam War, détente with the Soviet Union, and budding rapprochement with China, the Nixon and Ford administrations grappled with multiple other foreign policy crises, among them the Indo-Pakistan War in 1971, the Arab-Israeli War and a US-backed coup in Chile in 1973, and a brewing civil war (and proxy war) in newly independent Angola, beginning in 1975. Amid this political upheaval, aiding countries affected by natural hazards remained as critical as ever to American foreign affairs. "U.S. foreign assistance is central to U.S. foreign policy," Nixon stressed to Congress in 1971, and "we must be able to provide prompt and effective assistance to countries struck by natural disaster."[23]

When Tripp retired from government service in 1971, his successors continued the work he initiated.[24] Under their leadership, the foreign disaster relief coordinator's office organized US governmental, military, and voluntary responses to scores of international catastrophes. After several particularly extensive disasters, perceiving the need for additional support and more direct White House involvement, Nixon and Ford also designated a special relief coordinator, appointing either the administrator or deputy administrator of USAID to fill this role.[25]

With these officials at the reins, the US government and its partners in the voluntary sector responded to some truly extraordinary crises. They included the Ancash earthquake in Peru of May 1970 and the Bhola cyclone in East Pakistan in November that same year; Typhoon Gloring (Rita), which struck the Philippines in July 1972, and an earthquake that destroyed Nicaragua's capital, Managua, five months later; floods in the Punjab region of India and Pakistan in August 1973; Hurricane Fifi, which hit Honduras in September 1974; and major earthquakes in Guatemala and Italy, in February and May 1976.[26] Although these catastrophes captured the bulk of the US government's attention and resources, the federal disaster relief coordinator's office also steered hundreds of millions of dollars' worth of humanitarian assistance to disaster survivors all over the world, averaging roughly fifty relief operations per year.[27]

The administrators of USAID, John Hannah and Daniel Parker, evidently valued the foreign disaster relief coordinator's role in these aid efforts, for they steadily expanded the size and budget of this office—even while scaling back many other parts of USAID. Between 1968 and 1972, the staff of the foreign disaster relief coordinator's office more than doubled, growing to twelve personnel; by 1976, the number increased again, to include nine professional and five clerical staff.[28] In 1972, acting on earlier recommendations, USAID's leaders also established a Disaster Coordination and Information Exchange Center, providing a larger and better-equipped headquarters for foreign disaster relief operations.[29] These changes enabled the foreign disaster relief coordinator's office to focus not only on disaster response but also on the related tasks of disaster prevention, preparedness, and risk reduction. By the mid-1970s, these mitigation activities became central to the office's mission and operations, a vital complement to its traditional emphasis on humanitarian relief.

In addition to implementing these internal reforms, USAID's leaders worked with the US military and American voluntary sector to enhance their respective disaster response capabilities. Following the model they implemented in the Canal Zone in the late 1960s, USAID and the Defense Department established a standing Disaster Assistance Response Team in Okinawa in 1971, designed to respond more rapidly to catastrophes in East, South, and Southeast Asia.[30] As they had in Panama, USAID and the Defense Department also began maintaining stockpiles of disaster relief supplies elsewhere in the world, establishing them on US military bases in Guam in 1972, Livorno, Italy, in 1974, and Singapore in 1975.[31]

To assist the government's partners in the voluntary sector, meanwhile, the foreign disaster relief coordinator's office began convening annual disaster aid conferences with leaders of the ARC and the ACVFA affiliates involved in this field. Starting in 1972, staff of these organizations came together for a week every year to discuss best practices, new policies, and matters of shared concern in international disaster response.[32] Between these meetings, the foreign disaster relief coordinator maintained regular communications with the ARC and ACVFA's affiliates, sharing information about disasters and ongoing relief efforts and promoting the humanitarian work of these partner organizations in various other ways.[33]

Even as they made these assorted changes, however, USAID leaders in the early 1970s were simultaneously working toward a more ambitious goal: a comprehensive reform of US foreign aid legislation, intended to finally establish a more formal, legal basis for international disaster assistance. Although US officials had contemplated such a measure for some time, the immediate impetus for it arose in 1970, during the second year of Nixon's presidency. Despite the ongoing efforts to improve the coordination of US foreign disaster aid, the US responses to the Ancash earthquake in Peru and the Bhola Cyclone in East Pakistan that year proved chaotic and controversial, hampered by political, legal, and logistical obstacles. Criticized for these aid efforts at home and abroad, frustrated USAID leaders and the foreign disaster relief coordinator agreed "that a better organization was needed to cope with disasters."[34] By early 1971, they started seriously discussing how to address these ongoing challenges.

Over the next several years, taking decisive steps in this direction, USAID's administrators established two internal task forces on the "Enhancement of the A.I.D. Disaster Relief Function." Comprising members of the foreign disaster relief coordinator's office and other USAID personnel, these committees were charged with studying the enduring problems of foreign disaster aid and suggesting strategies to resolve them. These efforts resulted in a series of studies and reports, completed between 1972 and 1974, which proposed concrete measures to "strengthen A.I.D.'s disaster relief capability."[35]

These reports identified multiple problems with the US foreign disaster aid system in its current form. But the principal issue, both concurred, stemmed from the perceived disconnect between disasters and development, a flawed assumption that had led policymakers to neglect the former while privileging the latter. When USAID was established in 1961, the first of these reports explained, government officials imagined disaster aid "to be quite separate and distinct from AID's primary mission of long-term economic development." They regarded disaster relief as "at best tangential to AID's primary role," the task force observed, "a step-child of AID." Compounding this issue, the Foreign Assistance Act of 1961 had been "designed for economic development assistance, *not* for emergency humanitarian assistance." As a result, this historic aid legislation provided little clarity on the type of aid required in the case of sudden disasters: "saving life and alleviating suffering during an emergency by providing on a crash basis, the minimum necessary food, shelter, clothing, and medicines."[36]

Because they neglected to consider seriously the specific problem of natural hazards, neither the White House nor Congress ever explicitly authorized USAID to coordinate or administer foreign disaster aid. Of course, USAID had long since assumed responsibility for responding to catastrophes by "informal understandings," receiving "tacit approval within the Executive Branch and from the Congress" for such actions.[37] Still, as the authors of the second report reiterated, the Foreign Assistance Act did "not spell out an affirmative authority to conduct foreign disaster relief functions. . . . Except in a rather broad context," they complained, "it is difficult to articulate a fixed policy or identify a specific policy source to guide a disaster response."[38]

The lack of clear legislative authority and policy, these committees stressed, generated tremendous confusion. Having assumed an obligation to "respond to disasters anytime, anywhere," they observed, the US government "has wound up playing the world's fireman . . . and has spent hundreds of millions of dollars in so doing—all without a coherent policy in the matter."[39] Although the two task forces commended all that the foreign disaster relief coordinator's office had accomplished since its inception, affirming "the need and desirability of having a point of control for disaster coordination within the U.S. Government," they emphasized that this office alone was insufficient.[40] "Clearly," they insisted, "stronger steps are needed."[41]

To resolve these problems, task force members proposed a series of legislative reforms, calling on Congress "to include disaster relief as a specific priority objective of the Foreign Assistance Act." More precisely, they recommended that Congress explicitly establish the president's and USAID's authority to contribute foreign disaster aid, coordinate governmental and voluntary relief efforts, acquire and stockpile supplies, and conduct the various other tasks associated with US international disaster response. Noting that "disaster relief funding has become chaotic and unpredictable," they also advised legislators to establish a permanent Foreign Disaster Relief Fund, with sufficient appropriations to finance emergency relief, short-term recovery, and long-range disaster planning and prevention activities. "It is time," task force members concluded, "that Congress should regularize disaster relief and rehabilitation as a full-fledged part of America's assistance to less fortunate countries."[42] Fully supportive of these proposals, and declaring, "My interest in improving AID's disaster relief capability is well known," USAID Administrator Daniel Parker urged Congress to implement these reforms without delay.[43]

Parker's timing was auspicious, for US legislators had recently been grappling with similar issues as they pertained to disasters domestically. Over the preceding ten years, several momentous crises had rocked the United States, among them the Great Alaskan earthquake in 1964, Hurricane Betsy in 1965, Hurricane Camille in 1969, and the Super Outbreak of tornadoes in 1974. Stunned by the horrific scale and exorbitant cost of these catastrophes, Congress passed a slate of disaster management legislation, including the Disaster Relief Acts of 1966, 1970, and 1974 and the National Flood Insurance Act of 1968. Building on foundations established by the Disaster Relief Act of 1950, this legislation expanded the federal government's powers to respond to catastrophes within US states and territories. It also paved the way for the eventual creation of the Federal Emergency Management Association in 1979.[44]

Already taking steps to enhance the federal coordination of disaster assistance at home, Congress proved equally amenable to adopting similar reforms for catastrophes occurring abroad. Although the legislation Parker and his USAID colleagues proposed experienced several false starts, it eventually took form as Public Law 94-161: the International Development and Food Assistance Act of 1975. Introduced to the House of Representatives in July 1975, the bill made its way through both houses of Congress and was signed into law by President Gerald Ford on December 20, 1975.[45]

The enactment of this legislation marked a milestone in the history of US foreign disaster aid. Public Law 94-161 amended the Foreign Assistance Act of 1961 to include an entirely new chapter, dedicated to "International Disaster Assistance." Affirming the "willingness of the United States" to provide humanitarian aid in the case of foreign catastrophes, it empowered the president (or his appointed delegates in USAID) to furnish relief and short-term rehabilitation assistance to any people or country affected by "natural or manmade disasters." It also created a special fund expressly for this purpose, financed with an initial congressional appropriation of $25 million annually. Finally, it formalized the president's authority to appoint a special coordinator for international disaster assistance, with the understanding that USAID's administrator would typically fill this role.

Transforming long-standing precedents, tacit agreements, and unstructured policies into federal law, in sum, the act affirmatively established the US government's authority to provide international disaster aid, defined the state's role in coordinating and planning overseas disaster response and prevention, and allocated funds exclusively for these purposes.

As 1975 drew to a close, USAID's leaders achieved a long-sought objective: Congress and the Ford administration codified international disaster assistance in federal law. Just a few months later, reflecting both "the priority which the Administration gives to international disaster assistance" and the growing embrace of assistance activities beyond emergency relief, USAID's leadership renamed the foreign disaster relief coordinator's office the Office of Foreign Disaster Assistance (OFDA), the title it was to hold for nearly half a century.[46]

With these changes, the legal and bureaucratic apparatuses structuring US foreign disaster aid today, in the twenty-first century, were largely in place. Although the system of US international disaster assistance has evolved in critical ways since that time, the mid-1970s arguably mark its modern foundations, the origins of the United States' contemporary catastrophic diplomacy.

★　★　★

"The historical record is one of a substantial and growing US government contributions to international disaster assistance."[47] So began a report issued in December 1976 by the National Research Council's Committee on International Disaster Assistance, a commission of experts convened at USAID's request to assess the state of this humanitarian field. Between the dawn of the twentieth century and the mid-1970s, as the committee observed and as this book has traced, the US government's foreign disaster aid commitments had indeed expanded progressively. What officials once carried out as an ad hoc, reflexive humanitarian activity was finally an established, formal element of US foreign relations.

Although the members of this committee paused to reflect on the history of US foreign disaster aid, their primary concern was not with the past; they were far more interested in what lay ahead. Anticipating "the inevitable occurrences of future disasters," the commission's principal goal was to plan for those events, recommending steps to ensure "that future international disaster assistance operations can be made more effective and efficient."[48] Writing in the mid-1970s, the members of this committee were not the only ones contemplating the future of US foreign disaster aid. "As the world population continues to rise, more people will inhabit the valleys subject to flooding, the lands that will be shaken by earthquakes," the foreign disaster relief coordinator's office predicted in late 1975. As a result, they projected, "more rather than less disaster relief will be needed . . . in the future."[49] To maintain the political and humanitarian effectiveness of American assistance, these and other officials recognized, the US foreign disaster aid system had to continuously adapt, evolving in response to an ever-changing international context.

In the half-century since analysts made these remarks, the trends they forecast have largely come to fruition. Throughout the world—but particularly in the Global South—people have become far more vulnerable to natural hazards. The causes of this increased vulnerability, it bears emphasizing, were neither natural nor inevitable. Rather, they are attributable to a complex array of political and socioeconomic factors, including extreme poverty and global wealth inequality, urbanization and increased population density, and anthropogenic climate change. Catastrophes precipitated by natural hazards, in turn, have become more frequent, more intense, and more destructive, their harms disproportionately felt by poorer nations and marginalized communities. Well into the twenty-first century, human actions, decisions, and indecision have ensured that "natural" disasters remain a constant and omnipresent threat.[50]

Across these same five decades, the US government's foreign disaster assistance program has expanded; the mechanisms of disaster aid and motivations for providing it, moreover, have evolved in appreciable ways. Since the mid-1970s, multiple historic ruptures—among them the end of the Cold War, the 9/11 attacks and the ensuing War on Terror, and escalating political and ideological polarization within the United States—have successively altered the course of US international affairs and the aims of US foreign policy. As the United States' global objectives and interests shifted over the years, the State Department and US Foreign Service, USAID, the US Armed Forces, and the American voluntary organizations with which they partner all transformed in response, adapting to novel geopolitical contexts and fluctuating governmental priorities. As a result, the way these pillars of US humanitarian aid respond to international catastrophes changed as well, morphing into the system of US foreign disaster assistance that exists today.

What does this current system look like? Despite discernible differences, its basic architecture has not changed all that dramatically since the mid-1970s. Still acting under the authority of Chapter 9 of the Foreign Assistance Act of 1961—the section on "International Disaster Assistance" that Congress appended in 1975—the US government maintains an active program of humanitarian aid designed to "save lives, alleviate human suffering, and reduce the social and economic impact of disasters worldwide."[51] From 1964 until well into the twenty-first century, USAID's Office of Foreign Disaster Assistance (and its predecessor, the foreign disaster relief coordinator's office) remained the unit responsible "for leading and coordinating the US government's response to disasters overseas."[52] That changed only in 2020, when USAID's leadership dissolved OFDA and transferred its responsibilities to a new Bureau for Humanitarian Assistance (BHA). This bureau, which also absorbed the Office of Food for Peace, became the "lead federal coordinator for international disaster assistance."[53]

In recent years, personnel of the BHA and former OFDA have responded to an average of roughly sixty to seventy international disasters annually, including both rapid- and slow-onset crises. When sudden catastrophes occur in other nations, their staff orchestrate the delivery of food, water, and other relief supplies to the scene. They also routinely dispatch Disaster Assistance Response Teams to affected areas, sending these US humanitarian experts and technical advisers to assess the situation, identify urgent needs, and coordinate aid efforts on the scene. In addition to these emergency relief efforts, the BHA administers short-term recovery assistance projects in the weeks and months after disasters occur. It also organizes preparedness and risk-reduction programs, intended to help other nations mitigate the threats natural hazards pose in the first place.[54]

While the BHA focuses on disaster relief, recovery, and risk-reduction, other parts of USAID regularly participate in disaster aid operations as well. In other countries, USAID missions routinely engage in longer-term reha-bilitation and reconstruction assistance projects, often tied to their existing development programs. In Washington, many USAID offices and bureaus contribute specialized services, resources, and administrative assistance to the BHA in support of its disaster aid operations. The USAID administrator, finally, retains the position as the president's special coordinator for interna-tional disaster assistance, maintaining the precedent first established during the early 1970s.[55]

More generally, the three-pillared system of US foreign disaster assistance remains firmly intact. Although the State Department has delegated its disas-ter aid authority to USAID since the 1960s, both the State Department and US Foreign Service continue to play an important role in these efforts. All US ambassadors and other chiefs of mission are still charged with "responsibility for the conduct of [U.S. governmental] foreign disaster assistance within their jurisdiction." When catastrophes occur, these diplomats determine whether US aid is needed, whether the host nation would accept US assis-tance, and whether "it is in the interest of the U.S. Government to respond."[56]

In Washington, meanwhile, many State Department bureaus and offices assist with specific disaster-related issues, including the needs of refugees, help for US citizens in foreign countries, and multilateral cooperation with United Nations agencies and other international organizations. State Depart-ment regional bureaus also provide foreign policy guidance to USAID, both to assist the agency's personnel with their planning and to ensure foreign disaster aid does, indeed, serve the federal government's interests abroad.

While the State Department and USAID remain the government's desig-nated leads for foreign disaster assistance, the Defense Department and US Armed Forces still perform a crucial supporting role in these aid operations. If they happen to be nearby when a disaster erupts, military commanders and their assigned forces are authorized to respond immediately and to provide emergency, life-saving assistance. Most of the time, however, the US military conducts foreign disaster relief operations at the State Department and US-AID's request, in support of the BHA and other pertinent bureaus and offices.

This aid takes myriad forms. In any given year, the US military lends its immense logistical capabilities to multiple foreign disaster relief operations, using air force planes and navy ships to transport humanitarian supplies or personnel abroad. Members of the armed forces also contribute material aid designed to meet basic needs, such as food, water, clothing, bedding, and emergency shelter, often drawing these supplies from Defense Department stockpiles. Finally, American servicemen and women regularly provide emergency health care, communication, sanitation, and other essential services to nations affected by disaster.[57]

Today, the US government also maintains official partnerships with many American voluntary organizations active in international disaster assistance, upholding a long-standing tradition of state-private cooperation. As it has for more than a century, the American Red Cross remains a unique "federal instrumentality," mandated by its charter "to maintain a system of domestic and international relief in time of peace."[58] Over time, however, the ARC has become far less central to the US government's foreign disaster aid program than it once was. Although ARC leaders have always averred the organization's independence, moreover, the organization now operates with much greater autonomy than it did during the twentieth century, when it often behaved, for all intents and purposes, as a quasi-governmental entity. Even though it no longer functions as the US government's principal humanitarian auxiliary, the ARC still maintains a robust international disaster relief program. Each year, the organization contributes millions of dollars' worth of disaster assistance other countries, coordinating its efforts with the BHA, other US government agencies, and the International Federation of Red Cross and Red Crescent Societies in Geneva.[59]

As it has since 1946, the Advisory Committee on Voluntary Foreign Aid— now housed in USAID—remains the liaison between the US government and American private voluntary organizations operating overseas, including those involved in international disaster assistance.[60] In addition, USAID maintains formal partnerships with dozens of private voluntary organizations (though it has altered its traditional registration requirements). By partnering with the government, these organizations are eligible for many benefits, including ocean freight reimbursements, surplus federal property, grants of money and commodities, and contracts. These state resources supplement and subsidize the American voluntary sector's foreign disaster response capabilities, just as they have for decades.[61]

Since the mid-1970s, then, the central pillars of the US foreign disaster aid system have all transformed in notable ways. Budgets and personnel have ballooned, and the system itself has grown far larger and more complex. While continuing to privilege bilateral forms of disaster aid, the US government and its partners have expanded their participation in multilateral disaster assistance, cooperating with a dense network of international governmental and nongovernmental organizations. Conceptually, disaster management specialists have developed a more nuanced appreciation of the relationship between natural hazards and development—and, by extension, of the intertwined links between disaster relief and development assistance.

Even so, despite significant structural changes and the passage of several generations, clear continuities in foreign disaster response remain. Building on foundations laid across the first three quarters of the twentieth century, US diplomatic and development agencies, the US military, and their partners in the voluntary sector still cooperate in planning and executing the nation's official foreign disaster aid operations. The underlying motivations that compel US officials to provide international disaster aid, moreover, are not perceptibly different than they were fifty or even one hundred years ago. Today, US foreign disaster assistance efforts are driven by a complex mixture of diplomatic interests, strategic and economic objectives, and humanitarian concerns, just as they have been for well over a century. In moments of extreme global upheaval, catastrophic diplomacy remains a valuable and flexible instrument of US foreign policy.

Like its twentieth-century antecedents, today's US foreign disaster aid system is not beyond reproach or critique. For better and for worse, Americans involved in international disaster response are in a position to exercise considerable influence, power, and control over the affairs of other nations and their populations. Although many US officials working in this field act with altruistic aims and the best of intentions, their interactions with disaster survivors—whether consciously or not—are indelibly shaped by their own social and economic positions, political objectives, racial and cultural biases, and ideological assumptions. Like their twentieth-century counterparts, American diplomats and development workers juggle to balance US interests with the needs and desires of aid recipients. In a similar vein, although members of the US Armed Forces perform valuable humanitarian work in today's world, the militarized character of that assistance sometimes breeds animosity and resentment—not least when military personnel behave as occupiers rather than equal partners, prioritizing the enforcement of order over the distribution of life-saving aid. The government's reliance on the American voluntary sector, finally, merits scrutiny. Although nonstate aid workers can and do perform considerable good, they are sometimes less accountable to foreign nations than to their own donors and boards of directors. The tendency to outsource US foreign disaster assistance to private voluntary organizations via grants and contracts risks undermining democratic over-sight and local buy-in.

Perhaps, as the foregoing points suggests, the clearest line of continuity between the twentieth century and today is this: US responses to interna-tional catastrophes are, just as they always have been, complicated and deeply political acts. Both historically and in our present moment, they warrant a more critical eye.

In the mid-1970s, US analysts presaged the "the inevitable occurrences of future disasters."[62] Today, unfortunately, their grim predictions have come to fruition. Since that time, catastrophes triggered by natural hazards have upended the lives of hundreds of millions of people throughout the world. In the process, they have altered the course of local, national, regional, and global histories. Just as surely, the American responses to these disasters have marked critical chapters in modern US diplomatic history and contempo-rary foreign affairs. Yet, as this book has shown, there is nothing new—or natural—about these patterns. Since the earliest years of the twentieth cen-tury, catastrophes and foreign disaster aid have been a central and consistent facet of US international relations.

There is much to learn from studying the history of US foreign disaster as-sistance, and from analyzing the successes, the failures, and the complicated politics of earlier relief and recovery efforts. The history of American cata-strophic diplomacy demonstrates, for one, that the US government and mil-itary became involved in foreign assistance much earlier than we commonly perceive. It also highlights the pivotal role that voluntary organizations play in the conduct of US foreign policy, revealing their intimate ties with the state. Additionally, this history invites us to think more deeply about the intwined relationship between humanitarian relief and development assistance. Above all, it underscores the importance of natural hazards, disasters, humanitarian assistance, and environmental forces to US foreign relations. In the wake of sudden and momentous cataclysms, as the history of catastrophic diplomacy shows, twentieth-century Americans found repeated opportunities to exert influence and project power in other nations and empires.

Even as we contemplate and absorb these historical lessons, it would be a mistake to focus solely on the past. Like those analysts in the mid-1970s, it behooves all of us, in our present moment, to contemplate the future of global catastrophes and US foreign disaster aid.

Looking ahead to the remainder of the twenty-first century, it appears all but impossible *not* to predict "the inevitable occurrences of future disasters."

In coming years, climate change—together with its concomitant political, social, and economic effects—will exacerbate the grave threats natural hazards already pose to billions of people. In a world beleaguered by rising seas, scorched and desiccated lands, and overheated cities, the disasters precipitated by climate change and other hazards will almost certainly become more numerous, more destructive, and more devastating.

In the face of these looming global crises, the US government and its partners will, in all likelihood, continue engaging in both bilateral and multilateral disaster aid operations. Ideally, if history teaches us anything, future US responses to global catastrophes will become more efficacious and more ethical, less paternalistic and less politicized, than they so often were in the past.

In 1966, the ARC's director of international services, Sam Krakow, criticized the US government for prioritizing "the political implications of international disaster relief," chastising US officials for placing American diplomatic interests and strategic objectives over the needs of disaster victims around the world.[63] Though many decades have since passed, it is not too late to heed Krakow's message. The world is staring down the barrel of what the UN Human Rights Council has termed a looming "climate apartheid."[64] At this critical moment, it is time to focus less on American diplomatic interests and strategic objectives, and more on the human consequences and authentic humanitarian possibilities, of US foreign disaster assistance.

Notes

ABBREVIATIONS

acc/	Accession
ACVAFS	American Council of Voluntary Agencies for Foreign Service
ANRC	Records of the American National Red Cross
bx/	Box
cdf/	Central Decimal File
coll/	Collection
conf/	Confidential Files
corr/	Correspondence Files
CR	*Congressional Record*
DAF	Dennis A. Fitzgerald Papers
DDEPL	Dwight D. Eisenhower Presidential Library
DDE-RP	Dwight D. Eisenhower–Records as President
DOS	Department of State
FAF	Foreign Affairs Files
FAO	Archives of the Food and Agricultural Organization
FDRPL	Franklin D. Roosevelt Presidential Library
FDR-PP	Franklin D. Roosevelt–Papers as President
fi/	File
fo/	Folder
FOIA-ERR	Freedom of Information Act Electronic Reading Room
FRUS	*Papers Relating to the Foreign Relations of the United States*
HSTPL	Harry S. Truman Presidential Library
HST-PP	Papers of Harry S. Truman
HHPL	Herbert Hoover Presidential Library
HH-PP	Herbert Hoover Papers

INTRODUCTION

1. Samuel Krakow to James Collins, November 15, 1966, fo/890.21, bx/V-65, acc/200-85 -33, rg/ANRC, NARA-II.

2. Fassin and Pandolfini, *Contemporary States of Emergency*, 29–88; Button and Schuller, *Contextualizing Disaster*; Williamson and Courtney, "Disasters Fast and Slow."

3. Squires and Hartman, *There Is No Such Thing*; Kelman, *Disaster by Choice*; Remes and Horowitz, *Critical Disaster Studies*.

4. Steinberg, *Acts of God*; Rozario, *Culture of Calamity*; Kierner, *Inventing Disaster*.

5. Bankoff, "Time Is of the Essence"; Hewitt, *Regions of Risk.*

6. Bandopadhyay, *All Is Well*; Bankoff, *Cultures of Disaster*; Barry, *Rising Tide*; Clancey, *Earthquake Nation*; Coen, *Earthquake Observers*; Courtney, *Nature of Disaster in China*; Dauber, *Sympathetic State*; Davies, *Saving San Francisco*; Davis, *Late Victorian Holocausts*; Dyl, *Seismic City*; Fuller, *Famine Relief in Warlord China*; Garcia, *State of Disaster*; Grego, *Hurricane Jim Crow*; Healey, *Ruins of the New Argentina*; Horowitz, *Katrina*; Hutchinson, "Disasters and the International Order"; Keller, *Fatal Isolation*; Kierner, *Inventing Disaster*; Klinenberg, *Heat Wave*; Knowles, *Disaster Experts*; Mauch and Pfister, *Natural Disasters, Cultural Responses*; Mukherjee, *Hungry Bengal*; Parrinello, *Fault Lines*; Poster, "Hierarchy of Survival"; Remes, *Disaster Citizenship*; Rohland, *Changes in the Air*; Rozario, *Culture of Calamity*; Schemper, "Humanity Unprepared"; Schencking, *Great Kantō Earthquake*; Schwartz, *Sea of Storms*; Segalla, *Empire and Catastrophe*; Skilton, *Tempest*; Steinberg, *Acts of God*; Valencius, *Lost History of the New Madrid Earthquakes.*

7. Barnett, *Empire of Humanity*; Baughan, *Saving the Children*; Cabanes, *Great War and Origins of Humanitarianism*; Cohen, *In War's Wake*; Curtis, *Holy Humanitarians*; Davey, *Idealism beyond Borders*; Demmer, *After Saigon's Fall*; Farré, *Colis de guerre*; Fassin, *Humanitarian Reason*; Fiori et al., *Amidst the Debris*; Gatrell, *Making of the Modern Refugee*; Granick, *International Jewish Humanitarianism*; Heerten, *Biafran War and Postcolonial Humanitarianism*; Hong, *Global Humanitarian Regime*; Humbert, *Reinventing French Aid*; Hutchinson, *Champions of Charity*; Johnson, *Battle for Algeria*; Lipman, *In Camps*; Oppenheimer et al., "Resilient Humanitarianism?"; O'Sullivan, *NGO Moment*; Paulmann, *Dilemmas of Humanitarian Aid*; Porter, *Benevolent Empire*; Rodogno, *Night on Earth*; Romero, "Moving People"; Salvatici, *History of Humanitarianism*; Sobocinska, *Saving the World?*; Steinacher, *Humanitarians at War*; Tanielian, *Charity of War*; van Dijk, *Preparing for War*; Watenpaugh, *Bread from Stones*; Wieters, *NGO Care*; Wylie, Oppenheimer, and Crossland, *Red Cross Movement.*

8. Balogh, *Associational State*; Rosenberg, *Spreading the American Dream.*

9. Citino, *Envisioning the Arab Future*; Connelly, *Fatal Misconception*; Cullather, *Hungry World*; Ekbladh, *Great American Mission*; Engerman, *Price of Aid*; Field, *From Development to Dictatorship*; Frey, Kunkel, and Unger, *International Organizations and Development*; Helleiner, *Forgotten Foundations of Bretton Woods*; Immerwahr, *Thinking Small*; Lal, *African Socialism in Postcolonial Tanzania*; Latham, *Right Kind of Revolution*; Lorenzini, *Global Development*; Macekura, *Of Limits and Growth*; Macekura and Manela, *Development Century*; McVety, *Rinderpest Campaigns*; Meyerowitz, *War on Global Poverty*; Offner, *Sorting Out the Mixed Economy*; Simpson, *Economists with Guns*; Staples, *Birth of Development*; Thornton, *Revolution in Development*; Young, *Transforming Sudan.*

10. Sobocinska, "New Histories of Foreign Aid."

11. Bass, *Freedom's Battle*; Laderman, *Sharing the Burden*; Klose, *Cause of Humanity*; Klose, *Emergence of Humanitarian Intervention*; Rodogno, *Against Massacre*; Simms and Trim, *Humanitarian Intervention*; Laycock and Piana, *Aid to Armenia.*

12. Nunan, *Humanitarian Invasion.*

13. Barnett, *Humanitarianism and Human Rights*; Geyer, "Humanitarianism and Human Rights," in Klose, *Emergence of Humanitarian Intervention*, 31–55; Moyn, *Humane.*

CHAPTER 1

1. Kennan, *Tragedy of Pelée*, 3–5.

2. Scarth, *Catastrophe*; Zebrowski, *Last Days of St. Pierre*.

3. Alexander Scott to James Monroe, November 16, 1812, in Manning, *Diplomatic Correspondence*, 1159.

4. "Earthquake in Venezuela," *Weekly Register* (Baltimore), April 25, 1812.

5. 12th Cong., 1st sess., *Annals of Congress* (April 29 and May 4, 1812): 1348–52, 1378.

6. *Act for the Relief of the Citizens of Venezuela* (1812).

7. John Calhoun, speaking in 12th Cong., 1st sess., *Annals of Congress* (April 29, 1812): 1348.

8. Fitz, *Our Sister Republics*, 4, 80–155.

9. Manning, *Diplomatic Correspondence*, 1157–70.

10. John Rhea, speaking in 12th Cong., 1st sess., *Annals of Congress* (April 29, 1812): 1350.

11. "List of Acts and Resolutions of Congress Granting Relief to the People of Foreign Nations," 60th Cong., 2nd sess., *CR* (January 5, 1909): 453; "Acts of Congress Granting, or Ratifying Grant of, Relief to Sufferers from Floods, Fires, Earthquakes, and So Forth," 81st Cong., 2nd sess., *CR* (August 7, 1950): 11900–11902.

12. J. A. Barrenechea to A. P. Hovey, August 21, 1868, in *Papers Relating to Foreign Affairs*, 873.

13. "Genesis of U.S. Foreign Disaster Relief," 1970, fo/"earthquake reports," bx/1, ent /P-891, rg/286, NARA-II; Curti, *American Philanthropy Abroad*, 3–98.

14. Rozario, *Culture of Calamity*, 31–66; Kierner, *Inventing Disaster*.

15. Dauber, *Sympathetic State*, 17–25.

16. Davies, "Dealing with Disaster," 53.

17. Davies, "Emergence of a National Politics of Disaster"; Dauber, *Sympathetic State*, 25–49.

18. Steinberg, *Acts of God*, 3–24; Knowles, *Disaster Experts*, 21–109; Grego, *Hurricane Jim Crow*.

19. Jones, *American Red Cross*, 3–94; Irwin, *Making the World Safe*, 13–34.

20. *Act to Incorporate the American Red Cross* (1900).

21. Tyrrell, *Reforming the World*; Curtis, *Holy Humanitarians*; Laderman, *Sharing the Burden*, 16–47.

22. *List of Acts and Resolutions of Congress Granting Relief to the People of Foreign Nations* (1909).

23. Schwartz, *Sea of Storms*, 192–225.

24. "Japan's Latest Calamity," *Baptist Missionary Magazine*, November 1896, 546; "An Appeal on Behalf of Sufferers from the Yellow River Floods," in *Consular Reports*, 65.

25. Curtis, *Holy Humanitarians*, 123–70.

26. Louis Aymé to DOS, May 9, 1902, in Miller and Durham, *Martinique Horror*, 30.

27. "Survivors Tell of St. Pierre Horror," *NYT*, May 11, 1902.

28. "AWFUL," *Guthrie (OK) Daily Leader*, May 10, 1902; "Buried under a Storm of Fire from Heaven," *Appeal* (St. Paul, MN), May 17, 1902; "Relief Parties Reach St. Pierre," *NYT*, May 13, 1902.

29. "History's Greatest Disaster," *Frank Leslie's Popular Monthly*, July 1902, i–xvi; "The Martinique Pompeii," *Scribner's Magazine*, July 1902, 20a–20d; "Mont Pelée in Its Might," *McClure's Magazine*, August 1902, 359–68.

30. "Martinique Relief May Be Overdone," *NYT*, May 18, 1902.

31. William Corwine to Elihu Root, May 12, 1902, fi/434288, bx/3029, ent/25, rg/94, NARA-I.

32. Louis Klopsch to Elihu Root, May 12, 1902; and Louis Klopsch to Abby Baker, May 13, 1902, both fi/434288, bx/3029, ent/25, rg/94, NARA-I.

33. Miller and Durham, *Martinique Horror*, 32.

34. Theodore Roosevelt to Emile Loubet, May 10, 1902, Document 370, *FRUS*, 1902.

35. "Relief Measures of the Government," *NYT*, May 13, 1902.

36. "For the Relief of Citizens of the West Indies," 57th Cong., 1st sess., *CR* (May 12, 1902): 5330–34.

37. Louis Aymé to John Hay, in 57th Cong., 1st sess., *CR* (May 12, 1902): 5288.

38. Théophile Delcassé to Jules Cambon, May 11, 1902, Document 372, *FRUS*, 1902.

39. Theodore Roosevelt to Congress, 57th Cong., 1st sess., *CR* (May 12, 1902): 5304.

40. James Hemenway, speaking in 57th Cong., 1st sess., *CR* (May 12, 1902): 5331.

41. "For the Relief of Citizens of the French West Indies," 57th Cong., 1st sess., *CR* (May 12, 1902): 5331.

42. Oscar Underwood, speaking in 57th Cong., 1st sess., *CR* (May 12, 1902): 5331.

43. Various members of Congress, speaking in 57th Cong., 1st sess., *CR* (May 12, 1902): 5332.

44. *Act for the Relief of Citizens of the French West Indies* (1902).

45. "1902–Martinique–1903," *Charities*, August 1903, 114–16.

46. Major-General Brooks to Adjutant General, May 13, 1902; and Captain Crabbs to Adjutant General, May 23, 1902, both fi/434288, bx/3029, ent/25, rg/94, NARA-I.

47. H. C. Corbin, memorandum, May 12, 1902; and J. F. Weston to Secretary of War, May 14, 1902, both fi/434288, bx/3029, ent/25, rg/94, NARA-I; Log of *Dixie*, May 12, 1902–June 1, 1902, vol. 7, ent/118G, rg/24, NARA-I.

48. Kennan, *Tragedy of Pelée*, 3–5.

49. Gustav Schwab to Elihu Root, May 20, 1902, fi/434288, bx/3029, ent/25, rg/94, NARA-I.

50. Louis Aymé to DOS, May 24, 1902, ent/A1-85, rg/59, NARA-II.

51. "1902–Martinique–1903," 116.

52. Hugh Gallagher to Adjutant General, June 13, 1902, fi/434288, bx/3029, ent/25, rg/94, NARA-I; Log of *Dixie*, June 1, 1902.

53. Kennan, *Tragedy of Pelée*, 197.

54. Gallagher to Adjutant General, June 13, 1902.

55. Governor Heurre to Captain Gallagher, May 22, 1902, fi/434288, bx/3029, ent/25, rg/94, NARA-I; Jules Cambon to Theodore Roosevelt, May 15, 1902, Document 373; and President Loubet to President Roosevelt, May 22, 1902, Document 374, both *FRUS*, 1902.

CHAPTER 2

1. Rosenberg, *Spreading the American Dream*; Balogh, *Associational State*.

2. Theodore Roosevelt, Message to Congress, December 3, 1901, *FRUS*, 1901.

3. Baer, *One Hundred Years of Sea Power*, 27–48.

4. Millett, Maslowski, and Feis, *For the Common Defense*, 285, 292–300.

5. Werking, *Master Architects,* 88–142.

6. *Act to Incorporate the American Red Cross* (1900).

7. Jones, *American Red Cross,* 97–115.

8. *Act to Incorporate the American Red Cross* (1905).

9. Irwin, *Making the World Safe,* 35–208.

10. Theodore Roosevelt, in "President Asks Nation to Aid Stricken People," *Washington Post,* April 20, 1906.

11. Jones, *American Red Cross,* 116–36.

12. Foster, *Demands of Humanity,* 52–66.

13. Davies, *Saving San Francisco,* 42–62.

14. Steinberg, *Acts of God,* 25.

15. Rozario, *Culture of Calamity,* 67–100; Steinberg, *Acts of God,* 25–46; Davies, *Saving San Francisco;* Dyl, *Seismic City.*

16. John Hicks to DOS, August 21, 1906, ent/UD-9, rg/84, NARA-II.

17. "The Valparaiso Earthquake," *NYT,* August 20, 1906.

18. Some Quincy Citizens to the President, August 22, 1906, fi/241, NMFDS, rg/59, NARA-II.

19. John Doxsee to Theodore Roosevelt, August 20, 1906, fi/241, NMFDS, rg/59, NARA-II.

20. Theodore Roosevelt, in William Loeb to DOS, August 20, 1906, fi/241, NMFDS, rg/59, NARA-II.

21. John Hicks to DOS, September 12, 1906, ent/UD-9, rg/84, NARA-II.

22. Elihu Root, speech in Santiago, Chile, September 1, 1906, ent/UD-9, rg/84, NARA-II.

23. Hicks to DOS, September 12, 1906.

24. Theodore Roosevelt, Proclamation, August 25, 1906, fi/241, NMFDS, rg/59, NARA-II.

25. "Chile's Quake Victims Get Small Aid from US," *NYT,* August 25, 1906.

26. John Hicks to DOS, December 31, 1906, and May 5, 1907, both fi/241, NMFDS, rg/59, NARA-II.

27. Mabel Boardman to Robert Bacon, September 1, 1906, fi/241, NMFDS, rg/59, NARA-II.

28. "Suffering in Chile Very Great," manuscript, August 1906, fo/892.5, bx/I-59, rg /ANRC, NARA-II.

29. Boardman to Bacon, September 1, 1906; Hicks to DOS, September 14, 1906, ent/UD-9, rg/84, NARA-II.

30. Boardman to Bacon, September 1, 1906.

31. Hicks to DOS, December 31, 1906.

32. Miscellaneous correspondence from the US consulate in Valparaíso, Chile, August–December 1906, corr/848, ent/UD-914, rg/84, NARA-II.

33. Hicks to DOS, December 31, 1906.

34. Hicks to DOS, May 5, 1907.

35. "Secular News," *Christian Observer,* October 3, 1906.

36. Tilchin, "Jamaica Incident of 1907."

37. US Consul, Port Antonio, to DOS, January 16, 1907, fi/4001, NMFDS, rg/59, NARA-II.

38. "The Jamaican Disaster," *NYT*, January 17, 1907.

39. Theodore Roosevelt to King Edward VII, January 16, 1907; and Elihu Root to Governor of Jamaica, January 16, 1907, both fi/3892, NMFDS, rg/59, NARA-II.

40. ARC Executive Committee, Resolution, January 17, 1907, fo/893.5, bx/I-59, rg /ANRC, NARA-II.

41. US Vice Consul, Kingston, Jamaica, to DOS, January 26, 1907, fi/3892, NMFDS, rg/59, NARA-II.

42. *Act for the Relief of the Citizens of the Island of Jamaica* (1907).

43. John Fitzgerald, speaking in 59th Cong., 2nd sess., *CR* (January 17, 1907): 1296.

44. C. H. Davis to Secretary of the Navy, January 21, 1907, fi/4001, NMFDS, rg/59, NARA-II.

45. Miscellaneous correspondence, fi/24024, bx/837, ent/19, rg/80, NARA-I.

46. Theodore Wint to War Department, January 20 and 29, 1907, fi/1202124, bx/4741, ent/25, rg/94, NARA-I.

47. C. H. Davis to Secretary of the Navy, January 21, 1907; and C. H. Bourne, Memorandum, January 20, 1907, both fi/4001, NMFDS, rg/59, NARA-II.

48. J. L. Sticht to Commander, USS *Indiana*, January 17, 1907; and William Orrett to DOS, January 20, 1907, both fi/4001, NMFDS, rg/59, NARA-II.

49. C. H. Davis to Secretary of the Navy, January 21, 1907; and Alexander Swettenham to British Secretary of State, January 29, 1907, both fi/4001, NMFDS, rg/59, NARA-II.

50. Alexander Swettenham to British Secretary of State, January 18, 1907, fi/4001, NMFDS, rg/59, NARA-II.

51. C. H. Davis to Navy Department, January 18, 1907, ent/286, rg/45, NARA-I.

52. Davis to Secretary of the Navy, January 21, 1907.

53. A. S. Hoff to Commander, USS *Missouri*, January 19, 1907; and Howard Ames to C. H. Davis, January 20, 1907, both fi/4001, NMFDS, rg/59, NARA-II.

54. Alexander Swettenham to Admiral Davis, January 18, 1907, ent/286, rg/45, NARA-I.

55. C. H. Davis to Mayor of Kingston, January 19, 1907, fi/4001, NMFDS, rg/59, NARA-II.

56. Davis to Secretary of the Navy, January 21, 1907.

57. Tilchin, "Jamaica Incident of 1907," 392–405.

58. William Orrett to DOS, January 26, 1907, fi/3892, NMFDS, rg/59, NARA-II.

59. William Orrett to DOS, January 28, 1907, fi/4001, NMFDS, rg/59, NARA-II.

60. Theodore Roosevelt to DOS, April 24, 1907, fi/4001, NMFDS, rg/59, NARA-II.

61. Eugene Hale to DOS, January 22, 1907, fi/4001, NMFDS, rg/59, NARA-II.

CHAPTER 3

1. Dickie, *Catastrofe patriottica*; Parrinello, *Fault Lines*, 21–118.

2. LaGumina, *Great Earthquake*.

3. Theodore Roosevelt to Victor Emmanuel III, December 29, 1908, fo/895.4/2, bx /I-60, rg/ANRC, NARA-II.

4. Commander, *Culgoa*, to Charles Sperry, January 15, 1909, ent/500, rg/45, NARA-I.

5. Truman Newberry to George Logan, December 30, 1908, ent/500, rg/45, NARA-I.

6. Truman Newberry to Theodore Roosevelt, December 31, 1908, fi/17191, NMFDS, rg/59, NARA-II.

7. Miscellaneous correspondence, 1907–9, fi/"ZO–Great White Fleet and Atlantic Fleet," NHHC; Reckner, *Teddy Roosevelt's Great White Fleet*, 145–48.

8. Charles Sperry to Edith Sperry, January 4, 1909; and Charles Sperry to "My dear kid," January 6, 1909, both fo/"Family Correspondence," bx/6, coll/MSS-40923, LOC; Reginald Belknap to Lloyd Griscom, January 19, 1909, fo/895.4/08, bx/I-59, rg/ANRC, NARA-II.

9. Theodore Roosevelt, Message to Congress, 60th Cong., 2nd sess., *CR* (January 4, 1909): 451.

10. *Act for the Relief of Citizens of Italy* (1909).

11. Eugene Hale, speaking in 60th Cong., 2nd sess., *CR* (January 4, 1909): 452–53.

12. T. C. Belsito to Theodore Roosevelt, January 5, 1909, fi/17191, NMFDS, rg/59, NARA-II.

13. Joseph Bailey, speaking in 60th Cong., 2nd sess., *CR* (January 4, 1909): 452.

14. Henry Higginson to Robert Bacon, January 4, 1909, fi/17191, NMFDS, rg/59, NARA-II.

15. Louis Klopsch to Robert Bacon, December 31, 1908, fi/17191, NMFDS, rg/59; and Curtis Guild to Mabel Boardman, January 4, 1909, fo/894.4/02, bx/I-59, rg/ANRC, both NARA-II.

16. Minutes, ARC Annual Meeting, 1908, fo/[nameless], bx/I-13, rg/ANRC, NARA-II.

17. William Taft to Italian Red Cross, December 29, 1908, fi/17191, NMFDS, rg/59, NARA-II.

18. "American Sympathy for Quake Victims," *NYT*, December 30, 1908.

19. ARC headquarters to Arthur Warren, January 12, 1909, fo/895.4/7, bx/I-60, rg/ANRC; and Robert Bacon to G. M. Lane and Herbert Burrell, January 1, 1909, fi/17191, NMFDS, rg/59, both NARA-II.

20. Mabel Boardman to Curtis Guild, January 1, 1909, fo/894.4/02, bx/I-59, rg/ANRC, NARA-II.

21. "Subscriptions by States," manuscript, January 5, 1909, fo/895.4/21, bx/I-60, rg/ANRC, NARA-II.

22. Mabel Boardman to Lloyd Griscom, February 13, 1909, fo/894.4/02, bx/I-59; and Report of the Chairman of the ARC Central Committee, December 7, 1909, bx/I-60, both rg/ANRC, NARA-II.

23. Mimi Marseglia to Edward Breene, [1909], fi/17191, NMFDS, rg/59, NARA-II.

24. Arnoldo Raimondi to Philander Knox, March 17, 1909; and Odoardo Toscano to the President, April 21, 1909, both fi/17191, NMFDS, rg/59, NARA-II. See also Choate, *Emigrant Nation*, 189–91, 200–202; Iorizzo and Rossi, *Italian Americans*, 65–83; and Staiti, "Terremoto di Messina."

25. Minutes, American Relief Committee Meetings, January 3–8, 1909, fo/895.4, bx/I-59, rg/ANRC, NARA-II.

26. Bayard Cutting to DOS, January 23, 1909, bx/I-59, rg/ANRC, NARA-II.

27. American Relief Committee, "Earthquake in Sicily and Calabria," [1909], corr/848, ent/UD-35, rg/84, NARA-II.

28. Lloyd Griscom to Mabel Boardman, January 23, 1909, fo/895.4/02, bx/I-59, rg/ANRC, NARA-II.

29. "Report of the American Relief Committee, Rome, for the Earthquake Sufferers," [1909], fo/895.4/08, bx/I-59, rg/ANRC, NARA-II.

30. Minutes, American Relief Committee Meetings, January 3–8, 1909.

31. G. W. Logan to Secretary of the Navy, January 8, 13, and 19, 1909, ent/500, rg/45, NARA-I.

32. American Relief Committee, Report to ARC, March 30, 1909, fo/895.4, bx/I-59, rg /ANRC, NARA-II.

33. L. Mirabello to Rear-Admiral Sperry, January 14, 1909, fi/6072/593, bx/1024, ent/88, rg/24; and Wireless relays from USS *Connecticut*, December 30, 1908–January 20, 1909, bx/1, ent/UD-8B, rg/313, both NARA-I; Secretary of Archbishop of Messina, untitled manuscript, March 20, 1909, fo/895.4/08, bx/I-59, rg/ANRC, NARA-II.

34. Bayard Cutting, "Report on Catania and Syracuse," January 14, 1909, fi/17191, NMFDS, rg/59, NARA-II.

35. "Report of the Ladies' Committee," [1909], fo/895.4/08, bx/I-59, rg/ANRC, NARA-II.

36. Griscom to "My dear Rod," January 22, 1909, fo/ "Family Correspondence 1908–09," bx/2, coll/MSS-24208, LOC.

37. "The Relief Expedition in Calabria of H. Nelson Gay," [1909], fo/895.4/08, bx /I-59, rg/ANRC; and Bayard Cutting to Lloyd Griscom, January 14, 1909, fi/17191, NMFDS, rg/59, both NARA-II; Charles Sperry to "My dear girl" from Naples, January 11, 1909, fo/"Family Correspondence," bx/6, coll/MSS-40923, LOC.

38. Cutting to DOS, January 23, 1909.

39. Stuart Lupton to R. Douglas, March 22, 1909, corr/848, ent/UD-240, rg/84, NARA-II.

40. "Relief Expedition in Calabria of H. Nelson Gay," [1909].

41. "Report of the Ladies' Committee," [1909].

42. Bayard Cutting to Mabel Boardman, January 28, 1909, fo/895.4/7, bx/I-60, rg /ANRC, NARA-II.

43. William Henry Bishop to DOS, April 26, 1909, fi/17191, NMFDS, rg/59, NARA-II.

44. "Report of the Ladies' Committee," [1909].

45. Cutting to Griscom, January 14, 1909.

46. Cutting, "Report on Catania and Syracuse," January 14, 1909.

47. Charles Sperry to "My dear kid," January 17, 1909, fo/"Family Correspondence," bx/6, coll/MSS-40923, LOC.

48. Reginald Belknap to Lloyd Griscom, January 19, 1909, fo/895.4/08; and Lloyd Griscom to Mabel Boardman, January 23, 1909, fo/895.4/02, both bx/I-59, rg/ANRC, NARA-II.

49. Cutting to DOS, January 23, 1909.

50. Lloyd Griscom to Mabel Boardman, February 21, 1909, fo/895.4/08, bx/I-59, rg /ANRC, NARA-II.

51. Charles Sperry to "My dear boy," February 9, 1909, fo/"Family Correspondence," bx/6, coll/MSS-40923, LOC.

52. Ships' logs of *Celtic, Connecticut, Culgoa, Illinois, Scorpion,* and *Yankton,* January– February 1909, ent/118G, rg/24, NARA-I.

53. Griscom to Boardman, January 23, 1909.

54. Reginald Belknap to Lloyd Griscom, June 14, 1909, fo/895.4/08, bx/I-59, rg /ANRC, NARA-II.

55. Boardman to Griscom, February 13, 1909.

56. Griscom to Boardman, January 23, 1909.

57. "President to Send Lumber to Italy," *NYT*, January 17, 1909.

58. "ARC Aid to Italy—Statement of March 6, 1909," fo/895.4/2, bx/I-60, rg/ANRC, NARA-II.

59. Secretary of the Navy to DOS, January 22, 1909, fi/27162, bx/1394, ent/19, rg/80, NARA-I; Robert Bacon to Lloyd Griscom, January 28, 1909, fi/17191, NMFDS, rg/59, NARA-II.

60. Bruno Chimirri to Lloyd Griscom, January 26, 1909; and Lloyd Griscom to DOS, January 27, 1909, both fo/895.4/65, bx/I-60, rg/ANRC, NARA-II.

61. Griscom to Boardman, January 23, 1909.

62. Griscom to DOS, January 27, 1909.

63. "Convention pour la création du 'American Red Cross Orphanage,'" February 13, 1909; and Lloyd Griscom to George Davis, February 11 and 19, 1909; and "American Red Cross Orphanage at Palmi, Italy," manuscript [ca. 1911–13], all fo/895.4/65, bx/I-60, rg /ANRC, NARA-II.

64. Minutes, American Relief Committee Meeting, March 4, 1909, fo/895.4.

65. Reginald Belknap to Lloyd Griscom, March 24 and June 14, 1909, fo/895.4/08, bx/I-59, rg/ANRC, NARA-II.

66. Cutting to Boardman, January 28, 1909.

67. Lloyd Griscom to George Davis, June 12, 1909, fo/895.4/08, bx/I-59, rg/ANRC, NARA-II.

68. Belknap, *American House Building in Messina and Reggio.*

69. "Mr. Winthrop Chanler's Relief Expedition in Calabria," April 30, 1909; and Reginald Belknap to Lloyd Griscom, June 14, 1909, both fo/895.4/08, bx/I-59, rg/ANRC, NARA-II.

70. "Report of the American Relief Committee, Rome, for the Earthquake Sufferers," [1909].

71. Reginald Belknap to Lloyd Griscom, April 19, 1909, fi/17192, NMFDS, rg/59, NARA-II.

72. Reginald Belknap to Lloyd Griscom, June 14, 1909.

73. Reginald Belknap to Lloyd Griscom, March 2, 1909, fo/895.4/6, bx/I-60, rg /ANRC, NARA-II.

74. Reginald Belknap to Lloyd Griscom, March 8 and June 14, 1909, fo/895.4/08, bx /I-59, rg/ANRC, NARA-II.

75. Mabel Boardman to Robert de Forest, February 8, 1909, fo/895.4/21, bx/I-60; and Robert Bacon to Lloyd Griscom, February 16, 1909, fo/895.4, bx/I-59, both rg/ANRC, NARA-II.

76. Bicknell, *Pioneering with the Red Cross*, 111.

77. Reginald Belknap to Lloyd Griscom, March 8, 1909, fo/895.4/08; and Minutes, American Relief Committee Meeting, March 22, 1909, fo/895.4; both bx/I-59, rg/ANRC, NARA-II.

78. Reginald Belknap to Lloyd Griscom, March 24, 1909, fo/895.4/08, bx/I-59, rg /ANRC, NARA-II.

79. Belknap to Griscom, June 14, 1909.

80. Reginald Belknap to Lloyd Griscom, April 18 and 24, 1909, fi/17191, NMFDS, rg/59, NARA-II.

81. Reginald Belknap to Lloyd Griscom, April 19, 1909; and "The American Houses at Messina," memorandum, [1909], both fi/17191, NMFDS, rg/59, NARA-II.

82. Lloyd Griscom to DOS, June 4 and 12, 1909, fi/17191, NMFDS, rg/59, NARA-II; Belknap to Griscom, June 14, 1909.

83. Ponzio Vaglia to John Garrett, September 1, 1909, fi/17191, NMFDS, rg/59, NARA-II.

84. Belknap to Griscom, June 14, 1909.

85. Giovanni Giolitti to Lloyd Griscom, May 17, 1909, fi/17191, NMFDS, rg/59, NARA-II; Belknap, *American House Building in Messina and Reggio*, 242, 251.

86. "Italy Grateful to America," *NYT*, July 8, 1909.

87. "Griscom Praises People of Italy," *NYT*, June 27, 1909.

88. Francesco Giucciardini to John Leishman, January 3, 1910, corr/848, ent/UD-35, rg/84, NARA-II.

89. John Grillo to the President, May 2, 1910, fi/17191, NMFDS, rg/59, NARA-II.

90. Mabel Hill to Mr. Wilson, March 15, 1910, fo/895.4/6, bx/I-60, rg/ANRC, NARA-II.

91. Robert Bacon, quoted in Cutting to DOS, January 23, 1909.

92. Mabel Hill to Mr. Wilson, March 19, 1910, fo/895.4/6, bx/I-60, rg/ANRC, NARA-II.

93. George Dunn, Address at Dedication of ARC Orphanage at Palmi, October 9, 1913; and George Dunn to American Ambassador, October 12, 1913, both fo/895.4/65, bx/I-60, rg/ANRC, NARA-II.

94. Bicknell, *Pioneering with the Red Cross*, 133.

CHAPTER 4

1. Todd, *Story of the Exposition*, 103, 108–10.

2. "The Red Cross Exhibit at the World's Fair," *American Red Cross Magazine*, April 1915, 151–54.

3. Thompson, *Sense of Power*, 25–55.

4. International Relief Board membership lists, 1913–16, fo/118.11, bx/I-16, rg/ANRC, NARA-II.

5. Huntington Wilson, "Report to the Ninth International Red Cross Conference," 1912, fo/118.1, bx/I-16, rg/ANRC, NARA-II.

6. Huntington Wilson, "Report of the International Relief Board," 1910, fo/118.1, bx /I-16, rg/ANRC, NARA-II.

7. "Presidential Proclamations, Official Statements and Messages Pertaining to the American National Red Cross," 1882–1917, compiled August 13, 1934, off/124, WHCF /HST-PP, HSTPL.

8. Irwin, *Making the World Safe*, 33–34, 49.

9. William Taft, "Proclamation by the President of the United States," 1911, cdf/1910–29 /811.142/218, rg/59, NARA-II.

10. Ernest Bicknell to Mabel Boardman, September 21, 1909, fo/893.1, bx/I-59, rg /ANRC, NARA-II.

11. List compiled from various archival collections included in this book's bibliography.

12. ARC International Relief Board, 1910 Annual Report, fo/118.1, bx/I-16, rg/ANRC, NARA-II.

13. *American Red Cross Bulletin*, January 1912, 70; *American Red Cross Bulletin*, January 1913, 80.

14. Bicknell to Boardman, September 21, 1909.

15. Mabel Boardman to Robert Lansing, July 7 and October 2, 1914, cdf/1910–29 /893.48/6 and cdf/1910–29/893.48/19, rg/59, NARA-II.

16. Jones, *American Red Cross*, 137–56.

17. William Taft to Clément Armand Fallières, January 27, 1910, fo/894.2, bx/I-59, rg/ANRC; and Woodrow Wilson to Vittorio Emanuele, January 15, 1915, cdf/1910–29/865.48/9a, rg/59, both NARA-II.

18. Huntington Wilson to International Relief Board, May 12, 1910, fo/891.8, bx/I-59, rg/ANRC, NARA-II.

19. Miscellaneous correspondence, fi/27439, bx/1454, ent/19, rg/80, NARA-I.

20. J. B. Murdock to Secretary of the Navy, September 25, 1911, fi/27759-92, bx/1535, ent/19, rg/80, NARA-I.

21. Frank Upham to G. B. Ravndal, August 25, 1912, fo/896.5, bx/I-61, rg/ANRC, NARA-II.

22. *Register of the Department of State.*

23. ARC International Relief Board, 1910 Annual Report; Wilson, "Report to the Ninth International Red Cross Conference," 1912.

24. G. L. Monroe to George Davis, May 21, 1910, fo/891.8, bx/I-59, rg/ANRC, NARA-II.

25. C. Williams to DOS, September 12, 1911, cdf/1910–29/893.48el/13, rg/59, NARA-II.

26. David Thompson to DOS, August 30, 1909, fo/891.2, bx/I-59, rg/ANRC, NARA-II.

27. Philip Hanna, in Alvey Adee to Charles Magee, September 12, 1909, fo/891.2, bx /I-59, rg/ANRC, NARA-II.

28. A. J. Lespinasse to DOS, December 2, 1916, fo/892.1, bx/I-59, rg/ANRC, NARA-II.

29. Paul Jameson to Paul Reinsch, August 12, 1916, cdf/1910–29/893.48/60, rg/59, NARA-II.

30. Philip Hanna to DOS, October 24, 1909, fo/891.2/08, bx/I-59, rg/ANRC, NARA-II.

31. Clarence Miller to DOS, September 8 and December 8, 1909, both fo/891.2/08, bx/I-59, rg/ANRC, NARA-II.

32. Robert Bergh, in ARC International Relief Board, 1910 Annual Report.

33. Alvin Gilbert to E. T. Williams, September 4, 1911, cdf/1910–29/893.48el/11, rg/59, NARA-II.

34. Alvin Gilbert to DOS, September 15, 1911, cdf/1910–29/893.48el/15, rg/59, NARA-II.

35. John Davis to DOS, September 14, 1914, cdf/1910–29/893.48/21, rg/59, NARA-II.

36. William Kent to DOS, August 26, 1911, cdf/1910–29/893.48el/10; and C. Williams to DOS, September 12, 1911, cdf/1910–29/893.48el/13;, both rg/59, NARA-II.

37. Fleming Cheshire to DOS, September 3, 1915, cdf/1910–29/893.48/54, rg/59, NARA-II.

38. Fleming Cheshire to DOS, June 29 and July 23, 1914, cdf/1910–29/893.48/8 and cdf/1910–29/893.48/15, rg/59, NARA-II.

39. Robert Bacon, paraphrased in Mabel Boardman to Starr Murphy, February 2, 1910, fo/894.2, bx/I-59, rg/ANRC, NARA-II.

40. Charles Bryan to Philander Knox, May 24, 1909, fo/895.7, bx/I-60; and ARC International Relief Board, 1910 Annual Report, fo/118.1, bx/I-16, both rg/ANRC, NARA-II.

41. George Guthrie to DOS, January 30, 1914, fo/898.6, bx/I-62, rg/ANRC, NARA-II.

42. Jones, *American Red Cross*, 147–56; Remes, *Disaster Citizenship*, 54–104.

43. "The Mexican Flood and Relief Work," *American Red Cross Bulletin*, January 1910, 16.

44. Clarence Miller to DOS, September 8 and December 8, 1909, both fo/891.2/08, bx/I-59, rg/ANRC, NARA-II.

45. Monroe to Davis, May 21, 1910.

46. W. W. Rockhill to DOS, August 15, 1912, fo/896.5, bx/I-59, rg/ANRC, NARA-II.

47. Upham to Ravndal, August 25, 1912.

48. R. E. Chambers to Fleming Cheshire, December 11, 1914, cdf/1910–29/893.48/26, rg/59, NARA-II.

49. Fleming Cheshire to DOS, August 10, 1915, cdf/1910–29/893.48/50, rg/59, NARA-II.

50. H. O. T. Burkwell to Fleming Cheshire, July 21, 1914, cdf/1910–29/893.48/15; and Florence Drew to Fleming Cheshire, August 13, 1915, cdf/1910–29/893.48/51;, both rg/59, NARA-II.

51. G. L. Monroe to DOS, May 17, 1910, fo/891.8, bx/I-59, rg/ANRC, NARA-II.

52. Wilfred Port to G. B. Ravndal, August 23, 1912, fo/896.5, bx/I-61, rg/ANRC, NARA-II.

53. Thomas Page to DOS, January 25, 1915, cdf/1910–29/865.48/13, rg/59; and Thomas Page, "Report: Relief Work Done by Americans," March 9, 1915, corr/848, ent/UD-35, rg/84, both NARA-II.

54. John Leishman to DOS, February 12, 1910, fi/17192, NMFDS, rg/59, NARA-II.

CHAPTER 5

1. Jones, *American Red Cross*, 157–75; Irwin, *Making the World Safe*, 105–84.

2. Millett, Maslowski, and Feis, *For the Common Defense*, 316–79; Baer, *One Hundred Years of Sea Power*, 48–82.

3. Johnson, *Administration of United States Foreign Policy*, 59–68.

4. ARC to US Legation, Peking, October 6, 1917, cdf/1910–29/893.48/65, rg/59; and Director of Civilian Relief, ARC, to Miss Baldwin, March 3, 1918, fo/898.5, bx/II-704, rg/ANRC, both NARA-II.

5. Miscellaneous correspondence, fo/891.6 and fo/893.5, bx/II-703, rg/ANRC, NARA-II.

6. Roger Greene, "The Chihli Floods of 1917 and the Relief Work of the American Red Cross," December 5, 1918, fo/898.5/08, bx/II-705, rg/ANRC, NARA-II.

7. ARC, "Expenditures in China, 1906–1937," October 1937, fo/895.8/208, bx/III-1222, rg/ANRC, NARA-II.

8. Chang, *Fateful Ties*, 90–129.

9. P. R. Josselyn to Paul Reinsch, September 26, 1917, corr/848, ent/UD-897, rg/84; and Paul Reinsch to DOS, September 28 and October 2, 1917, cdf/1910–29/893.48/65 and cdf/1910–29/893.48/74, rg/59, all NARA-II.

10. ARC to US Legation, Peking, October 6 and November 2, 1917, cdf/1910–29/893.48 /65 and cdf/1910–29/893.48/79a, rg/59, NARA-II.

11. Reinsch to DOS, September 28, 1917. See also F. L. Belin to Paul Reinsch, October 1, 1917, corr/848, ent/UD-897, rg/84; and Special Order 191, September 15, 1917, bx/2, ent /NM-94/5964, rg/395, both NARA-II.

12. Cornebise, *United States 15th Infantry Regiment in China*.

13. Paul Reinsch to DOS, October 15, 1917, cdf/1910–29/893.48/80, rg/59; and Roger Greene to Henry Davison, December 7, 1917, fo/898.5/08, bx/II-705, rg/ANRC, both NARA-II.

14. Memorandum of Meeting, F. L. Belin and Hsiung Hsi-Ling, October 2, 1917, fo/898.5/04, bx/II-705, rg/ANRC, NARA-II.

15. Belin to Reinsch, October 1, 1917.

16. US Consul-General, Tientsin, "Organized Relief Work in Tientsin," September 29, 1917, corr/848, ent/UD-897, rg/84, NARA-II.

17. P. R. Josselyn to Paul Reinsch, October 20, 1917, corr/848, ent/UD-897, rg/84, NARA-II.

18. Greene, "Chihli Floods of 1917."

19. F. L. Belin to Paul Reinsch, October 13, 1917, cdf/1910–29/893.48/80, rg/59, NARA-II.

20. Roger Greene to Paul Reinsch, October 19, 1917, corr/848, ent/UD-10, rg/84, NARA-II.

21. Roger Greene to Paul Reinsch, October 14, 1917, corr/848, ent/UD-10, rg/84; and Minutes, ARC Committee Meeting, October 12, 1917, fo/898.5, bx/II-705, rg/ANRC, both NARA-II.

22. "The Tientsin Refugee Camp," *Peking Times*, February 6, 1918, in fo/898.5, bx/II-705, rg/ANRC, NARA-II.

23. Roger Greene to Henry Davison, December 7, 1917, fo/898.5, bx/II-705, rg/ANRC, NARA-II.

24. C. H. Morrow, Report on Operation of ARC Flood Relief Camp of Tientsin, [1918], cdf/1910–29/893.48/97, rg/59, NARA-II.

25. "Statement of Policy of the American Red Cross Committee," [1917], corr/848, ent /UD-897, rg/84, NARA-II.

26. Greene to Davison, December 7, 1917.

27. Reinsch to DOS, October 26, 1917, cdf/1910–29/893.48/75, rg/59, NARA-II.

28. ARC to US Legation, Peking, November 2, 1917.

29. Reinsch to DOS, October 26, 1917.

30. Roger Greene, "The Peking-Tungchow Highway," December 5, 1918, fo/898.5/08, bx/II-705, rg/ANRC, NARA-II.

31. Reinsch to DOS, June 17, 1918.

32. Greene, "Peking-Tungchow Highway."

33. Greene, "Chihli Floods of 1917."

34. William Fee to DOS, January 9, 1918, corr/848, ent/UD-401, rg/84; and Alvin Struse to Frank Persons, March 5, 1918, fo/891.4/508, bx/II-703, rg/ANRC, both NARA-II.

35. Dosal, *Doing Business with the Dictators*, 37–94; Colby, *Business of Empire*, 79–145.

36. Walter Thurston to DOS, December 26, 1917, cdf/1910–29/814.48/5; and Robert Lansing to US Legation, Guatemala, December 26, 1917, cdf/1910–29/814.48/7, both rg/59, NARA-II.

37. Walther Thurston to DOS, December 28, 1917, cdf/1910–29/814.48/9; and Manuel Estrada Cabrera to Woodrow Wilson, December 30, 1917, cdf/1910–29/814.48/17, both rg/59, NARA-II.

38. USS *Cincinnati* to Opnav, Washington, January 5, 1918, cdf/1910–29/814.48/22, rg/59, NARA-II; Miscellaneous Correspondence [January 1918], fo/"*USS Cincinnati*," bx/1066, ent/520, rg/45, NARA-I.

39. ARC News Service, press release, January 2, 1918, fo/891.4, bx/II-703, rg/ANRC; and ARC to US Legation, Guatemala, December 28, 1917, cdf/1910–29/814.48/10, rg/59, both NARA-II.

40. DOS to US Legation, Guatemala, December 30, 1917; and Alfred Clark to Manuel Estrada Cabrera, December 30, 1917, both corr/848, ent/UD-27, rg/84, NARA-II.

41. John Merrill to DOS, January 5, 1918, cdf/1910–29/814.48/23, rg/59, NARA-II.

42. Walter Thurston to DOS, January 5, 1918, cdf/1910–29/814.48/27, rg/59, NARA-II.

43. Judge Fouille to Secretary of War, January 8, 1918, cdf/1910–29/814.48/37, rg/59; and Samuel Heald to Chester Harding, February 4, 1918, fo/891.4/08, bx/II-703, rg/ANRC, both NARA-II.

44. DOS to US Legation, Guatemala, January 9, 1918, cdf/1910–29/814.48/27, rg/59, NARA-II.

45. Walter Thurston to DOS, January 9, 1918, cdf/1910–29/814.48/44, rg/59, NARA-II.

46. Thurston to DOS, January 12, 1918, cdf/1910–29/814.48/30, rg/59, NARA-II.

47. Thurston to DOS, January 9, 1918.

48. Dosal, *Doing Business with Dictators*, 37.

49. S. W. Heald to Alfred Clark, January 27, 1918, cdf/1910–29/814.48/90, rg/59, NARA-II.

50. Samuel Heald to Chester Harding, February 4, 1918, fo/891.4/08, bx/II-703, rg/ANRC, NARA-II.

51. Meeting Minutes, ARC Department of Civilian Relief, [1918], fo/891.4, bx/II-703, rg/ANRC, NARA-II.

52. Frank Parsons to Alfred Clark, January 17, 1918, cdf/1910–29/814.48/35, rg/59, NARA-II.

53. Frank Polk to ARC, January 18, 1918, cdf/1910–29/814.48/43a, rg/59, NARA-II.

54. John O'Connor to Frank Persons, February 26, 1918, fo/891.4/08, bx/II-703, rg/ANRC, NARA-II.

55. John O'Connor to Frank Persons, February 5, 1918, cdf/1910–29/814.48/90, rg/59, NARA-II.

56. O'Connor to Persons, February 5, 1918.

57. Alvin Struse to Frank Persons, February 12 and March 5, 1918, fo/891.4/5 and fo/891.4/92; both bx/II-703, rg/ANRC, NARA-II.

58. O'Connor to Persons, February 5, 1918.

59. H. Apfel, "Report of the Camp Department," June 5, 1918, fo/891.4/68; and John O'Connor to Executive Committee, ARC Relief Committee, June 11, 1918, fo/891.4/08, both bx/II-703, rg/ANRC, NARA-II.

60. Apfel, "Report of the Camp Department"; O'Connor to Persons, February 5, 1918.

61. Louise O'Connor to Executive Committee, June 1, 1918, fo/891.4/08, bx/II-703, rg/ANRC, NARA-II.

62. Struse to Persons, March 5, 1918.

63. Alvin Struse, "Report of the Medical Relief of the American Red Cross," 1918, fo/891.4/508, bx/II-703, rg/ANRC, NARA-II.

64. Edward Stuart to Frank Persons, February 11, 1918, cdf/1910–29/814.48/77, rg/59, NARA-II.

65. John O'Connor to Frank Persons, February 26, 1918, fo/891.4/08; and Edward Stuart, "Recommendations Regarding a Permanent Water Supply," June 1, 1918, fo/891.4/51, both bx/II-703, rg/ANRC, NARA-II.

66. Frank Persons to DOS, March 18, 1918, fo/891.4/92, bx/II-703, rg/ANRC, NARA-II.

67. Frank Persons to Alfred Clark, March 18, 1918, fo/891.4/92, bx/II-703, rg/ANRC, NARA-II.

68. William Leavell to DOS, February 11, 1918, cdf/1910–29/814.48/69, rg/59, NARA-II.

69. O'Connor to Persons, February 26, 1918.

70. Alfred Clark to Frank Persons, April 10, 1918, fo/891.4/5, bx/II-703, rg/ANRC, NARA-II.

71. Meeting Minutes, ARC Department of Civilian Relief, [1918].

72. Alvin Struse to Frank Persons, April 2, 1918, fo/891.4; and H. Apfel, "Report of the Camp Department," June 5, 1918, fo/891.4/68, both bx/II-703, rg/ANRC, NARA-II.

73. Stuart, "Recommendations Regarding a Permanent Water Supply."

74. John O'Connor to Frank Persons, April 30, 1918, fo/891.4/08, bx/II-703, rg/ANRC, NARA-II.

75. Struse to Persons, April 2, 1918.

76. John O'Connor to Frank Persons, May 14, 1918, fo/891.4/08, bx/II-703, rg/ANRC, NARA-II.

77. "En honor de la Cruz roja americana," *Diario de Central América*, May 13, 1918, in cdf/1910–29/814.48/106, rg/59, NARA-II.

78. O'Connor to Executive Committee, ARC Relief Committee, June 11, 1918.

79. O'Connor to Executive Committee, ARC Relief Committee, June 11, 1918.

CHAPTER 6

1. Clancey, *Earthquake Nation*, 212–34; Schencking, *Great Kantō Earthquake*.

2. Cabanes, *Great War and the Origins of Humanitarianism*, 189–247; Rodogno, *Night on Earth*, 152–68; Irwin, *Making the World Safe*, 141–84.

3. List compiled from various archival collections included in this book's bibliography.

4. Lester Schnare to ARC China Central Committee, October 1, 1922, fo/DR-47, bx /II-719, rg/ANRC, NARA-II.

5. Schencking, "Giving Most and Giving Differently," 734.

6. Schencking, "Giving Most and Giving Differently"; Borland and Schencking, "Objects of Concern, Ambassadors of Gratitude."

7. Davidann, *Cultural Diplomacy in U.S.-Japanese Relations*, 1–102; Auslin, *Pacific Cosmopolitans*, 130–68.

8. Cyrus Woods to DOS, September 24, 1923, cdf/1910–29/894.48B/200, rg/59, NARA-II.

9. Charles Burnett to Cyrus Woods, September 21, 1923, cdf/1910–29/894.48B/200, rg/59, NARA-II.

10. Nelson Johnson to MacMurray, September 14, 1923, bx/3, coll/MSS-27912, LOC.

11. E. F. Witsell to Officer, American Embassy Relief, September 15, 1923, cdf/1910–29/894.48B/200, rg/59, NARA-II.

12. Cyrus Woods to DOS, September 5, 1923, corr/848, ent/UD-36, rg/84, NARA-II.

13. Schencking, *Great Kantō Earthquake*.

14. Vice-Minister of Home Affairs to Charles Burnett, February 1924, cdf/1910–29/894.48b/517, rg/59, NARA-II.

15. Cyrus Woods to DOS, September 2 and 3, 1923, corr/848, ent/UD-36, rg/84, NARA-II.

16. Woods to DOS, September 24, 1923.

17. Burnett to Woods, September 21, 1923.

18. Reports of American Embassy Relief Committee, September 15 and 19, 1923, cdf/1910–29/894.48B/200, rg/59, NARA-II.

19. John Doty and W. W. Johnston, Diary, September 1–12, 1923, coll/FO-908/6, TNA; W. I. Eisler to US Shipping Board, September 21, 1923, corr/848, ent/UD-948, rg/84, NARA-II.

20. E. C. Creager to Erle Dickover, September 10, 1923, corr/848, ent/UD-472, rg/84, NARA-II.

21. Erle Dickover to Cyrus Woods, September 8, 1923, corr/848, ent/UD-472, rg/84; and Erle Dickover to DOS, October 1, 1923, cdf/1910–29/894.48B/219, rg/59, both NARA-II.

22. Erle Dickover to DOS, October 30, 1923, ent/UD-948, rg/84, NARA-II; Tommy Tompkins to Arthur Bassett, September 12, 1923; and W. W. Peter to Ernest Bicknell, October 5, 1923, both fo/898.6/02, bx/II-707, rg/ANRC, NARA-II.

23. Charles Hughes to Hanihara Masanao, September 29, 1923, cdf/1910–29/894.48B /160a, rg/59, NARA-II.

24. DOS to US Embassy, Tokyo, September 3, 1923, corr/848, ent/UD-36, rg/84, NARA-II.

25. H. H. Tebbitts to Commanding General Philippine Department, September 3, 1923; and S. Heinzelman to Adjutant General of the Army, September 5, 1923, both bx/929, fi/400.38, ent/PI-17/37-A, rg/407, NARA-II.

26. Henry Rogers, "My Experience with American Red Cross Japan Relief Mission," October 1923, fo/898.6/08, bx/II-707, rg/ANRC; and H. G. Fairbanks, "Diary of Co. B, 1st

En., 14th Engrs (PS), September 23, 1923, fo/"14th Engrs Diary," bx/432, ent/NM-84/310, rg/165, both NARA-II.

27. Adjutant General to Frank McCoy, September 6, 1923, bx/929, fi/400.38, ent/PI-17 /37-A, rg/407, NARA-II.

28. James Fieser to DOS, September 8, 1923, cdf/1910–29/894.48B/504, rg/59, NARA-II.

29. James Fieser to ARC Regional Division Managers, September 3, 1923, fo/898.6, bx /II-707, rg/ANRC, NARA-II.

30. Calvin Coolidge, Proclamation, September 4, 1923, fo/898.6/08, bx/II-707, rg /ANRC, NARA-II.

31. William Philips to Elizabeth Lavelle, September 19, 1923, cdf/1910–29/894.48B/86, rg/59; and Calvin Coolidge to Evangeline Booth, fo/898.6/21, bx/II-707, rg/ANRC, both NARA-II.

32. John Payne to DOS, September 14, 1923, cdf/1910–29/894.48B/112, rg/59, NARA-II.

33. Edward Farley to Calvin Coolidge, September 4, 1923, fo/898.6/02, bx/II-707, rg /ANRC, NARA-II; Edward Farley to Edwin Denby, September 4, 1923, fi/7266-345:1, bx/332, ent/19, rg/80, NARA-I.

34. Ernest Swift to Mr. Lockhart, November 23, 1923, cdf/1910–29/894.48B/286, rg/59, NARA-II.

35. James Fieser to Division Managers, September 10, 1923, fo/898.6/031, bx/II-707, rg /ANRC, NARA-II.

36. Woods to DOS, September 24, 1923.

37. Hughes to Hanihara, September 29, 1923.

38. William Philips to Cyrus Woods, September 4, 1923, cdf/1910–29/894.48B/19a, rg/59, NARA-II.

39. Commander, USS *Huron*, to Chief of Naval Operations, October 1, 1923, corr/848, ent/UD-36, rg/84, NARA-II.

40. Frank McCoy, "Report of the American Relief Mission to Japan," November 9, 1923, bx/1857, fi/400.38, ent/PI-17/37-C, rg/407, NARA-II; Edwin Denby to Calvin Coolidge, September 4, 1923, fi/7266-345; and Report, "Naval Vessels in and en Route to Japanese Waters in Connection with Relief Work," October 6, 1923, fi/7266-345:118, both ent/19, rg/80, NARA-I.

41. Nelson Johnson to DOS, September 10, 1923; and Erle Dickover to DOS, October 1, 1923, both corr/848, ent/UD-472, rg/84, NARA-II.

42. Bundy (Manila) to Adjutant General, September 6, 1923, bx/929, fi/400.38, ent /PI-17/37-A, rg/407; and Frank McCoy to Cyrus Woods, October 17, 1923, cdf/1910–29 /894.48B/275, rg/59, both NARA-II.

43. ARC to US Embassy, Tokyo, September 12, 1923, cdf/1910–29/894.48B/67m, rg/59, NARA-II; Eisler to US Shipping Board, September 21, 1923.

44. Cyrus Woods to DOS, September 6, 1923, cdf/1910–29/894.48B/51, rg/59, NARA-II.

45. D. W. Hand, "Report of Commanding Officer . . . American Relief Mission," October 31, 1923, corr/848, ent/UD-36, rg/84, NARA-II.

46. Arthur McCollum to Director of Naval Intelligence, December 8, 1923, fo/c-9-e, fi/16546-B, bx/449, ent/98, rg/38, NARA-I.

47. Woods to DOS, September 6, 1923.

48. Cyrus Woods, Notice to Americans, September 5, 1923, corr/848, ent/UD-36, rg/84, NARA-II.

49. Woods to DOS, September 6, 1923.

50. Cyrus Woods to DOS, September 10, 1923, corr/848, ent/UD-36, rg/84, NARA-II.

51. Charles Hughes to Cyrus Woods, September 6, 1923, cdf/1910–29/894.48B/52j, rg/59, NARA-II.

52. Frank McCoy to Adjutant General, September 22, 1923, bx/929, fi/400.38, ent /PI-17/37-A, rg/407, NARA-II; Woods to DOS, September 24, 1923.

53. Commander, USS *Huron*, to Chief of Naval Operations, October 1, 1923.

54. Cyrus Woods to DOS, September 15, 1923, cdf/1910–29/894.48B/80, rg/59, NARA-II.

55. McCoy, "Report of the American Relief Mission to Japan."

56. McCoy to Woods, October 17, 1923.

57. Frank McCoy to Commander, USS *Merritt*, September 13, 1923, corr/848, ent /UD-36, rg/84; and G. W. Read to Adjutant General of the Army, October 4, 1923, bx/929, fi/400.38, ent/PI-17/37-A, rg/407, both NARA-II.

58. H. Haragachi to John Weeks, September 17, 1923, bx/929, fi/400.38, ent/PI-17/37-A, rg/407, NARA-II.

59. Tokugawa Iesato, speech, October 3, 1923, fo/"Japan Relief Mission," bx/75, coll /MSS-31989, LOC.

60. Count Yamamoto to US President and Secretary of State, via Japanese Embassy, September 9, 1923, cdf/1910–29/894.48B/47, rg/59, NARA-II.

61. E. A. Anderson to Secretary of the Navy, September 18, 1923, cdf/1910–29/894.48B /170, rg/59, NARA-II.

62. Cyrus Woods to DOS, September 22, 1923, corr/848, ent/UD-36, rg/84, NARA-II.

63. Cyrus Woods to DOS, September 17, 1923, fo/898.6, bx/II-707, rg/ANRC, NARA-II.

64. Kobayashi Seizō to Commander, US Asiatic Fleet, September 18, 1923, cdf/1910–29 /894.48B/170, rg/59, NARA-II.

65. William Philips to John Payne, September 19, 1923, fo/898.6/02, bx/II-707, rg /ANRC, NARA-II.

66. William Philipps to US Embassy, Tokyo, September 19, 1923, cdf/1910–29 /894.48B/121c, rg/59, NARA-II; Eisler to US Shipping Board, September 21, 1923.

67. Anderson to Secretary of the Navy, September 18, 1923.

68. Cyrus Woods to DOS, September 22, 1923, cdf/1910–29/894.48B/226, rg/59, NARA-II.

69. Cyrus Woods to ARC, October 5, 1923, cdf/1910–29/894.48B/170, rg/59, NARA-II.

70. Frank McCoy to Adjutant General, October 10, 1923, bx/929, fi/400.38, ent/PI-17 /37-A, rg/407; and Howard Ramsey, "Relief Work for Japanese Earthquake Victims," January 1924, cdf/1910–29/894.48B/372, rg/59, both NARA-II.

71. George Read to Adjutant General, September 21, 1923, bx/929, fi/400.38, ent /PI-17/37-A, rg/407; and "Feature Story on American Field Hospital," translated article from Tokyo-*Nichi-Nichi*, September 30, 1923, in cdf/1910–29/894.48B/275, rg/59, both NARA-II.

72. E. L. Munson, "Final Report on the Medical and Hospital Activities," October 13, 1923, corr/848, ent/UD-36, rg/84, NARA-II.

73. Frank McCoy to John Pershing, October 26, 1923, fo/"General Correspondence," bx/17, coll/MSS-31989, LOC; Howard Ramsey to ARC, October 10, 1923, cdf/1910–29 /894.48B/275, rg/59, NARA-II.

74. "Speech of Acceptance of the Japanese Minister of Foreign Affairs," October 13, 1923, in McCoy, "Report of the American Relief Mission to Japan."

75. Translation of Report by Baron Hirayama, [October 1923], cdf/1910–29/894.48B /515, rg/59, NARA-II.

76. Hand, "Report of Commanding Officer . . . American Relief Mission."

77. D. H. Blake, "The Japanese Earthquake of September 1st, 1923," October 1924, fo/898.6/08, bx/II-707, rg/ANRC, NARA-II.

78. James Fieser to US Embassy, Tokyo, November 12, 1923, cdf/1910–29/894.48B/261, rg/59, NARA-II.

79. Jefferson Caffery to DOS, November 13, 1923, cdf/1910–29/894.48B/263; and Jefferson Caffery to DOS, November 20, 1923, cdf/1910–29/894.48B/272; and James Fieser to Jefferson Caffery, November 30, 1923, cdf/1910–29/894.48B/302b, all rg/59, NARA-II.

80. James Fieser to Jefferson Caffery, December 26, 1923, cdf/1910–29/894.48B/332, rg /59, NARA-II.

81. Fieser to Caffery, December 26, 1923.

82. Jefferson Caffery to DOS, December 19, 1923, cdf/1910–29/894.48B/326, rg/59, NARA-II.

83. Jefferson Caffery to DOS, February 13, 1924, cdf/1910–29/894.48B/385, rg/59, NARA-II.

84. Translation of report by Baron Hirayama, [October 1923]; Jefferson Caffery to DOS, October 11, 1924, corr/848, ent/UD-36, rg/84, NARA-II.

85. Ernest Swift to John Payne, January 24, 1924, fo/898.6/2; and D. H. Blake, "The Japanese Earthquake of September 1st, 1923," October 1924, fo/898.6/08, both bx/II-707, rg/ANRC, NARA-II.

86. "Roster of Personnel American Relief Mission," [1923], in McCoy, "Report of the American Relief Mission to Japan."

87. John Weeks to Speaker of the House of Representatives, April 21, 1924, bx/929, fi/400.38, ent/PI-17/37-A, rg/407, NARA-II; *Act for the Relief of Sufferers from Earthquake in Japan* (1925).

88. Vice-Minister of Home Affairs to Charles Burnett, February 1924, cdf/1910–29 /894.48b/517, rg/59, NARA-II; LRCS, "The Japanese Disaster and the American Red Cross," March 1, 1924, Bulletins and Information Circulars of the League of Red Cross Societies, 1919–1939, IFRC.

89. Woods to DOS, September 24, 1923.

90. Hirayama Seishin to Frank McCoy, October 27, 1923, fo/"General Correspondence," bx/17, coll/MSS-31989, LOC. Memorandum of conversation, Baron Tanaka and Frank McCoy, November 6, 1923, cdf/1910–29/894.48B/372, rg/59, NARA-II.

91. Ijūin Hikokichi to Leonard Wood, October 21, 1923, fo/"General Correspondence," bx/17, coll/MSS-31989, LOC; Hanihara Masanao to DOS, October 30, 1923, cdf /1910–29/894.48B/237, rg/59, NARA-II.

92. Translated editorial from *Kokumin*, October 12, 1923, in Cyrus Woods to DOS, October 15, 1923, cdf/1910–29/894.48B/2353, rg/59; and Translated editorial from *Jiji*, September 28, 1923, in Cyrus Woods to DOS, October 1, 1923, corr/848, ent/UD-36, rg/84, both NARA-II.

93. H. Nishimura to Cyrus Woods, October 21, 1923; and Inomatsu Mita to Cyrus Woods, October 31, 1923, both corr/848, ent/UD-36, rg/84, NARA-II.

94. Acting Naval Attaché, "Attitude of Japanese toward United States," November 2, 1923, fo/c-9-e, fi/16546-B, bx/449, ent/98, rg/38, NARA-I.

95. McCoy to Woods, October 17, 1923.

96. Woods to DOS, September 24, 1923.

97. Charles Burnett, "Comments on the Earthquake and Fire in Tokyo," in Jefferson Caffery to DOS, January 31, 1924, corr/848, ent/UD-36, rg/84, NARA-II.

98. Ramsey, "Relief Work for Japanese Earthquake Victims."

99. Schencking, "Giving Most and Giving Differently," 749–57.

100. Erle Dickover to DOS, September 12, 1923, corr/848, ent/UD-36; and Erle Dickover to DOS, October 11, 19, and 30, 1923, corr/848, ent/UD-472, all rg/84, NARA-II.

101. Dickover to DOS, October 11 and 30, 1923.

102. Nelson Johnson to Erle Dickover, November 13, 1923, corr/848, ent/UD-472, rg/84, NARA-II.

103. McCollum to Director of Naval Intelligence, December 8, 1923.

104. "Intelligence Report on Japanese Earthquake Disaster," September 26, 1923, fo /c-9-e, fi/16546-B, bx/449, ent/98, rg/38, NARA-I.

105. Charles McBrayer, "Hospital Order #5," October 10, 1923, bx/1857, fi/400.38, ent /PI-17/37-C, rg/407, NARA-II.

106. Barney Meeden to Adjutant General, December 20, 1923, bx/1364, fi/350.05, ent /PI-17/37-C, rg/407, NARA-II.

107. P. P. Bishop to Chief of Staff, [1923], bx/929, fi/400.38, ent/PI-17/37-A, rg/407, NARA-II.

108. Cyrus Woods to DOS, October 30, 1923, cdf/1910–29/894.48B/279, rg/59, NARA-II.

109. Jefferson Caffery to DOS, November 5, 1923, cdf/1910–29/894.48B/246, rg/59, NARA-II.

110. Jefferson Caffery to DOS, November 20, 1923.

111. Jefferson Caffery to DOS, January 18, 1924, cdf/1910–29/894.48B/361, rg/59, NARA-II.

112. Jefferson Caffery to DOS, February 15, 1924, corr/848, ent/UD-36, rg/84, NARA-II.

113. Blake, "Japanese Earthquake of September 1st, 1923."

CHAPTER 7

1. Ernest Bicknell, "The American Red Cross in Far Places," radio transcript, April 15, 1925, fo/"Bicknell," bx/3, ent/P-105, rg/ANRC, NARA-II.

2. Ernest Bicknell, untitled speech, 1928, fo/"Bicknell," bx/3, ent/P-105, rg/ANRC, NARA-II.

3. List compiled from various archival collections included in this book's bibliography.

4. James Fieser to ARC Central Committee, April 15, 1927, fo/FDR-25, bx/II-709, rg /ANRC, NARA-II.

5. Ernest Bicknell to DOS, August 14, 1924, cdf/1910–29/893.48P36/3, rg/59, NARA-II.

6. Frank Kellogg to US Embassy, Havana, October 24, 1926, cdf/1910–29/837.48/23, rg/59, NARA-II.

7. Bicknell, untitled speech, 1928.

8. Ernest Swift, Memorandum of Conference with ARC members, October 25, 1926, fo/FDR-21, bx/II-709, rg/ANRC, NARA-II.

9. Frank Kellogg to US Embassy, Havana, October 25, 1926, corr/848, ent/UD-13, rg/84, NARA-II.

10. Robert Skinner to Charles MacFarland, May 1, 1928, corr/848, ent/UD-26, rg/84, NARA-II.

11. C. C. Thurber to Robert Skinner, April 26, 1928, corr/848, ent/UD-26, rg/84, NARA-II.

12. Enoch Crowder to DOS, November 5 and 15, 1926, corr/848, ent/UD-13, rg/84, NARA-II.

13. Henry Baker to James Fieser, October 31, 1926, fo/FDR-21, bx/II-709, rg/ANRC, NARA-II.

14. Enoch Crowder to DOS, October 23, 1926, fi/H4–9/EF19, bx/855, ent/19, rg/80, NARA-I.

15. Bicknell, untitled speech, 1928.

16. Ernest Swift to Henry Baker, September 24, 1928, fo/FDR-40, bx/II-709, rg/ANRC, NARA-II; Bicknell, "American Red Cross in Far Places"; Bicknell, untitled speech, 1928.

17. K. C. Melhorn to Minister of Interior, August 20, 1928, fo/FDR-40, bx/II-709, rg /ANRC, NARA-II.

18. R. L. Lovejoy to Commandant, Seventh Naval District, November 26, 1926; and Commander, USS *Milwaukee*, to Chief of Naval Operations, October 28, 1926, both fi /H4–9/EF19, bx/855, ent/19, rg/80, NARA-I.

19. Bicknell, untitled speech, 1928.

20. Joseph Grew to DOS, April 13, 1928, fo/FDR-35, bx/II-709, rg/ANRC, NARA-II.

21. George Cross to William Blocker, January 29, 1926; and William Blocker to DOS, January 20, 1926, both fo/FDR-7, bx/II-709, rg/ANRC, NARA-II.

22. John Cremer to James Fieser, August 23, 1928, fo/FDR-40, bx/II-709, rg/ANRC, NARA-II.

23. US Legation, Riga, "American Foreign Service Report," September 29, 1924, cdf/1910–29/893.48P36/9, rg/59, NARA-II.

24. List compiled from various archival collections included in this book's bibliography.

25. William Baxter to ARC Chapter Chairmen, September 13, 1930; and Ernest Swift to James Fieser, September 19, 1930, both fo/FDR-65, bx/II-713, rg/ANRC, NARA-II.

26. ARC, "Managua Earthquake: Official Report of the Relief Work in Nicaragua," October 1931, fo/FDR-73, bx/II-714, rg/ANRC, NARA-II.

27. Veeser, *World Safe for Capitalism*; Tillman, *Dollar Diplomacy by Force*; McPherson, *Invaded*, 34–50, 68–72, 159–93.

28. Roorda, *Dictator Next Door*, 6–62.

29. Gobat, *Confronting the American Dream*, 205–66; McPherson, *Invaded*, 13–21, 53–58, 73–90, 213–37.

30. Matthew Hanna to DOS, March 18, 1931, fo/"2d.Brig.-Misc.," bx/14, ent/43A, rg/127, NARA-I.

31. Roorda, *Dictator Next Door*, 54–59; Charles Curtis to DOS, September 15, 1930, corr/848, ent/UD-16, rg/84, NARA-II.

32. Charles Curtis to DOS, September 14, 1930, corr/848, ent/UD-16, rg/84, NARA-II.

33. William Beaulac, "Report on the Work of the American Red Cross to Relieve Distress in Nicaragua," [1931], cdf/1930–39/817.48/193, rg/59, NARA-II; F. B. Garrett to Major General Commandant, April 11, 1931, fi/Nicaragua-Earthquake, MCHRB.

34. James McClintock to Judge Payne, September 6, 1930, fo/FDR-65, bx/II-713; and James McClintock to Judge Payne, April 20, 1931, fo/FDR-73, bx/II-714, both rg/ANRC, NARA-II.

35. Swift to Fieser, September 19, 1930.

36. Henry Stimson to Matthew Hanna, March 31, 1931, cdf/1930–39/817.48/11, rg/59, NARA-II.

37. ARC News Service, press release, November 16, 1930, fo/FDR-65, bx/II-713, rg /ANRC, NARA-II; ARC, "Managua Earthquake: Official Report."

38. Swift to Fieser, September 19, 1930; Commander, USS *Grebe*, to Governor of Virgin Islands, September 15, 1930; and Commander, USS *Gilmer*, to Chief of Naval Operations, September 20, 1930, both fi/H4–9/EF63, bx/855, ent/19, rg/80, NARA-I.

39. Memorandum, "Assistance the Navy Has Thus Far Rendered," April 1, 1931, cdf/1930–39/817.48/61, rg/59, NARA-II; Ships' logs of the *Chaumont, Lexington, Relief*, and *Rochester*, March and April 1931, ent/118G, rg/24, NARA-I; Memorandum of telephone message, Secretary Jahncke for Herbert Hoover, March 31, 1931, coll/1008-8, FAF /HH-PP, HHPL.

40. Theodore Roosevelt Jr. to Secretary of War, September 4, 1930, fi/651, bx/1537, ent/82A, rg/26, NARA-I; Ernest Swift to Charles Curtis, September 4, 1930, corr/848, ent/UD-16, rg/84, NARA-II.

41. Baxter to ARC Chapter Chairmen, September 13, 1930; Swift to Fieser, September 19, 1930.

42. Saint Clair Smith to Chief of Naval Operations, April 21, 1931, fi/Nicaragua-Earthquake, MCHRB.

43. Office of Naval Intelligence, "Report of Landing Force (Aircraft) at Managua," 1931, fi/A-1-Z-20781, bx/181, ent/98, rg/38, NARA-I; USS *Lexington* to USMC Headquarters, April 2, 1931, fo/"2d.Brig.-Patrol Rpts.," bx/15, ent/43A, rg/127, NARA-I.

44. J. T. Trippe to Secretary of Navy, April 1, 1931, fi/H1-5/EF49, bx/831, ent/19, rg/80, NARA-I.

45. Van Vleck, *Empire of the Air*.

46. Curtis to DOS, September 15, 1930.

47. James Fieser to Herbert Hoover, September 6, 1930, coll/998-9, FAF/HH-PP, HHPL.

48. James McClintock to Herbert Hoover, March 31, 1931, coll/1008-8, FAF/HH-PP, HHPL.

49. Curtis to DOS, September 14, 1930; Ernest Swift, "The Red Cross Flies to Managua," *Red Cross Courier*, September 1931, 454.

50. Joseph Cotton to Charles Curtis, September 10, 1930, corr/848, ent/UD-16, rg/84, NARA-II.

51. Charles Curtis to DOS, September 7, 1930, corr/848, ent/UD-16, rg/84, NARA-II; Swift to Fieser, September 19, 1930.

52. Charles Curtis to DOS, September 22, 1930, corr/848, ent/UD-16, rg/84, NARA-II.

53. ARC, "Managua Earthquake: Official Report."

54. Swift to Fieser, September 19, 1930.

55. Thomas Watson to Colonel [Cutts], September 18, 1930, fo/"Correspondence, September 1930," bx/1, coll/3615, MCAB.

56. Lucius Johnson, "Report on Relief Work in the Santo Domingo Disaster," 1930, fo/"Reports," bx/3, coll/3615, MCAB.

57. Curtis to DOS, September 22, 1930.

58. Matthew Hanna to DOS, April 3, 1931, cdf/1930–39/817.48/53, rg/59, NARA-II.

59. ARC, "Managua Earthquake: Official Report."

60. British Legation, Managua, to H. A. Grant Watson, April 9, 1931, fi/25, coll/FO-809/2, TNA; Johnson, "Report on . . . the Santo Domingo Disaster"; ARC, "Managua Earthquake: Official Report"; Beaulac, "Report on the Work of the American Red Cross . . . in Nicaragua."

61. Frederic Bradman to Chief of Naval Operations, March 20, 1933, cdf/1930–39/817.48/240, rg/59, NARA-II.

62. Matthew Hanna to José María Moncada, April 14, 1931, corr/848, ent/UD-45, rg/84, NARA-II; Garrett to Major General Commandant, April 11, 1931.

63. Barry, *Rising Tide*.

64. Ernest Swift to Rafael Trujillo, September 15, 1930; and Thomas Watson and Barney Morgan, "Report of Relief Work Carried On in Santo Domingo," 1930, both fo/FDR-65, bx/II-713, rg/ANRC, NARA-II.

65. Watson and Morgan, "Report of Relief Work . . . in Santo Domingo."

66. ARC, "Managua Earthquake: Official Report."

67. Matthew Hanna to DOS, April 5, 1931, cdf/1930–39/817.48/73, rg/59, NARA-II.

68. ARC, "Managua Earthquake: Official Report."

69. Thomas Watson to Ernest Swift, October 10, 1930, fo/FDR-65, bx/II-713, rg/ANRC, NARA-II.

70. Thomas Watson to George Richards, October 18, 1930, fo/"Correspondence, October 1930," bx/1, coll/3615, MCAB.

71. Watson and Morgan, "Report of Relief Work . . . in Santo Domingo."

72. Matthew Hanna to José María Moncada, April 30, 1931, corr/848, ent/UD-45, rg/84, NARA-II.

73. José María Moncada to Matthew Hanna, April 29, 1931, corr/848, ent/UD-45, rg/84, NARA-II.

74. Watson to [Cutts], September 18, 1930.

75. Watson and Morgan, "Report of Relief Work . . . in Santo Domingo."

76. Matthew Hanna to DOS, April 21 and 29, 1931, cdf/1930–39/817.48/139 and cdf/1930–39/817.48/142, rg/59, NARA-II.

77. Hanna to Moncada, April 30, 1931.

78. ARC, "Managua Earthquake: Official Report."

79. Hanna to DOS, April 5, 1931; Hanna to Moncada, April 30, 1931.

80. Watson to [Cutts], September 18, 1930.

81. Matthew Hanna to DOS, April 10, 1931, cdf/1930–39/817.48/114, rg/59, NARA-II.

82. Watson and Morgan, "Report of Relief Work . . . in Santo Domingo."

83. Hanna to DOS, April 5, 1931.

84. George Hepburn to Matthew Hanna, July 25, 1931, ent/UD-45, rg/84, NARA-II.

85. Matthew Hanna to Ernest Swift, June 6, 1931, fo/FDR-73, bx/II-714, rg/ANRC, NARA-II.

86. Watson to [Cutts], September 18, 1930.

87. Thomas Watson to Rafael Trujillo, September 16, 1930; and Thomas Watson to Ernest Swift, September 10, 1930; and Thomas Watson to George Richards, October 18, 1930, all fo/"Correspondence," bx/1, coll/3615, MCAB.

88. ARC, "Managua Earthquake: Official Report."

89. Matthew Hanna to Ernest Swift, August 22, 1931; and Matthew Hanna to José María Moncada, August 29, 1931, both fo/FDR-73, bx/II-714, rg/ANRC, NARA-II.

90. Statement prepared for Herbert Hoover by DOS, September 22, 1930, coll/998-9, FAF/HH-PP, HHPL.

91. Ernest Swift to Thomas Watson, September 24, 1930, fo/"Correspondence, September 1930," bx/1, coll/3615, MCAB.

92. Rafael Trujillo to Herbert Hoover, September 20, 1930, corr/848, ent/UD-16, rg/84, NARA-II.

93. José María Moncada to Matthew Hanna, June 18, 1931, fo/FDR-73, bx/II-714, rg /ANRC, NARA-II.

94. Matthew Hanna to Francis White, June 2, 1931, corr/848, ent/UD-45, rg/84, NARA-II.

95. Swift to Trujillo, September 15, 1930; Swift to Fieser, September 19, 1930.

96. ARC, "Managua Earthquake: Official Report."

97. Juan Bautista Sacasa to Henry Stimson, April 6, 1931, cdf/1930–39/817.48/193, rg/59, NARA-II.

98. DOS, Division of Current Information, Memorandum of Press Conference, April 6, 1931, fo/11, bx/16, Frances White Letters Collection, HHPL.

99. "Historic Responsibility for the Magnitude of the Disaster," translated article from Diario de Occidente (León, Nicaragua), April 25, 1931, cdf/1930–39/817.48/147, rg/59, NARA-II.

100. "American Marines Responsible for the Destruction of Managua," translated article from La Prensa (Mexico City), May 9, 1931, cdf/1930–39/817.48/163, rg/59, NARA-II.

101. US Legation in El Salvador to DOS, April 14, 1931, cdf/1930–39/817.48/126, rg/59, NARA-II.

102. "Did the Forces of Occupation in Managua Prevent the Great Disaster from Being Kept within Bounds?," translated article from El Gráfico (Mexico City), April 24, 1931, cdf/1930–39/817.48/150, rg/59, NARA-II.

103. Transcript of interview with Vicente Lombardo Toledano, in El Universal (Mexico City), April 29, 1931, cdf/1930–39/817.48/149, rg/59, NARA-II.

104. "The Health Envoy to Nicaragua Has Just Returned to Mexico," translated article from *Excélsior* (Mexico City), May 29, 1931, cdf/1930–39/817.48/166, rg/59, NARA-II.

105. US Legation, El Salvador, to DOS, April 7 and 14, 1931, cdf/1930–39/817.48/126 and cdf/1930–39/817.48/136, rg/59, NARA-II.

106. Hanna to Moncada, April 14, 1931; Matthew Hanna to DOS, April 27, 1931, cdf/1930–39/817.48/147, rg/59, NARA-II.

107. DOS, Division of Current Information, Memorandum of Press Conference, April 6, 1931; "Swift Praises Marines' Job in Nicaragua," *Washington (DC) Herald*, April 30, 1931, cdf/1930–39/817.48/146, rg/59, NARA-II.

108. Matthew Hanna to DOS, April 4, 1931, cdf/1930–39/817.48/66, rg/59, NARA-II; Monthly Reports of 2nd Marine Brigade, March 1931–January 1932, bx/4 and bx/8, ent /113C, rg/127, NARA-I.

CHAPTER 8

1. Dauber, *Sympathetic State*.

2. Edwin Cunningham to DOS, June 30, 1932, cdf/1930–39/893.48/574, rg/59, NARA-II.

3. Courtney, *Nature of Disaster in China*, 3.

4. Nelson Johnson to DOS, August 14, 1931, cdf/1930–39/893.48/268; and Edmund Clubb, "The Central China Flood," September 26, 1931, cdf/1930–39/893.48/478, both rg/59, NARA-II; Rear Admiral, HMS *Bee* to Commander-in-Chief, China Station, September 4, 1931, ADM-116/2843, TNA.

5. David Brown to John Payne, September 10, 1931, fo/FDR-74, bx/II-714, rg/ANRC, NARA-II.

6. Courtney, *Nature of Disaster in China*; Pietz, *Engineering the State*, 41–102; Pietz, *Yellow River*.

7. Kuo Min News Agency, press release, August 14, 1931, corr/848, ent/UD-603, rg/84, NARA-II; Ministry of Foreign Affairs, Republic of China, "Memorandum," August 15, 1931, cdf/1930–39/893.48/361, rg/59, NARA-II.

8. Kuo Min News Agency, press release, August 23, 1931, cdf/1930–39/893.48/408, rg/59, NARA-II; Edwin Cunningham to US Legation, Peiping, August 27, 1931, corr/848, ent/UD-10, rg/84, NARA-II.

9. Willys Peck to DOS, August 15, 1931, cdf/1930–39/893.48/274, rg/59, NARA-II.

10. Walter Adams to DOS, September 8, 1931, fo/FDR-74, bx/II-714, rg/ANRC, NARA-II.

11. Chang, *Fateful Ties*, 130–67; Tsui, *China's Conservative Revolution*.

12. Nelson Johnson to DOS, August 31, 1931, cdf/1930–39/893.48/326, rg/59, NARA-II.

13. Walter Adams to US Legation, Peiping, August 22, 1931, cdf/1930–39/893.48/415, rg/59, NARA-II.

14. M. M. Taylor to Nelson Johnson, October 3, 1931, fo/"General Correspondence," bx/16, MSS-27912, LOC.

15. Nelson Johnson to DOS, August 15, 1931, cdf/1930–39/893.48/271, rg/59, NARA-II.

16. Ernest Bicknell, "Memorandum" August 15, 1931, fo/FDR-74, bx/II-714, rg/ANRC, NARA-II.

17. Ernest Swift to David Brown, September 18, 1931, fo/FDR-74, bx/II-714, rg/ANRC, NARA-II.

18. ARC, "Expenditures in China, 1906–1937," October 1937, fo/895.8/208, bx/III-1222, rg/ANRC, NARA-II.

19. ARC, *Report of the American Red Cross Commission to China.*

20. ARC, *Report of the American Red Cross Commission to China,* 30.

21. Walter Adams to US Legation, Peiping, August 11, 1931, corr/848, ent/UD-603, rg/84, NARA-II.

22. Walter Adams to DOS, August 6, 1931, cdf/1930–39/893.48/265, rg/59, NARA-II.

23. Adams to DOS, September 8, 1931.

24. Editorial, Peiping *Leader,* September 8, 1931, in Clubb, "Central China Flood."

25. Memorandum of conversation, John Payne, Ernest Swift, and Stanley Hornbeck, August 19, 1931, cdf/1930–39/893.48/445, rg/59; and William Castle to US Consul, Hankow, August 19, 1931, corr/848, ent/UD-10, rg/84, both NARA-II.

26. DOS to John Payne, August 19, 1931, cdf/1930–39/893.48/277, rg/59, NARA-II.

27. Ernest Swift to Stanley Hornbeck, August 21, 1931, cdf/1930–39/893.48/296, rg/59, NARA-II.

28. Adams to DOS, September 8, 1931.

29. Walter Adams to DOS, October 19, 1931, corr/848, ent/UD-10, rg/84, NARA-II.

30. Walter Adams to DOS, October 8, 1931, fo/FDR-74, bx/II-714, rg/ANRC, NARA-II.

31. Division of Far Eastern Affairs, DOS, "Flood in China and Question of Relief," August 16, 1931, cdf/1930–39/893.48/443; and William Castle to US Legation, Peiping, August 17, 1931, cdf/1930–39/893.48/271, both rg/59, NARA-II.

32. Commander-in-Chief, Asiatic Fleet, to Secretary of the Navy, August 19, 1931, fi /H4-9/EF16, bx/855, ent/19, rg/80, NARA-I; Willys Peck to US Legation, Peiping, August 24, 1931, corr/848, ent/UD-10, rg/84, NARA-II.

33. McDonald, *Food Power,* 78–79.

34. Memorandum of conversation, Shen Tchang and Willys Peck, August 10, 1931, corr/848, ent/UD-603, rg/84, NARA-II.

35. T. L. Wang, "Memorandum," August 15, 1931, cdf/1930–39/893.48/361, rg/59, NARA-II.

36. Nelson Johnson to DOS, August 14, 1931, coll/997-4, FAF/HH-PP, HHPL.

37. Division of Far Eastern Affairs, "Flood in China and Question of Relief."

38. William Borah to David Brown, [August 1931], cdf/1930–39/893.48/279, rg/59, NARA-II.

39. James Stone to DOS, August 19, 1931, coll/997-4, FAF/HH-PP, HHPL.

40. Nelson Johnson to DOS, September 6, 1931, cdf/1930–39/893.48/345, rg/59, NARA-II.

41. T. V. Soong, in Edwin Cunningham to DOS, June 30, 1932, cdf/1930–39/893.48/574, rg/59, NARA-II.

42. Stone to DOS, August 19, 1931.

43. Chinese Ministry of Foreign Affairs to Willys Peck, in Edwin Cunningham to DOS, August 27, 1931, coll/997-4, FAF/HH-PP, HHPL.

44. William Castle to Nelson Johnson, August 27, 1931, cdf/1930–39/893.48/314, rg/59, NARA-II.

45. Walter Newton to J. H. Franklin, August 28, 1931, coll/997-4, FAF/HH-PP, HHPL.

46. T. V. Soong to Federal Farm Board, in Willys Peck to DOS, September 3, 1931, cdf/1930–39/893.48/339, rg/59, NARA-II.

47. Draft note from Chinese Minister of Foreign Affairs to US Minister, in Willys Peck to DOS, September 14, 1931, cdf/1930–39/893.48/354, rg/59, NARA-II.

48. Walter Adams to DOS, October 13, 1931, cdf/1930–39/893.48/484, rg/59, NARA-II.

49. Clubb, "Central China Flood."

50. Nelson Johnson, Memorandum of interview with Chiang Kai-Shek, November 4, 1931, fo/"General Correspondence," bx/16, MSS-27912, LOC.

51. [No name], Preliminary Report to National Flood Relief Commission, August 4, 1932, cdf/1930–39/893.48/582, rg/59, NARA-II.

52. Marine Detachment, US Legation, Peiping, November 1931 intelligence report, December 11, 1931, fi/"China Legation, 1928–1936 (Oversized)," MCHRB.

53. Perkins (China) to DOS, April 27, 1932, cdf/1930–39/893.48/554, rg/59, NARA-II.

54. T. V. Soong, paraphrased in Nelson Johnson to DOS, May 9, 1932, cdf/1930–39/893.48/543, rg/59, NARA-II.

55. Memorandum of conversation, John Simpson and Stanley Hornbeck, July 28, 1932, cdf/1930–39/893.48/575, rg/59, NARA-II.

56. Union of Farmers' Associations of Kiangsu, Chekiang, Anhwei, Kiangsi, and Hupeh to Willys Peck, October 4, 1932, cdf/1930–39/893.48/637, rg/59, NARA-II.

57. McDonald, *Food Power.*

58. ARC, "Disasters in the American Republics from 1930 to 1941," [ca. 1942], fo/891.08, bx/III-1221, rg/ANRC, NARA-II.

59. Pérez, Jr., *Cuba under the Platt Amendment,* 301–32; Schmitz, *Thank God They're on Our Side,* 73–84.

60. Sumner Welles to DOS, September 3, 1933, corr/848, ent/UD-13, rg/84; and Ernest Bicknell to Ernest Swift, September 5, 1933, fo/FDR-87, bx/II-715, rg/ANRC, both NARA-II.

61. James Fieser to Franklin Roosevelt, September 5, 1933, fo/1933, bx/1, off/124, FDR-PP, FDRPL.

62. Fieser to Roosevelt, September 5, 1933; Cordell Hull to US Embassy, Cuba, September 5, 1933, ent/UD-13, rg/84, NARA-II.

63. John Payne, paraphrased in James McClintock to Ernest Bicknell, September 21, 1933, fo/FDR-87, bx/II-715, rg/ANRC, NARA-II.

64. Sumner Welles to DOS, September 12, 1933, ent/UD-13, rg/84, NARA-II.

65. McClintock to Bicknell, September 21, 1933.

66. Antoinette Hardisty to Ernest Swift, September 23, 1933, FDR-87, bx/II-715, rg/ANRC, NARA-II.

67. Maurice Reddy, paraphrased in James McClintock to William Phillips, September 25, 1933, cdf/1930–39/837.48/27, rg/59, NARA-II.

68. Sumner Welles to DOS, September 30, 1933, cdf/1930–39/837.48/30, rg/59, NARA-II.

69. Henry Roosevelt to Franklin Roosevelt, December 20, 1933, fo/1933, bx/1, off/159 FDR-PP, FDRPL.

70. Schmitz, *Thank God They're on Our Side,* 78–84, 214–18.

71. Reginald Carey to DOS, September 16, 1933, corr/848, ent/UD-855, rg/84, NARA-II.

72. Reginald Carey, confidential memorandum, September 29, 1933, corr/848, ent /UD-855, rg/84, NARA-II.

73. Memorandum of telephone conversation, Josephus Daniels and William Phillips, September 26, 1933, corr/848, ent/UD-41, rg/84, NARA-II.

74. Franklin Roosevelt to Abelardo Rodriguez, September 26, 1933, fo/1933–37, bx/1, off/146, FDR-PP, FDRPL; Josephus Daniels to José Puig Casauranc, September 26, 1933, corr/848, ent/UD-41, rg/84, NARA-II.

75. Josephus Daniels to DOS, September 29, 1933, fo/FDR-88, bx/II-715, rg/ANRC, NARA-II.

76. "The Assistance of Texas," translated article from *El Nacional* (Mexico City), October 6, 1933, corr/848, ent/UD-41, rg/84, NARA-II.

77. Daniels to DOS, September 29, 1933.

78. Cordell Hull to George Dern, June 13 and 21, 1934, bx/2408, fi/400.38, ent/PI-17 /37-H, rg/407, NARA-II.

79. Julius Lay to DOS, June 21, 1934, bx/2408, fi/400.38, ent/PI-17/37-H, rg/407, NARA-II.

80. Julius Lay to DOS, June 13, 1934, fo/FDR-93, bx/II-715, rg/ANRC, NARA-II.

81. ARC, "Disasters in the American Republics from 1930 to 1941."

82. Clyde Buckingham to C. H. Whelden, February 8, 1952, fo/890, bx/IV-1659, rg /ANRC, NARA-II.

83. James McClintock to DOS, August 20, 1934, fo/FDR-97, bx/III-715, rg/ANRC, NARA-II; John Payne to Franklin Roosevelt, January 24, 1934, fo/1933–41, bx/12, off/48h, FDR-PP, FDRPL.

84. Ernest Swift to Cary Grayson, June 5, 1935, fo/FDR-98, bx/III-1222, rg/ANRC, NARA-II.

85. US Legation, China, press release, August 21, 1935, corr/848, ent/UD-10, rg/84, NARA-II; ARC, "Expenditures in China, 1906–1937."

86. ARC, "Disasters in the American Republics from 1930 to 1941"; Pavilack, *Mining for the Nation*, 109–15.

87. Gil, "Disasters as Critical Junctures."

88. Office of Naval Intelligence, "Earthquake in Chile," 1939, fi/H-5-C 22725, bx/928, ent/98, rg/38, NARA-I; Wesley Frost, "Political Complications Resulting from the Earthquake," February 4, 1939, cdf/1930–39/825.48/61, rg/59, NARA-II.

89. Cordell Hull to Franklin Roosevelt, [February 1939], fo/"Jan-June 1939," bx/7, off/20, FDR-PP, FDRPL; Norman Armour, radio address, February 6, 1939, fo/FDR-123.7, bx/III-1228, rg/ANRC, NARA-II.

90. Norman Davis to ARC Chapter Chairmen, February 2, 1939, fo/FDR-123; and George Smith, "Mission of Mercy, American Red Cross," February 20, 1939, fo/FDR-123.08, both bx/III-1228, rg/ANRC, NARA-II.

91. Stone to Adjutant General, February 2, 1939, bx/2403, fi/400.38, ent/PI-17/37-H, rg/407, NARA-II.

92. Memorandum of telephone conversation, Grant Mason and Edward Sparks, January 26, 1939, cdf/1930–39/825.48/38, rg/59, NARA-II.

93. British Embassy, Chile, to Viscount Halifax, January 30, 1939, coll/FO-371/22739 /633, TNA; Franklin Wolfe to Commanding General, Panama Canal Department, February 18, 1939, bx/2403, fi/400.38, ent/PI-17/37-H, rg/407; and Ernest Swift to ARC Chairman, March 1, 1939, fo/800.08, bx/III-680, rg/ANRC, both NARA-II.

94. Division of American Republics, DOS, to Sumner Welles, January 27, 1939, cdf/1930–39/825.48/26, rg/59, NARA-II.

95. Memorandum of conversation, Norman Armour and Edward Sparks, January 31, 1939, cdf/1930–39/825.48/40, rg/59, NARA-II.

96. Division of American Republics to Welles, January 27, 1939; Memorandum of conversation, Edward Sparks and Nelson Armour, February 1, 1939, cdf/1930–39/825.48 /41, rg/59, NARA-II.

97. Edward Sparks to Sumner Welles, February 1, 1939, cdf/1930–39/825.48/41, rg/59, NARA-II.

98. Cecil Lyon, "Earthquake in Chile," [1939], bx/42, corr/848, ent/UD-2554, rg/84, NARA-II.

99. Ernest Swift to Bonabes, Comte de Rougé, March 7, 1939, fo/FDR-123.7, bx/III -1228, rg/ANRC, NARA-II.

100. Sergio Huneeus, radio address, February 6, 1939, fo/FDR-123.7; and Mario Illanes, letter to San Francisco *Call-Bulletin*, February 23, 1939, fo/FDR-123.93, both bx/III-1228, rg/ANRC, NARA-II.

101. Wesley Frost to DOS, February 8, 1939, cdf/1930–39/825.48/66, rg/59, NARA-II.

102. "Los aviones de bombardeo," *El Callao* (Callao, Peru), February 3, 1939; and "Pan-americanism in Action," translated article from *La Crónica* (Lima, Peru), February 7, 1939, both cdf/1930–39/825.48/75, rg/59, NARA-II.

103. Memorandum of telephone conversation, Ernest Swift and George Smith, January 27, 1939, fo/FDR-123, bx/III-1228, rg/ANRC; and Cordell Hull to US Embassy, Chile, February 8, 1939, bx/42, corr/848, ent/UD-2554, rg/84; and Ralph Wooten to Adjutant General, February 9, 1939, bx/2403, fi/400.38, ent/PI-17/37-H, rg/407, all NARA-II.

104. Edward Sparks to Sumner Welles, January 30, 1939, cdf/1930–39/825.48/22 1/2, rg/59, NARA-II; Ernest Swift to Norman Davis, April 15, 1939, fo/FDR-123.08; and "Further Aid for Chile," *Red Cross Courier*, May 1939, fo/FDR-123.07, both bx/III-1228, rg /ANRC, NARA-II.

105. "For the Relief of the Sufferers from the Earthquake in Chile," H.R. 5031, 76th Cong., 1st sess., *CR* (March 14, 1939): 2755; "Relief of the Sufferers from the Earthquake in Chile," 76th Cong., 1st sess., *CR* (July 5, 1939): 8622–25.

CHAPTER 9

1. Sherry, *In the Shadow of War*, 15–187; Sparrow, *Warfare State*; Wertheim, *Tomorrow, the World*.

2. McCleary, *Global Compassion*, 36–59; Zunz, *Philanthropy in America*, 137–68; Mather, "Citizens of Compassion," 35–93.

3. Clyde Buckingham to C. H. Whelden, February 8, 1952, fo/890, bx/IV-1658, rg /ANRC, NARA-II.

4. List compiled from various archival collections included in this book's bibliography.

5. Buckingham to Whelden, February 8, 1952.

6. L. M. Mitchell to Mr. Allen, May 20, 1942, fo/FDR-142, bx/III-1228, rg/ANRC, NARA-II.

7. Sparrow, *Warfare State*.

8. Sparrow, *Warfare State*, 237.

9. Baer, *One Hundred Years of Seapower*, 182.

10. Van Vleck, *Empire of the Air*, 91, 104–5.

11. Johnson, *Sorrows of Empire*, 151; Immerwahr, *How to Hide an Empire*, 18.

12. Sanders, *America's Overseas Garrisons*, 62–285; Vine, *Base Nation*, 17–44.

13. Millett, Maslowski, and Feis, *For the Common Defense*, 440–53.

14. Sherry, *In the Shadow of War*, 130–31.

15. Leffler, *Preponderance of Power*, 25–140; Hogan, *Cross of Iron*, 1–264.

16. Johnson, *Administration of United States Foreign Policy*, 73–88.

17. *Act for the Acquisition of Buildings and Grounds in Foreign Countries* (1946).

18. Leffler, *Preponderance of Power*, 141–81; Hogan, *Cross of Iron*, 23–68.

19. *Office of Foreign Relief and Rehabilitation Operations*.

20. Reinisch, "Internationalism in Relief"; Salvatici, *History of Humanitarianism*, 116–40.

21. Harold Stein to John Steelman, May 8, 1945, bx/1420, off/426, WHCF/HST-PP, HSTPL.

22. Hogan, *Marshall Plan*.

23. Draft of statement by President Truman to ACVAFS Delegates, February 20, 1947, bx/1428, off/426, WHCF/HST-PP, HSTPL.

24. Franklin Roosevelt, Proclamation 2530: "Red Cross War Fund Campaign" (December 12, 1941), and Proclamation 2576: "Red Cross War Fund Campaign" (February 23, 1943), in *Cumulative Supplement to the Code of Federal Regulations*, 281, 326.

25. Dulles, *American Red Cross*, 339–524; Jones, *American Red Cross*, 263–65; Irwin, *Making the World Safe*, 199–208.

26. *Act to Amend the Act of January 5, 1905, to Incorporate the American National Red Cross* (1947); Congressional Research Service, *Congressional Charter of the American National Red Cross*.

27. McCleary, *Global Compassion*, 36–59; Porter, *Benevolent Empire*, 75–100; Mather, "Citizens of Compassion," 35–93; Wieters, *NGO CARE*, 1–42.

28. Davies, Taft, and Warren, *Voluntary War Relief during World War II*.

29. Harry Truman, Executive Order 9723, 1 *Fed. Reg.* 5345 (May 14, 1946).

30. Harry Truman to DOS, May 14, 1946, in Charles Taft to Boards of Directors of Agencies Engaged or Interested in Voluntary Foreign Aid, July 11, 1946, fo/"ACVFA /Administration and Operation," bx/I-171, coll/MSS-42218, LOC.

31. Charter of the Advisory Committee on Voluntary Foreign Aid, effective May 14, 1946, fo/"ACVFA Charter," bx/52, USGF, coll/MC-655, RUSC.

32. Charles Taft, memorandum: "Advisory Committee on Voluntary Foreign Aid," August 28, 1946, fo/"ACVFA/Administration and Operation," bx/I-171, coll/MSS-42218, LOC.

33. Section 117(c), *Economic Cooperation Act of 1948*; Section 535, *Mutual Security Act of 1952*.

34. Section 416, *Agricultural Act of 1949*.

35. "A Report on the Future of the Advisory Committee on Voluntary Foreign Aid," February 1969, bx/I-175, coll/MSS-42218, LOC; McCleary, *Global Compassion*, 60–82; Mather, "Citizens of Compassion," 94–168.

36. ACVFA, Budgetary Statement for Fiscal Year 1950, June 24, 1949, bx/I-171, coll /MSS-42218, LOC; McCleary, *Global Compassion*, 63.

37. Charles Taft, in Charlotte Owen to Agencies of ACVAFS, July 17, 1946, fo/"ACVFA Established by Executive Order," bx/52, USGF, coll/MC-655, RUSC.

CHAPTER 10

1. See, e.g., Lancaster, *Foreign Aid*, 63–65.

2. Harry Truman, Inaugural Address, 81st Cong., 1st Session, *CR* (January 20, 1949): 477–79.

3. Hogan, *Marshall Plan*; Cullather, *Hungry World*, 43–107; Ekbladh, *Great American Mission*, 77–152; Engerman, *Price of Aid*, 21–87.

4. *Federal Disaster Relief Act* (1950); Robert, *Disasters and the American State*, 47–69.

5. List compiled from various archival collections included in this book's bibliography.

6. Maurice Bernbaum to DOS, August 13, 1949, cdf/1945–49/822.48/8-1349, rg/59, NARA-II.

7. Milton Wells to DOS, October 17, 1949, cdf/1945–49/814.48/10-1749, rg/59, NARA-II.

8. Harold Tittman to DOS, May 25, 1950, cdf/1950–54/823.49/5-2550, rg/59, NARA-II; Covert, "Planeamiento urbano post-desastre."

9. Charles Bolen to Chief of Naval Operations, August 18, 1949, bx/794, fi/400.38, ent /NM-3/363-C, rg/407; and Maurice Bernbaum to DOS, August 10, 1949, cdf/1945–49 /822.48/8-1049, rg/59; and H. E. Russell to Thomas Dinsmore, October 20, 1949, fo /DR-573, bx/IV-1686, rg/ANRC; and B. M. Bryan to Russell Brock, May 26, 1950, cdf/1950–54/823.49/5-2950, rg/59, all NARA-II.

10. Maurice Bernbaum to DOS, August 12, 1949, cdf/1945–49/822.48/8-1249; and John Simmons to DOS, September 5, 1949, cdf/1945–49/822.48/9-549, both rg/59, NARA-II; British Embassy, Lima to K. C. Younger, June 9, 1950, coll/FO-371/22739/633, TNA.

11. Harold Tittman to DOS, May 29, 1950, cdf/1950–54/823.49/5-2950, rg/59, NARA-II.

12. Douglass Ballentine to DOS, November 16, 1949, cdf/1945–49/814.48/11-1049, rg/59, NARA-II.

13. Dean Acheson to Herbert Hoover, August 31, 1949, cdf/1945–49/822.48/8-1849, rg/59, NARA-II.

14. Richard Patterson to Medical Director, Office of the Defense Secretary, December 7, 1949, cdf/1945–49/814.48/12-749; and Douglas Ballentine to DOS, December 9, 1949, 45–49cdf814.48/12-949, both rg/59, NARA-II.

15. Maurice Bernbaum to DOS, August 12, 1949; Philip Glick to Edward Miller, June 1, 1950, cdf/1950–54/823.49/6-150, rg/59, NARA-II.

16. Acheson to Hoover, August 31, 1949.

17. Herbert Hoover to President Truman, August 18, 1949, cdf/1945–49/822.48/8-1849, rg/59, NARA-II.

18. Tittman to DOS, May 29, 1950.

19. Richard Patterson to DOS, October 26, 1949, fo/"Correspondence 1949," bx/40, conf/"State Department Correspondence," WHCF/HST-PP, HSTPL; Richard Patterson to President Truman, October 27, 1949, cdf/1945–49/814.48/10-2749; and Richard Patterson, paraphrased in Edward Jamison to Mr. Daniels, November 2, 1949, cdf/1945–49/814.48/11-249, both rg/59, NARA-II.

20. Ballentine to DOS, November 16, 1949.

21. Richard Patterson, Memorandum for President Truman: "Report on Guatemalan Flood," November 1949, fo/"Correspondence 1949," bx/40, conf/"State Department Correspondence," WHCF/HST-PP, HSTPL.

22. Edward Winsall to M. Milson, September 1, 1950, fo/"Disaster-Earthquake -1950–1951," bx/19749, IFRC.

23. Loyd Steere to Department of Agriculture, August 29, 1950, cdf/1950–54/891.49/8 -2950, rg/59, NARA-II.

24. Loy Henderson to DOS, September 30, 1950, cdf/1950–54/890d.49/9-2950; and A. R. Preston to DOS, October 5, 1950, cdf/1950–54/890d.49/10-550, both rg/59, NARA-II.

25. Clare Timberlake to DOS, November 8, 1950, cdf/1950–54/891.49/11-850, rg/59, NARA-II.

26. Henderson to DOS, September 30, 1950.

27. Henderson to DOS, September 30, 1950.

28. Memorandum of conversation, ARC leaders and DOS personnel, October 2, 1950, cdf/1950–54/891.49/10-250, rg/59, NARA-II.

29. Secretary of Defense to DOS, September 15, 1950, cdf/1950–54/891.49/9-1550, rg/59, NARA-II.

30. Memorandum of conversation, Gaile Golub and Frank Collins, October 13, 1950, cdf/1950–54/891.49/10–1350, rg/59; and James Webb to US Embassy, India, October 4, 1950, fi/571, bx/106, ent/UD-2710A, rg/84, both NARA-II.

31. Dean Acheson to US Embassy, India, October 19, 1950, fi/571, bx/106, ent/UD-2710A, rg/84; and Loy Henderson to DOS, October 30, 1950, cdf/1950–54/891.49/10-3050, rg/59; and Dean Acheson to George Marshall, November 13, 1950, cdf/1950–54/891.49/11 -350, rg/59, all NARA-II.

32. Roland Harriman to Dean Acheson, December 13, 1950, cdf/1950–54/891.49/12 -1350, rg/59, NARA-II.

33. Secretary of Defense to DOS, December 1, 1950, cdf/1950–54/891.49/12-150, rg/59, NARA-II.

34. Dean Acheson to US Embassy, India, December 30, 1950, cdf/1950–54/891.49/12 -3050, rg/59, NARA-II.

35. Loy Henderson to DOS, October 11, 1950, fi/571, bx/106, ent/UD-2710A, rg/84, NARA-II.

36. Mortimer Cooke to Zed Crawford, July 17, 1948, fo/898.5, bx/IV-1663, rg/ANRC, NARA-II.

37. Mary Ligthle to Gaile Galub, July 19, 1949, fo/898.5, bx/IV-1663, rg/ANRC, NARA-II.

38. Section 202, *Foreign Economic Assistance Act of 1950*.

39. George Marshall to Dean Rusk, October 9, 1950; and James Nicholson to Frank Cleverley, May 15, 1950; both fo/898.5, bx/IV-1663, rg/ANRC, NARA-II.

40. Gaile Galub to Thomas Dinsmore, October 4, 1950, fo/898.5, bx/IV-1663, rg /ANRC, NARA-II.

41. Miscellaneous correspondence, folders 898.6 and 898.6/08, bx/III-1222, rg/ANRC, NARA-II

42. W. de St. Aubin to James Nicholson, March 14, 1952, fo/898.6, bx/IV-1664, rg /ANRC, NARA-II.

43. Myron Cowen to DOS, February 7, 1953, cdf/1950–54/840.49/2-753, rg/59, NARA-II.

44. Ramone Eaton to James Hagerty, February 6, 1953, bx/474, off/113-H, WHCF /DDE-RP, DDEPL.

45. John Evans, Interim Reports of Committee on Flood Relief, February 11 and 19, 1953, cdf/1950–1954/840.49/2-1153 and cdf/1950–1954/840.49/2-1953, rg/59, NARA-II.

46. Freeman Matthews, Memorandum for President Eisenhower, February 5, 1953, bx/474, off/113-H, WHCF/DDE-RP, DDEPL.

47. John Foster Dulles, Interim Report of Committee on Flood Relief, February 13, 1953, bx/474, off/113-H, WHCF/DDE-RP, DDEPL.

48. Evans, Interim Reports of Committee on Flood Relief. Miscellaneous correspondence, fi/"East Coast Floods . . . Awards to Five Americans," August 1953, coll/HO-286 /12, TNA.

49. Cowen to DOS, February 7, 1953.

50. John Evans, Final Report of the Inter-Agency Working Group on Flood Relief, May 1953, cdf/1950–54/840.49/5-2053, rg/59, NARA-II.

51. Memorandum of conversation, Mr. Speekenbrink, Mr. van Kretschmar, and Mr. Evans, February 25, 1953, bx/3, ent/A1-1537B, rg/59, NARA-II.

52. *Reorganization Plan No. 7 of 1953.*

53. Norbert Anschuets to DOS, October 21, 1953, fo/"Earthquake 1953," bx/3, ent/UD -359, rg/469, NARA-II.

54. Leland Barrows to Harold Stassen, April 15, 1953; and W. F. Stettner to Dennis Fitzgerald, August 24, 1953, both fo/"Earthquake 1953," bx/3, ent/UD-359, rg/469, NARA-II.

55. North American Council Information Service, "Note for the Press on NATO Aid to Greece," [1953]; and Harry Turkel to US Embassy, Greece, August 20, 1953, both fo/"Earthquake 1953," bx/3, ent/UD-359, rg/469, NARA-II.

56. FOA/Washington to USRO/Paris, December 19, 1953, bx/3, ent/UD-359, rg/469, NARA-II; "Resumé of Activity Re. Relief Needs in Greece," August 1953, fo/"Greece–Earthquakes," bx/107, MRC-DF, coll/MC-655, RUSC.

57. Minutes, Meeting of FOA Mission to Greece, August 31, 1953, fo/"Disaster Areas," bx/5, ent/UD-1220; and Memorandum of telephone conversation, N. Carter de Paul and J. I. Eiker, August 28, 1953, bx/3, ent/UD-359, both rg/469, NARA-II.

58. Harry Turkel to US Embassy, Greece, August 20, 1953, fo/"Earthquake 1953," bx/3, ent/UD-359, rg/469, NARA-II.

59. Anschuets to DOS, October 21, 1953.

60. Gerald Morgan to Overton Brooks, September 14, 1953, bx/471, off/113-D-1, WHCF/DDE-RP, DDEPL; Daniel Hopkinson to Dennis Fitzgerald, September 16, 1953, fo/"Earthquake 1953," bx/3, ent/UD-359, rg/469, NARA-II.

61. Turkel to US Embassy, Greece, August 20, 1953.

62. *Agricultural Trade and Development Assistance Act of 1954.*

63. Section 416, *Agricultural Act of 1949.*

64. For additional details, see Irwin, "Raging Rivers and Propaganda Weevils."

65. Miscellaneous correspondence, fo/1-1-54, bx/245, fi/370.1, ent/NM-3/363-E, rg/407, NARA-II; "Danube Flood Forces Evacuation of 24,000 from Linz," *NYT*, July 12, 1954; "Homeless Return as Danube Falls," *NYT*, July 13, 1954.

66. ACVAFS, "Report on July 28 Meeting . . . on Austrian Flood," July 29, 1954, fo /DR-620, bx/IV-1709, rg/ANRC, NARA-II.

67. John Foster Dulles to Certain American Diplomatic Missions, July 16, 1954, cdf/1950–54/840.49/7-1654, rg/59, NARA-II.

68. Selma Freedman to Carl McCardle, July 15, 1954, cdf/1950–54/840.49/7-1554, rg/59, NARA.

69. ARC, "Confidential Summary of Danube Project," [August 1954], fo/DR-620.08, bx/IV-1710, rg/ANRC, NARA-II.

70. Dwight Eisenhower, Statement, July 29, 1954, bx/476, off/113-K, WHCF/DDE-RP, DDEPL.

71. Walter Walmsley to DOS, July 17, 1954, cdf/1950–54/840.49/7-1754, rg/59, NARA-II.

72. John Foster Dulles to HICOG Bonn, HICOG Berlin, and US Consulate, Geneva, August 18, 1954, cdf/1950–54/862b.49/8-1754; and HICOG Bonn to DOS, August 19, 1954, cdf/1950–54/862b.49/8-1954, both rg/59, NARA-II.

73. Memoranda of telephone conversations, Dennis Fitzgerald and various US government officials, July 21–29, 1954, fo/"Telephone Conversations (4)," bx/21, coll /DAF, DDEPL.

74. Henry Parkman to DOS, August 5, 1954, cdf/1950–54/862b.49/8-554, rg/59, NARA-II.

75. Christian Ravndal to DOS, August 10, 1954, cdf/1950–54/864.49/8-1054, rg/59; and Czechoslovak Minister of Foreign Affairs to US Embassy, Czechoslovakia, August 16, 1954, fo/571, bx/28, corr/848, ent/UD-2378A, rg/84; and John Foster Dulles to HICOG Bonn and HICOG Berlin, August 12, 1954, cdf/1950–54/862b.49/8-1254, rg/59, all NARA-II.

76. Eleanor Dulles to Mr. Lewis, October 7, 1954, cdf/1950–54/862a.49/10-754, rg/59, NARA-II; Dwight Eisenhower, Statement, October 30, 1954, bx/476, off/113-K, WHCF /DDE-RP, DDEPL.

77. LRCS, Press Communiqué, February 25, 1955, fo/DR-620, bx/IV-1709, rg/ANRC, NARA-II; LRCS, Press Communiqué, April 1, 1955, fo/"General 3," bx/A0967, IFRC.

78. Arthur Bryan to Murray Snyder, February 14, 1955, bx/476, off/113-K, WHCF /DDE-RP, DDEPL.

79. Henry Dunning, "Final Report on Distribution of American Danube Flood Relief," [1955], fo/"General 3," bx/A0967, IFRC.

80. John Foster Dulles, "Final Report on the President's Flood Relief Program," [1955], fo/"Danube Flood Relief Program," bx/1, ent/A1-1175D, rg/59, NARA-II.

81. Herbert Hoover Jr. to Dwight Eisenhower, February 3, 1955, bx/476, off/113-K, WHCF/DDE-RP, DDEPL.

82. McDonald, *Food Power.*

83. Memorandum of conversation, Roy Davis and Henry Hoyt, October 13, 1954, cdf/1950–54/838.49/10-1354; and Stanley Baranson to Eugene Clay, October 24, 1954, cdf/1950–54/838.49/10-2454, both rg/59, NARA-II.

84. USOM personnel to Stanley Baranson, January 4, 1955, fo/"Project 16 Disaster-Reports Special," bx/3, ent/UD-1229, rg/469, NARA-II.

85. Deck logs of *Saipan*, October 1954, ent/118B, rg/24; and Mr. Whitaker to Mr. Jamison, October 27, 1954, cdf/1950–54/838.49/10-2754, rg/59, both NARA-II.

86. Commander, Caribbean Sea Frontier, to DOS, October 17, 1954, cdf/1950–54 /838.49/10-1754; and Thurston Morton to Edward Boland, November 19, 1954, cdf /1950–54/838.49/11-1954, both rg/59, NARA-II.

87. Howard Ross, Report on ARC Operations in Caribbean, October–December 1954, fo/"Disaster, Flood, 1954–56," bx/19682, IFRC.

88. John Foster Dulles, Memorandum for President Eisenhower, November 1, 1954, cdf/1950–54/838.49/11-154, rg/59; and Roy Davis to Stanley Baranson, October 19, 1954, fo/"Information on Hurricane Hazel," bx/8, ent/P-262, rg/469, both NARA-II.

89. "Draft Agreement for Distribution of Emergency Aid Supplies," [1954], fo/"Project 16 Disaster–Reports Special," bx/3, ent/UD-1229; and William Shaw, "Report of Distribution of Food Supplies Shipped by FOA/W," March 24, 1955, fo/"Disaster Relief for Haiti," bx/7, ent/P-260, both rg/469, NARA-II.

90. Roy Davis to DOS, October 13, 1954, cdf/1950–54/838.49/10-1354, rg/59, NARA-II.

91. Mr. Newbegin to Mr. Holland and Mr. Cale, December 14, 1954, cdf/1950–54 /838.49/12-1454, rg/59, NARA-II.

92. Stanley Baranson, Statement regarding Fondation Magloire, October 1954, cdf/1950–54/838.49/10-2854, rg/59, NARA-II.

93. Milton Barall to DOS, October 26, 1954, cdf/1950–54/838.49/10-2654, rg/59, NARA-II.

94. Phillip Douglas to Sherman Adams, November 16, 1954, cdf/1950–54/838.49 /11-1654, rg/59, NARA-II.

95. William Shaw, Statement concerning Relations between FOA and Fondation Magloire, October 1954, cdf/1950–54/838.49/10-2854, rg/59, NARA-II.

96. Milton Barall to DOS, October 26 and 28, 1954, cdf/1950–54/838.49/10-2654 and cdf/1950–54/838.49/10-2854, rg/59, NARA-II.

97. Mr. Hoyt to Mr. Newbegin, December 14, 1954, cdf/1950–54/838.49/11-1454, rg/59, NARA-II.

98. Barall to DOS, October 28, 1954.

99. Barall to DOS, October 28, 1954. Mr. Newbegin to Mr. Holland and Mr. Cale, December 14, 1954, cdf/1950–54/838.49/12-1454, rg/59, NARA-II.

100. Shaw, "Report of Distribution of Food Supplies Shipped by FOA/W."

101. List compiled from various archival collections included in this book's bibliography.

CHAPTER 11

1. Dwight Eisenhower, Executive Order 10610, 10 *Fed. Reg.* 3179 (May 11, 1955).

2. Dwight Eisenhower to John Foster Dulles, April 15, 1955, fo/"International Cooperation Administration, Organization," bx/50, USGF, coll/MC-655, RUSC.

3. ICA, "Fact Sheet," [1955], fo/"International Cooperation Administration, Organization," bx/50, USGF, coll/MC-655, RUSC.

4. Joseph Stokes, "The International Cooperation Administration," *World Affairs* 119, no. 2 (1956): 35–37.

5. Operations Coordinating Board, "Foreign Disaster Relief Operations," August 22, 1958, fo/890.011, bx/IV-1660, rg/ANRC, NARA-II.

6. List compiled from various archival collections included in this book's bibliography.

7. Section 401, *Mutual Security Act of 1954*; Section 8, *Mutual Security Act of 1956*; Section 301, *Mutual Security Act of 1958*.

8. Roland Harriman to Oveta Hobby, January 19, 1955, fo/050, bx/IV-166, rg/ANRC, NARA-II.

9. ICA Advisory Committee on Voluntary Foreign Aid, Staff Logs, 1956–1958, fo/"Staff Logs," bx/I-176, coll/MSS-42218, LOC; Miscellaneous correspondence, bx/107, MRC-DF, coll/MC-655, RUSC.

10. Charles Taft, Statement to the Senate Foreign Relations Committee, February 24, 1955, fo/"Miscellany," bx/I-175, coll/MSS-42218, LOC; McCleary, *Global Compassion*, 60–82; Mather, "Citizens of Compassion," 169–242.

11. Dennis [US Army representative], in Minutes, Meeting on Caribbean Disaster Coordination, March 17, 1955, fo/890.011, bx/IV-1660, rg/ANRC, NARA-II.

12. John Hollister to John Foster Dulles, August 29, 1956, fo/"ICA Matters 1954–56," bx/5, Subject Series, John Foster Dulles Papers, DDEPL.

13. Operations Coordinating Board, "Foreign Disaster Relief Operations," September 14, 1956, rev. August 22, 1958; and Elmer Staats, memorandum to Operations Coordinating Board, July 23, 1958, both fo/890.011, bx/IV-1660, rg/ANRC, NARA-II.

14. Headquarters Far East Command, "Disaster Relief in Japan," March 16, 1955; and Headquarters Caribbean Command, "Disaster Relief in Latin America," September 1, 1955; and Headquarters US European Command, "Plans Operations and Training: Civil Disasters," April 25, 1956, all fo/890.011, bx/IV-1660, rg/ANRC, NARA-II.

15. John Foster Dulles, DOS Instruction CA-11026, June 16, 1958, bx/58, ent/UD-180, rg /469, NARA-II.

16. Joint State–Defense–ICA Circular on Emergency Relief Assistance in Foreign Disasters, [1958], fo/890.011, bx/IV-1660, rg/ANRC, NARA-II.

17. Christian Herter to all Chiefs of Mission and USOM [1959], bx/11, ent/P-368, rg/286, NARA-II.

18. Samuel Krakow to James Nicholson, March 13, 1958, fo/890.011, bx/IV-1660, rg /ANRC, NARA-II.

19. Robert Lewis to Jack Henry, September 18, 1959, fo/890, bx/IV-1659, rg/ANRC, NARA-II.

20. Selden Chapin to DOS, August 14, 1956, cdf/1955–59/888.49/8-1456, rg/59, NARA-II.

21. Roland Jacobs to DOS, October 7, 1959, fo/"Madagascar–Disaster," bx/2, ent/P-211, rg/469, NARA-II.

22. Francis White to DOS, September 19, 1955, cdf/1955–59/812.49/9-1955, rg/59, NARA-II.

23. Florence Kirlin to Lyndon Johnson, September 27, 1955, cdf/1955–59/812.49/9-2255, rg/59, NARA-II.

24. Louis Blanchard, memorandum for the file, September 29, 1955, cdf/1955–59 /812.49/9-2955, rg/59, NARA-II.

25. Memorandum of conversation, Captain Dissette and Cecil Lyon, September 29, 1955, cdf/1955–59/812.49/9-2955, rg/59, NARA-II.

26. John Foster Dulles to US Embassy, Mexico, September 30, 1955, cdf/1955–59 /812.49/9-3055, rg/59, NARA-II.

27. Memorandum of conversation, Gordon Gray, Cecil B. Lyon, and Captain Dissette, September 20, 1955, cdf/1955–59/812.49/9-2055; and Memorandum of conversation, Charles Thomas and Henry F. Holland, September 30, 1955, cdf/1955–59/812.49/9-3055, both rg/59, NARA-II.

28. Manuel Tello to Herbert Hoover Jr., October 24, 1955, cdf/1955–59/812.49/10-2455, rg/59, NARA-II.

29. "Summary of United States Relief to Mexican Disaster Area," [1955], bx/476, off/113-P, WHCF/DDE-RP, DDEPL; Cecil Lyon to Mr. Linebaugh, October 12, 1955, cdf/1955–59/812.49/10-1255, rg/59; and Deck logs of *Saipan*, *Siboney*, and *Oglethorpe*, September and October 1955, ent/118B, rg/24, both NARA-II.

30. Harris Austin, "ARC Relief Operation, Tampico, Mexico," 1955, fo/891.2, bx /IV-1661, rg/ANRC, NARA-II.

31. Herbert Hoover Jr. to A. J. Goodpaster, October 22, 1955, cdf/1955–59/812.49 /10-2255, rg/59, NARA-II.

32. Joseph Schutz to DOS, October 27, 1955, cdf/1955–59/812.49/10-2755, rg/59, NARA-II.

33. Francis White to Ellsworth Bunker, November 4, 1955, cdf/1955–59/812.49/11-455, rg/59, NARA-II.

34. John Foster Dulles, memorandum for President Eisenhower, October 19, 1955, off/113-P, WHCF/DDE-RP, DDEPL.

35. Douglas Dillon, Memorandum for President Eisenhower, August 11, 1959, cdf /1955–59/893.49/8-1159, rg/59, NARA-II.

36. Samuel Krakow to Robert Lewis, August 11, 1959, fo/909, bx/IV-1780, rg/ANRC; and Paul Sturm to DOS, August 24, 1959, cdf/1955–59/893.49/8-2459, rg/59; and Deck logs of *Thetis Bay*, August 1959, ent/118C, rg/24, all NARA-II.

37. Paul Sturm to DOS, September 25, 1959, cdf/1955–1959/893.49/9-2559; and William Macomber to Byron Johnson, October 27, 1959, cdf/1955–59/893.49/10-2759, both rg/59, NARA-II.

38. William Macomber to Charles Bennett, August 12, 1959, cdf/1955–59/893.49/8-1259, rg/59; and John Wilson to Mr. Nearman et al., September 18, 1959, fo/890, bx/IV-1659, rg /ANRC, both NARA-II.

39. Lewis Smith, "Report on Typhoon Sarah's Visit to Korea," October 16, 1959, fo /"Report on Typhoon Sarah," bx/108, ent/UD-1276, rg/469, NARA-II.

40. Douglas MacArthur to DOS, October 1, 1959, cdf/1955–59/894.49/10-159; and Mr. Parsons to DOS, October 1, 1959, cdf/1955–59/894.49/10-159, both rg/59, NARA-II.

41. Parsons to DOS, October 1, 1959; Christian Herter to Douglas MacArthur, October 1, 1959, fi/"571-Typhoon Vera," bx/81, ent/UD-2828A, rg/84, NARA-II.

42. Joseph Donelan to DOS, October 30, 1959, cdf/1955–59/894.49/10-3059; and Coburn Kidd to DOS, February 23, 1960, cdf/1955–59/894.49/2-2360, both rg/59, NARA-II.

43. Robert Gordon to Ramone Eaton, October 16, 1959, fo/898.6, bx/IV-1664, rg /ANRC; and Joan Kain, Memorandum on Typhoon Vera, September 30, 1959, bx/11, ent/P-368, rg/286, both NARA-II.

44. Douglas MacArthur to Japanese Minister of Foreign Affairs, November 12, 1959; and Report: "Use of Public Law 480 Flour and Grain Stocks for Relief of Nagoya," 1960, both fi/"571-Typhoon Vera," bx/211, ent/UD-2828, rg/84, NARA-II.

45. Raymond Moyer to ICA/Washington, November 11, 1959, fo/"Korea (Disasters)," bx/2, ent/P-211, rg/469, NARA-II.

46. Press Release, "USOM Proposes ROK-US Measures for Prompt Action in Typhoon Area," October 22, 1959, fo/750c, bx/14, ent/p-319, rg/469, NARA-II.

47. Raymond Moyer to ICA/Washington, December 7, 1959, fo/"Disasters–Typhoon Sarah," bx/108, ent/UD-1276, rg/469, NARA-II.

48. H. J. McCool to H. E. Timmis, February 25, 1960, fo/750c, bx/14, ent/p-319, rg/469; and Records of the Combined Economic Board, U.S.–ROK, October 1959 to May 1960, fo/"Typhoon Sarah Damage," bx/6, ent/P-590, rg/286, both NARA-II.

49. Everett Drumright to DOS, July 8, 1959, cdf/1955–1959/893.49/7-859, rg/59, NARA-II.

50. W. T. M. Beale to DOS, June 29, 1959, cdf/1955–1959/893.49/7-159, rg/59, NARA-II.

51. Douglas Dillon to DOS, July 1, 1959, cdf/1955–1959/893.49/7-159, rg/59, NARA-II.

52. William Macomber to Hubert Humphrey, August 28, 1959, cdf/1955–1959/893.49 /8-2159, rg/59, NARA-II.

53. Henry Dunning to Alfred Gruenther, March 16, 1960, fo/DR-922, bx/IV-1782, rg /ANRC; and David Nes to DOS, March 23, 1960, cdf/1960–63/871.49/3-2360, rg/59, both NARA-II.

54. Segalla, *Empire and Catastrophe*, 108–64; Williford, "Seismic Politics."

55. Charlie Yost to DOS, March 1, 1960, cdf/1960–63/871.49/3-160, rg/59, NARA-II.

56. Livingston Merchant to US Embassy, Morocco, March 1, 1960, cdf/1960–63 /871.49/3-160; and Christian Herter to US Embassy, Morocco, March 10, 1960, cdf /1960–63/871.49/3-1060, both rg/59, NARA-II.

57. Joan Kain to American Friends Service Committee et al., March 8, 1960, fo /DR-922, bx/IV-1782, rg/ANRC, NARA-II.

58. Christian Herter to US Embassy, Morocco, March 3, 1960, cdf/1960–63/871.49 /3-360, rg/59, NARA-II.

59. ARC, Foreign Disaster Relief Bulletin, January 23, 1958, fo/897.3, bx/IV-1663, rg /ANRC, NARA-II.

60. Joan Kain, Memorandum of a Conversation . . . concerning Algerian Floods, February 4, 1958, fo/"Algeria (Disasters)," bx/1, ent/P-211, rg/469; and Merritt Cootes to DOS, March 7, 1958, cdf/1955–59/851s.49/3-758, rg/59, both NARA-II.

61. Herter to US Embassy, Morocco, March 3, 1960.

62. Charlie Yost to DOS, March 5, 1960, cdf/1960–63/871.49/3-560, rg/59, NARA-II.

63. Nes to DOS, March 23, 1960.

64. Charlie Yost to DOS, April 15, 1960, cdf/1960–63/871.49/4-1560, rg/59, NARA-II.

65. Avaria and Segovia, *Identidad terremoteada.*

66. Walter Howe to DOS, May 21 and June 1, 1960, cdf/1960–63/825.49/5-2160 and cdf/1960–63/825.49/6-160; and Henry Villard to DOS, June 4, 1960,

cdf/1960–63/825.49/6-460, all rg/59, NARA-II. Thanks to Magdalena Gil for the modern estimates.

67. Walter Howe to DOS, May 28, 1960, cdf/1960–63/825.49/5-2860, rg/59, NARA-II.

68. Rabe, *Eisenhower and Latin America*, 117–52.

69. DOS, Press Release, May 26, 1960, off/113-T, WHCF/DDE-RP, DDEPL.

70. John Calhoun to A. J. Goodpaster, May 26, 1960; and DOS, "Relief and Reconstruction Assistance in Chile," June 9, 1960, both off/113-T, WHCF/DDE-RP, DDEPL; "After Action Report, Joint Task Force, Chilean Disaster Relief," in William Krieg to DOS, September 15, 1960, cdf/1960–63/825.49/9-1560, rg/59, NARA-II.

71. Dwight Eisenhower to Mike Mansfield, June 1, 1960; and Bureau of the Budget to Robert Merriam, June 9, 1960, both off/113-T, WHCF/DDE-RP, DDEPL.

72. Memoranda of telephone conversations, Dennis Fitzgerald and various US government officials, May 27–June 13, 1960, fo/"Telephone Conversations," bx/28 and bx/29, coll/DAF, DDEPL.

73. Christian Herter to Michael Mansfield, May 27, 1960, cdf/1960–63/825.49/5-2660, rg/59, NARA-II.

74. Dwight Eisenhower, Press statement, May 27, 1960, fo/"Chile–1960 Earthquake –Correspondence," bx/1, ent/UD-WW-862, rg/286, NARA-II.

75. Minutes, Meeting on Chilean Relief, June 3, 1960, off/113-T, WHCF/DDE-RP, DDEPL; George Elsey to Henry Dunning and Raymond Schaeffer, June 29, 1960, fo /DR-924.02, bx/IV-1782, rg/ANRC, NARA-II.

76. ARC, Press Release, June 5, 1960, fo/"Disaster, Earthquake," bx/19750, IFRC; ARC, "Progress Report on Chilean Relief," June 21, 1960, fo/DR-924.08, bx/IV-1783, rg/ANRC, NARA-II.

77. Robert Merriam to A. J. Goodpaster, June 23, 1960, fo/ "Chile (3)," bx/2, International Series, OSS/WHOF; and Alfred Gruenther to Dwight Eisenhower, June 24, 1960, off/113-T, WHCF/DDE-RP, both DDEPL.

78. Eduardo Frei, "Deeply Appreciated Diplomatic Action," translated article from *El Mercurio* (Valparaíso, Chile), June 22, 1960, off/113-T, WHCF/DDE-RP, DDEPL.

79. Walter Howe to DOS, May 30, 1960, cdf/1960–63/825.49/5-3060, rg/59, NARA-II. DOS, "Relief and Reconstruction Assistance in Chile."

80. Walter Dustmann to DOS, June 2, 1960, cdf/1960–63/825.49/6-260, rg/59, NARA-II.

81. DOS, "Relief and Reconstruction Assistance in Chile."

82. Howe to DOS, May 28, 1960.

83. Roy Rubottom to Douglas Dillon, June 16, 1960, fo/"Chile Disaster 1960," bx/1, ent /P-188, rg/59, NARA-II.

84. Roy Rubottom to Livingston Merchant, June 1, 1960, fo/"Chile Earthquake," bx/7, ent/A1-3137, rg/59, NARA-II.

85. Memorandum of conversation, Jorge Schneider, Horacio Suárez, et al., June 8, 1960, cdf/1960–63/825.49/6-860; and Minutes, Meeting of Chilean Reconstruction Committee, cdf/1960–63/825.49/6-1060, both rg/59, NARA-II.

86. White House, Draft Press Release: "Reconstruction Aid to Chile," June 23, 1960, fo/"Chile (3)," bx/2, International Series, OSS/WHOF, DDEPL.

87. Memorandum of conversation, Walter Müller, Jorge Schneider, et al., June 23, 1960, cdf/1960–63/825.49/6-2360, rg/59, NARA-II.

88. John Dreier, Statement to Organization of American States, July 8, 1860, fo/"COAS Special Committee, Chilean Disaster," bx/22, ent/p-163, rg/59, NARA-II.

89. Memorandum of conversation, Walter Müller, Jorge Schneider, and Roy Rubottom, June 28, 1960, cdf/1960–63/825.49/6-2860, rg/59, NARA-II.

90. Memorandum of conversation, Douglas Dillon, Walter Müller, et al., June 29, 1960, cdf/1960–63/825.49/6-2960, rg/59, NARA-II.

91. Memorandum of telephone conversation, Dennis Fitzgerald and Mr. Holmer, June 21, 1960, fo/"Telephone Conversations, June," bx/29, coll/DAF, DDEPL.

92. Roy Rubottom to Douglas Dillon, June 18, 1960, cdf/1960–63/825.49/6-1860, rg/59, NARA-II.

93. *An Act to Provide for Assistance in the Development of Latin America and in the Reconstruction of Chile* (1960).

94. Memorandum of conversation, Walter Howe, Eduardo Figueroa, et al., November 11, 1960, cdf/1960–63/825.49/11-1460, rg/59, NARA-II.

95. Walter Howe to DOS, November 25, 1960, fo/"Chile–Eco. & Non-Mil.," bx/1, ent/P-188, rg/59, NARA-II.

96. Newson Ryan to DOS, December 21, 1960, fo/"Chile Reconstruction," bx/1, ent /P-188, rg/59, NARA-II.

97. John Hollister, paraphrased in memorandum of telephone conversation, Dennis Fitzgerald and Mr. Lathram, May 8, 1957, fo/"Reading File 1/57–6/57," bx/36, coll/DAF, DDEPL.

98. George Elsey to Henry Dunning, January 9, 1959, fo/890.4, bx/IV-1660, rg/ANRC, NARA-II.

CHAPTER 12

1. Eric Kocher to Dimitri Dejanikus, December 2, 1963, fi/571-Skopje, bx/94, ent /UD-3350, rg/84, NARA-II.

2. Alexander Johnpell to DOS, August 31, 1963, snf/1964–66/SOC-10-Yugoslavia, rg/59, NARA-II.

3. Taffet, *Foreign Aid as Foreign Policy*; Cullather, *Hungry World*, 108–204; Field, *From Development to Dictatorship*; Citino, *Envisioning the Arab Future*, 212–50; Engerman, *Price of Aid*, 191–272.

4. Walt Rostow, "The Strategy of Foreign Aid, 1961," [ca. 1960–61], fo/"Foreign Aid: General, December 1960–February 1961," bx/297, NSF-S/JFK-PP, JFKPL.

5. John Kennedy, Executive Order 10915, 26 *Fed. Reg.* 781 (January 26, 1961).

6. John Kennedy, "Special Message on Foreign Aid," March 22, 1961, fo/"Foreign Aid: General, 1961: March–April," bx/297, NSF-S/JFK-PP, JFKPL.

7. List compiled from various archival collections included in this book's bibliography.

8. John Kennedy to Congress, "Message from the President . . . requesting the appropriation of $600 million," March 14, 1961, fo/"COAS–Special Committee, Chilean Disaster," bx/22 ent/p-163, rg/59, NARA-II; *Inter-American Social and Economic Cooperation Program Appropriation* (1961).

9. Memorandum of Discussion, Milton Barall with Ernesto Pinto, June 23, 1961, cdf/1960–63/825.49/6-2361, rg/59, NARA-II.

10. USAID, "Deficiencies in Administration of the Earthquake . . . Program for Chile," June 1964, bx/21, fi/CO-49,WHCF/LBJ-PP, LBJPL.

11. "Red Cross Gift to Assist Chile," *South Pacific Mail* (Santiago, Chile), December 30, 1960, in fo/DR-924.65, bx/IV-1784, rg/ANRC, NARA-II.

12. Edgar Zimmerman to Robert Edson and Robert Pierpont, March 20, 1961; and Robert Edson to Robert Gordon, April 12, 1961; and Charles Cole to Alfred Gruenther, December 21, 1961, all fo/DR-924.65, bx/IV-1784, rg/ANRC, NARA-II.

13. Clyde Buckingham to Robert Shea, May 29, 1962, fo/900.02, bx/IV-1793, rg/ANRC, NARA-II

14. Guillermo Gonzalez Mansilla to Alfred Gruenther, November 17, 1962; and Winifred Manns to Edson, June 3, 1963, both fo/DR-924.65, bx/IV-1784, rg/ANRC, NARA-II.

15. *Act for International Development of 1961.*

16. John Kennedy, Executive Order 10973, 26 *Fed. Reg.* 10469 (November 7, 1961).

17. Samuel Krakow to John Wilson, November 20, 1961, fo/890.407, bx/IV-1661, rg /ANRC, NARA-II.

18. Chapter 5, *Act for International Development of 1961.*

19. Section 216, *Act for International Development of 1961.*

20. *Agricultural Trade and Development Assistance Act of 1954.*

21. John Foster Dulles, DOS Instruction CA-11026, June 16, 1958, bx/58, ent/UD-180, rg/469, NARA-II; FDRC, "U.S. Foreign Disaster Relief," September 23, 1975, fo/"1975 AID/FDRC Disaster Conference," bx/101, MRC-DF, coll/MC-655, RUSC.

22. DOS Delegation of Authority 104, November 3, 1961, in M.S.-1563.1, "Agency for International Development Manual: Foreign Disaster Emergency Relief," September 14, 1964, fo/890.011, bx/IV-1660, rg/ANRC, NARA-II.

23. Newman Williams to DOS, November 27, 1961, fo/6.7, bx/8, ent/UD-07D-81, rg/59, NARA-II.

24. Andrew Lynch to DOS, November 28 and December 7, 1961, fo/6.7, bx/8, ent /UD-07D-81, rg/59, NARA-II; George McGovern, Memorandum for President Kennedy, November 28, 1961, bx/89, fi/DI-5-CO, WHCF/JFK-PP, JFKPL.

25. Andrew Lynch to DOS, December 14, 1961, fo/6.7, bx/8, ent/UD-07D-81, rg/59, NARA-II.

26. Sam Krakow to Frank Coffin, December 22, 1961, fo/DR-952, bx/IV-1787, rg /ANRC, NARA-II.

27. Robert Pierpont to Fowler Hamilton, January 24, 1962, fo/DR-952, bx/IV-1787, rg /ANRC, NARA-II.

28. Krakow to Wilson, November 20, 1961.

29. Samuel Krakow to Ramone Eaton, December 7, 1961, fo/DR-952, bx/IV-1787; and Samuel Krakow to Ramone Eaton, March 19, 1962, fo/900.02, bx/IV-1794, both rg/ANRC, NARA-II.

30. Ramone Eaton to Alfred Gruenther, April 18, 1962, fo/897.1, bx/IV-1663, rg/ANRC, NARA-II.

31. List compiled from various archival collections included in this book's bibliography.

32. Julius Holmes to DOS, September 3, 1962, cdf/1960–63/888.49/9-362, rg/59; and USAID Task Force on Earthquake Rehabilitation, "Report on Iran's Earthquake Disaster," September 12, 1962, fo/"Earthquake of September 1962," bx/1, ent/P-491, rg/286, both NARA-II.

33. Ansari, "Myth of the White Revolution"; Alvandi, *Age of Aryamehr.*

34. Offiler, *US Foreign Policy and the Modernization of Iran*, 13–48.

35. Robert Komer, "Our Policy on Iran," October 18, 1962, fo/"Iran General 1962," bx/116a, NSF-C/JFK-PP, JFKPL.

36. Lyndon Johnson, paraphrased in Frederick Reinhardt to DOS, September 5, 1962, cdf/1960–63/888.49/9-562, rg/59, NARA-II.

37. Dean Rusk to Lyndon Johnson, September 5, 1962, cdf/1960–63/888.49/9-562, rg/59, NARA-II.

38. Holmes to DOS, September 3, 1962.

39. Percy Liles and Vernon Scott, "Environmental Sanitation in Iran's Earthquake," fo/"Earthquake Disaster Sep. 1, 1962," bx/4, ent/P-512, rg/286, NARA-II.

40. USAID Task Force on Earthquake Rehabilitation, "Report on Iran's Earthquake Disaster"; Miscellaneous reports, 1962, fo/"Earthquake Disaster Sep. 1, 1962," bx/4, ent /P-512, rg/286, NARA-II.

41. Richard Reuter, Memorandum for President Kennedy, September 4, 1962, bx/89, fi/DI-3-CO, WHCF/JFK-PP, JFKPL.

42. Robert Shea to Robert Everetts, September 26, 1962, fo/DR-962, bx/IV-1790, rg /ANRC, NARA-II.

43. Samuel Krakow, Memorandum for the Record, September 4, 1962, DR-962, bx /IV-1790, rg/ANRC, NARA-II; A. W. Mustafa to A. H. Boerma, October 10, 1962, Project 802, bx/16WFP187, FP-5/3, rg/16, FAO.

44. Commander, US European Command, to DOS, September 3, 1962, cdf/1960–63 /888.49/9-362, rg/59; and DOS, Press Release, September 6, 1962, fo/DR-962, bx/IV-1790, rg/ANRC, both NARA-II.

45. Julius Holmes to DOS, September 5, 1962, cdf/1960–1963/888.49/9-562, rg/59, NARA-II.

46. Mohammad Reza Pahlavi, paraphrased in Julius Holmes to DOS, October 1, 1962, cdf/1960–63/888.49/10-162, rg/59, NARA-II.

47. Julius Holmes to DOS, October 1 and 4, 1962, cdf/1960–63/888.49/10-162 and cdf/1960–63/888.49/10-462, rg/59, NARA-II.

48. USAID Mission to Iran, "Proposed Plan for Village Reconstruction and Development in U.S. Designated Earthquake Areas," December 5, 1962, fo/"Earthquake of 1962," bx/1, ent/P-491; and Stephen Tripp, "Iran: Characteristics of the Disaster," [ca. 1964], fo/"Iran Earthquake Aug 1962," bx/1, ent/UD-WW-862, both rg/286, NARA-II.

49. Robert Galloway, "Self-Help Construction in U.S. Sponsored Earthquake Villages," October 27, 1962, fo/"Earthquake of September 1962," bx/1, ent/P-491, rg/286, NARA-II; Minutes, Meeting of USAID, CARE, FAO, and WFP representatives, February 13, 1963, Project 802, bx/16WFP187, FP-5/3, rg/16, FAO.

50. Herman Hendricks and Roger Sandage, Report on Trip to the Earthquake Area, November 21, 1962, bx/4, ent/P-512, rg/286, NARA-II.

51. USAID Mission to Iran, "Reconstruction and Rehabilitation U.S. Designated Earth-quake Areas," October 1962, fo/"Earthquake of September 1962," bx/1, ent/P-491, rg/286, NARA-II.

52. William Brubeck to McGeorge Bundy, January 21, 1963, fo/"Iran General 1962," bx/116a, NSF-C/JFK-PP, JFKPL.

53. Stephen Tripp, "Disaster Information Memo: Iran" [ca. 1964], fo/"Iran Earthquake Aug 1962," bx/1, ent/UD-WW-862, rg/286, NARA-II.

54. Eric Kocher to DOS, August 12, 1963, snf/1964–66/SOC-10-Yugoslavia, rg/59; and Stephen Tripp, "Disaster Information Memo: Yugoslavia–1963 Skopje Earthquake," [1966], fo/DR-909.08, bx/IV-1791, rg/ANRC, both NARA-II.

55. Močnik, "United States–Yugoslav Relations, 1961–80"; Rakove, *Kennedy, Johnson, and the Non-Aligned World*, 62–93.

56. Eric Kocher to DOS, July 26 and August 1, 1963, snf/1964–66/SOC-10-Yugoslavia, rg/59, NARA-II.

57. Richard Davis to DOS, August 13 and 19, 1963, snf/1964–66/SOC-10-Yugoslavia, rg/59, NARA-II.

58. Dean Rusk to US Embassy, France, July 31, 1963; and Robert Cleveland to DOS, August 6, 1963, both snf/1964–66/SOC-10-Yugoslavia, rg/59, NARA-II.

59. Richard Davis to DOS, August 13, 1963, snf/1964–66/SOC-10-Yugoslavia, rg/59, NARA-II; Kocher to Dejanikus, December 2, 1963; Tripp, "Disaster Information Memo: Yugoslavia–1963 Skopje Earthquake."

60. Eric Kocher to DOS, July 29 and August 10, 1963, fi/571-Skopje, bx/94, ent/UD -3350, rg/84, NARA-II.

61. Eric Kocher to DOS, July 31, 1963, snf/1964–66/SOC-10-Yugoslavia, rg/59, NARA-II.

62. George Ball to Robert McNamara, August 23, 1963, snf/1964–66/SOC-10 -Yugoslavia, rg/59, NARA-II.

63. Foreign Minister of Yugoslavia, quoted in Eric Kocher to DOS, July 30, 1963; and Memorandum of conversation, Veljko Mićunović, Richard Davis, and Nicholas Andrews, August 15, 1963, both snf/1964–66/SOC-10-Yugoslavia, rg/59, NARA-II.

64. Kocher to DOS, August 12, 1963; "Relief Aid to Tito," *Washington (DC) Daily News*, August 13, 1963.

65. R. G. Cleveland to DOS, August 17, 1963, snf/1964–66/SOC-10-Yugoslavia, rg/59, NARA-II.

66. Section 620(f), *Foreign Assistance Act of 1962*.

67. Richard Davis to DOS, August 13, 1963, snf/1964–66/SOC-10-Yugoslavia, rg/59, NARA-II; George Ball to US Embassy, Yugoslavia, August 24, 1963; and George Ball, Memorandum for President Kennedy, October 17, 1963, both fo/"Yugoslavia, General, 1963," bx/211, NSF-C/JFK-PP, JFKPL.

68. Ball to US Embassy, Yugoslavia, August 24, 1963.

69. Eric Kocher to DOS, August 28, 1963; and Dean Rusk to US Embassy, Yugoslavia, October 19, 1963, both fi/571-Skopje, bx/94, ent/UD-3350, rg/84, NARA-II; Shirley Robertson, Report on Emergency Feeding Programme for Children of Skopje, November 25, 1963, fo/DR-909, bx/IV-1791, rg/ANRC, NARA-II.

70. Eric Kocher to DOS, October 11, 1963, fi/571-Skopje, bx/94, ent/UD-3350, rg/84, NARA-II.

71. Robert Cleveland to DOS, October 1, 1963, snf/1964–66/SOC-10-Yugoslavia, rg/59, NARA-II.

72. Eric Kocher to DOS, September 13, 1963, snf/1964–66/SOC-10-Yugoslavia, rg/59, NARA-II.

73. Wolfe, "'Revolution Is a Force.'"

74. LRCS Relief Bureau, Circular 204, October 10, 1963, fo/893.1, bx/IV-1662; and Stephen Tripp, "Disaster Information Memo: Hurricane Flora–1963," [August 1966], fo /DR-912, bx/IV-1792, both rg/ANRC, NARA-II; Fidel Castro to Food and Agricultural Organization, October 12, 1963, Project 811, bx/16WFP190, FP-5/3, rg/16, FAO.

75. Rabe, *Most Dangerous Area in the World*, 56–98; Arthus, "Challenge of Democratizing the Caribbean."

76. "Haiti: Proposed Short Range Plan of Action, August 1 to October 30, 1963," July 1963, fo/"Haiti 6/63–11/63," bx/394, RADF, NSF/JFK-PP, JFKPL.

77. CIA Office of Current Intelligence, "The Current Situation in Haiti," November 5, 1963, Document CIA-RDP79T00429A001200050009-3, General CIA Records, FOIA-ERR; "Haiti: Proposed Short Range Plan of Action, November 1, 1963 to April 30, 1964," [November 1963], fo/"Haiti 6/63–11/63," bx/394, RADF, NSF/JFK-PP, JFKPL.

78. George Ball to US Embassy, Haiti, October 4, 1963, snf/1964–66/SOC-10-Haiti, rg/59, NARA-II.

79. Edward Curtis to DOS, October 26, 1963, snf/1964–66/SOC-10-Haiti, rg/59, NARA-II.

80. Edward Curtis to DOS, October 10, 1963, snf/1964–66/SOC-10-Haiti, rg/59, NARA-II.

81. Edward Curtis to DOS, October 9, 1963, snf/1964–66/SOC-10-Haiti, rg/59, NARA-II.

82. Deck logs of *Lake Champlain* and *Thetis Bay*, October 1963, ent/118D, rg/24, NARA-II; Curtis to DOS, October 9, 1963.

83. William Dabney, Memorandum for the Record, October 18, 1963, fo/893.3, bx /IV-1660, rg/ANRC, NARA-II.

84. Dean Rusk to US Embassy, Haiti, October 10, 1963, snf/1964–66/SOC-10-Haiti, rg/59, NARA-II.

85. Edward Curtis to DOS, October 22 and 25, 1963, snf/1964–66/SOC-10-Haiti, rg/59, NARA-II; ACVFA, Memorandum on Haiti, October 10, 1963; and Voluntary Foreign Aid Service, "U.S. Navy Airlift to Haiti," October 15, 1963, both fo/"Disasters–Hurricane Flora–Haiti," bx/1, ent/UD-WW-862, rg/286, NARA-II.

86. Edward Curtis to DOS, October 15, 1963; and George Ball to US Embassy, Haiti, October 30, 1963, both fo/"Haiti: Contingency Planning," bx/104a, NSF-C/JFK-PP, JFKPL; Robert Pierpont to Robert Shea, October 21, 1963, fo/DR-912, bx/IV-1792, rg /ANRC, NARA-II.

87. Edward Curtis to DOS, October 25, 1963, fo/"Haiti: Contingency Planning," bx/104a, NSF-C/JFK-PP, JFKPL; George Ball to Food and Agricultural Organization, December 18, 1923, Unnumbered Project, bx/16WFP220, FP-5/3, rg/16, FAO.

88. Curtis to DOS, October 26, 1963; "Haiti: Proposed Short Range Plan of Action, November 1, 1963 to April 30, 1964."

89. Rabe, *Most Dangerous Area in the World*; LeoGrande and Kornbluh, *Back Channel to Cuba*, 42–78.

90. Defense Intelligence Agency, "Special Intelligence Summary: Hurricane Damage in Cuba," October 11, 1963, fo/"Cuba, Hurricane Flora," bx/050, NSF-C/JFK-PP, JFKPL.

91. Commander, Atlantic Command to RUECW/JCS, October 9, 1963, fo/"Cuba, Hurricane Flora," bx/050, NSF-C/JFK-PP, JFKPL.

92. CIA Directorate of Intelligence, "Intelligence Memorandum: Interim Assessment of Hurricane Damage in Cuba," October 22, 1963, fo/"Cuba, Hurricane Flora," bx/050, NSF-C/JFK-PP, JFKPL.

93. Gordon Chase to McGeorge Bundy, October 7, 1963, fo/"Cuba, Hurricane Flora," bx/050, NSF-C/JFK-PP, JFKPL.

94. ARC to Cuban Red Cross, October 7, 1963, fo/893.1, bx/IV-1662, rg/ANRC, NARA-II.

95. Cruz Roja Cubana to ARC, October 9, 1963, fo/893.1, bx/IV-1662, rg/ANRC, NARA-II.

96. Editorial from Cuban newspaper *Hoy*, October 11, 1963, recorded in 88th Cong., 2nd sess., *CR* (January 14, 1964): 425.

97. Fidel Castro, radio broadcast, October 11, 1963, recorded in 88th Cong., 2nd sess., *CR* (January 14, 1964): 424.

98. Thomas Hughes to DOS, October 24, 1963, fo/"Cuba: Intelligence," bx/050a, NSF-C/JFK-PP, JFKPL.

99. Thomas Robinson to A. H. Boerma, October 25, 1963; and Shirley Robertson to A. L. Cardinaux, January 13, 1964, both Project 811, bx/16WFP190, FP-5/3, rg/16, FAO; Thomas Hughes to DOS, [October 1963], fo/"Cuba: Intelligence," bx/050a, NSF-C /JFK-PP, JFKPL.

100. Thomas Hughes to DOS, October 15, 1963, fo/"Cuba, Hurricane Flora," bx/050, NSF-C/JFK-PP, JFKPL.

101. George Ball, Proposed Press Release: "Private Disaster Relief for Cuba," October 16, 1963, fo/"Cuba, Hurricane Flora," bx/050, NSF-C/JFK-PP, JFKPL; Joan Kain, "Cuba–American Friends Service Committee," October 17, 1963, fo/"Disasters–Hurricane Flora–Cuba," bx/1, ent/UD-WW-862, rg/286, NARA-II.

102. Gordon Chase to McGeorge Bundy, October 18, 1963, NSF-C/JFK-PP, JFKPL.

103. American Friends Service Committee, International Service Division, "Report of Relief Mission for Cuban Hurricane Victims," 1963, fo/"Disasters–Hurricane Flora–Cuba," bx/1, ent/UD-WW-862, rg/286, NARA-II.

104. Gordon Chase to McGeorge Bundy, October 28, 1963; and Unknown author, "To Be Used To Answer Press Inquiries," [October 1963], both bx/050, NSF-C/JFK-PP, JFKPL.

105. Briefing Paper for President's Press Conference, October 31, 1963, fo/"Disasters–Hurricane Flora–Cuba," bx/1, ent/UD-WW-862, rg/286, NARA-II.

106. American Friends Service Committee, "Report of Relief Mission for Cuban Hurricane Victims."

107. Eric Kocher to DOS, October 5, 1963, fo/"Yugoslavia, General, 1963," bx/211, NSF-C/JFK-PP, JFKPL.

108. Draft of President Kennedy's Welcome Statement, October 17, 1963, fo /"Yugoslavia, General, 1963," bx/211, NSF-C/JFK-PP, JFKPL.

109. McGeorge Bundy, "Offer of Emergency Shelter Relief for Skopje," October 17, 1963, fo/"Yugoslavia, General, 1963," bx/211, NSF-C/JFK-PP, JFKPL.

110. National Security Action Memorandum 267: Disaster Assistance for Skopje, in McGeorge Bundy to Secretary of Defense and USAID Administrator, October 18, 1963, snf/1964–66/SOC-10-Yugoslavia, rg/59, NARA-II.

111. Commander, US Army Europe, to US Embassy, Yugoslavia, November 6, 1963, fi/571-Skopje, bx/94, ent/UD-3350, rg/84, NARA-II.

112. Commander, US Army Europe to Commanding General of 7th Army et al., November 7, 1963, fi/571-Skopje, bx/94, ent/UD-3350, rg/84, NARA-II.

113. Commander, US Army Europe, "Operation Home Run Report 6," December 14, 1963, fi/571-Skopje, bx/94, ent/UD-3350, rg/84; and William Tyler to Mr. Johnson, December 12, 1963, snf/1964–66/SOC-10-Yugoslavia, rg/59, both NARA-II.

114. Edgar Zimmerman to Robert Edson, February 27, 1964, fo/DR-909, bx/IV-1791, rg/ANRC; and US Embassy, Yugoslavia to DOS, February 29, 1964, fo/"Disasters –Yugoslavia Skopje–1963," bx/1, ent/UD-WW-862, rg/286, both NARA-II.

115. Eric Kocher to DOS, November 9, 1963, fi/571-Skopje, bx/94, ent/UD-3350, rg/84; and Edgar Zimmerman to Robert Edson, November 20, 1963, fo/DR-909, bx/IV-1791, rg /ANRC, both NARA-II.

116. Sec. 639, *Foreign Assistance Act of 1965*; Louis Boochever to DOS, March 10, 1966, snf/1964–66/SOC-10-Yugoslavia, rg/59, NARA-II.

117. Rusk, Memorandum to President Johnson, June 13, 1966, bx/232, NSF-C/LBJ-PP, LBJPL.

118. Irwin Tobin to US Embassy, Turkey, September 8, 1966, snf/1964–66/SOC-10 -Yugoslavia, rg/59, NARA-II.

CHAPTER 13

1. Chapter 9, *Foreign Assistance Act of 1961*, as amended by the *International Development and Food Assistance Act of 1975*.

2. FDRC, "U.S. Foreign Disaster Relief," September 23, 1975, fo/"1975 AID/FDRC Disaster Conference," bx/101, MRC-DF, coll/MC-655, RUSC.

3. William McCahon to Harry Shooshan, November 15, 1963, fo/"Executive," bx/88, fi /DI-1963, WHCF/JFK-PP, JFKPL; Marie-Louise Sigerist to Samuel Krakow, January 24, 1964, fo/898.5, bx/IV-1664, rg/ANRC, NARA-II.

4. Irwin Ross, "America's 'Mr. Catastrophe,'" *Reader's Digest*, October 1968, 141–44; FDRC, "U.S. Foreign Disaster Relief," September 23, 1975.

5. USAID Manual Orders 1562.1, 1563.1, 1563.2, 1564.1, and 1565.1, "Foreign Disaster Emergency Relief," October 9, 1964, fo/890.011, bx/IV-1660, rg/ANRC, NARA-II.

6. FDRC and ACVFA, "Summary Report: Foreign Disaster Emergency Relief Operations, 1964," February 5, 1965, fo/"Annual Disaster Reports," bx/101, MRC-DF, coll /MC-655, RUSC.

7. USAID Manual Order 1563.1, "Foreign Disaster Emergency Relief."

8. Robert Edson to Robert Shea, January 7, 1965, fo/900.02, bx/V-98, acc/200-85-33, rg/ANRC, NARA-II.

9. Samuel Krakow to Henrik Beer, January 15, 1965, fo/900.02, bx/V-98, acc/200-85-33, rg/ANRC, NARA-II.

10. Minutes, Meeting of Stephen Tripp and Miss Von Thurn, September 16, 1964, fo/"Haiti–Hurricane, Floods," bx/108, MRC-DF, coll/MC-655, RUSC.

11. Gerald Fleischer and Anthony Mason, "A Systems Analysis Case Study: Emergency Disaster Relief Operations of the Agency for International Development," [ca. 1966], fo/"Disaster, Relief," bx/I-174, coll/MSS-42218, LOC; The White House Conference on International Cooperation National Citizens' Commission, "Report of the Committee on Disaster Relief," 1965, bx/102, MRC-DF, coll/MC-655, RUSC; DOS Policy Planning Council, "Angry Nature: Strengthening Cooperation against Common Hazards," July 24, 1968, fo/"1.B. Disaster Relief," bx/8, ent/P-115, rg/286, NARA-II.

12. US Department of Defense Directive 5100.46: "Responsibilities for Foreign Disaster Relief Operations," October 15, 1964, fo/"1974 FDRC Annual Conference," bx/101, MRC-DF, coll/MC-655, RUSC.

13. Stephen Tripp, Peru Earthquake Disaster Relief Memo, June 15, 1970, fo/"DR-0 Earthquake Reports," bx/1, ent/P-891, rg/286, NARA-II.

14. Krakow to Beer, January 15, 1965.

15. James MacCracken to Howard Kresge, August 30, 1966, fo/"Correspondence and Memoranda, 1965–69," bx/I-174, coll/MSS-42218, LOC; Bernard Confer, "Testimony before the Subcommittee on Foreign Aid," March 19, 1968, fo/"Relation of ACVAFS to Governments," bx/5, Board of Directors Files, coll/MC-655, RUSC.

16. Ed O'Brien to Howard Kresge, January 31, 1968, fo/"Correspondence Prior to 1972," bx/101, MRC-DF, coll/MC-655, RUSC.

17. Samuel Krakow, in Ross, "America's 'Mr. Catastrophe,'" 144.

18. List compiled from various archival collections included in this book's bibliography.

19. Ross, "America's 'Mr. Catastrophe,'" 142.

20. Stephen Tripp, in Irwin Ross, "Stephen R. Tripp: Mr. Catastrophe," *Sunday Denver Post*, September 15, 1968, 14–15; and in Ross, "America's 'Mr. Catastrophe,'" 144.

21. Ross, "America's 'Mr. Catastrophe,'" 141.

22. "Foreign Aid: Mr. Catastrophe," *Time*, January 26, 1968, 19.

23. Richard Nixon, Message to Congress, April 21, 1971, in *Public Papers of the Presidents of the United States*, 564–77.

24. Alexanderina Shuler, "AID's 'Mr. Disaster' Earns Some Relief," *Front Lines* 9, no. 15 (1971): 1, 6.

25. Richard Nixon to Maurice Williams, March 28, 1974, fo/"5/73–5/74," bx/12, fi/FG-11–4/A, Subject Files, WHCF/RMN-PM, RMNPL; Daniel Parker to Brent Scowcroft, March 10, 1975, fo/ 'Disaster Relief PHA,' bx/7, ent/P-115, rg/286, NARA-II.

26. List compiled from various archival collections included in this book's bibliography.

27. Minutes, conference on "improving cooperative efforts in foreign disaster relief," November 13, 1972, fo/900.02/04, bx/V-99, acc/200-85-34; and OFDA, "Summary of U.S. Government Foreign Disaster Assistance since 1964," in *Foreign Disaster Assistance: The United States Government's Role* [ca. 1980], fo/800.02, bx/V-31, acc/200-86-7, both rg /ANRC, NARA-II.

28. Samuel Krakow, Memorandum for the record, November 13, 1972, fo/900.02/04, bx/V-65, acc/200-85-34; and National Research Council Report, "Committee on

International Disaster Assistance," December 13, 1976, fo/890.1, bx/V-65, acc/200-85-33, both rg/ANRC, NARA-II.

29. FDRC, "U.S. Foreign Disaster Relief," September 23, 1975.

30. PHA Working Paper, "Strengthening AID's Program for Disaster Relief," 1975, fo /"Disaster Relief PHA," bx/7, ent/P-115, rg/286, NARA-II.

31. FDRC, "U.S. Foreign Disaster Relief," September 23, 1975.

32. Minutes, Belmont Conferences, 1972, 1973, and 1974, bx/V-99, acc/200-85-34, rg /ANRC, NARA-II.

33. "Cooperation between Voluntary Agencies and the Office of the Foreign Disaster Relief Coordinator," August 29, 1975, fo/"AID and Voluntary Agencies," bx/50, USGF, coll/MC-655, RUSC.

34. Jarold Kieffer, "Plan of Action to Enhance the A.I.D. Role in Coordinating U.S. Responses to Natural and Other Disasters Abroad," October 12, 1974, fo/"Disaster Relief PHA," bx/7, ent/P-115, rg/286; and Tim McClure, paraphrased in minutes, conference on "improving cooperative efforts in foreign disaster relief," November 13, 1972, fo/900.02/04, bx/V-99, acc/200-85-34, rg/ANRC, both NARA-II.

35. Jarold Kieffer, "Strengthening AID's Disaster Relief Capability," April 11, 1972, fo/"Establishment of the PHA," bx/7, ent/P-115, rg/286, NARA-II.

36. Kieffer, "Strengthening AID's Disaster Relief Capability."

37. Kieffer, "Plan of Action."

38. Office of Management Planning, "A Study of Disaster Relief Activities of the Agency for International Development," June 17, 1974, fo/"Establishment of the PHA," bx/7, ent/P-115, rg/286, NARA-II.

39. Kieffer, "Plan of Action."

40. Office of Management Planning, "Study of Disaster Relief Activities."

41. Kieffer, "Plan of Action."

42. Kieffer, "Plan of Action."

43. Daniel Parker, "Action Plan for Enhancing AID's Role in Coordinating the U.S. Response to Disasters," January 8, 1975, fo/"Disaster Relief PHA," bx/7, ent/P-115, rg/286, NARA-II.

44. Knowles, *Disaster Experts*, 256–79; Robert, *Disasters and the American State*, 47–80.

45. Chapter 9, *Foreign Assistance Act of 1961*, as amended by the *International Development and Food Assistance Act of 1975*.

46. USAID Office of Public Affairs, Press Release, May 28, 1976, fo/"AID Correspondence," bx/50, USGF; and Minutes, meeting of voluntary disaster response agencies with AID/FDA Director, June 16, 1976, fo/"Disasters, Meetings," bx/101, MRC-DF, both coll /MC-655, RUSC.

47. National Research Council Report, "Committee on International Disaster Assistance."

48. National Research Council, *The U.S. Government Foreign Disaster Assistance Program*, v, 83.

49. FDRC, "U.S. Foreign Disaster Relief," September 23, 1975.

50. UN General Assembly Resolution A/Res/69/283, "Sendai Framework for Disaster Risk Reduction 2015–2030," June 3, 2015; Kelman, *Disaster by Choice*; Remes and Horowitz, *Critical Disaster Studies*.

51. Dina Esposito and Jeremy Konyndyk, "Merging USAID's Disaster Offices Means Answering Hard Questions," *Devex*, March 20, 2018.

52. USAID, Office of Foreign Disaster Assistance website, 2019, https://perma .cc/8U2V-AJWW; Olson, "The Office of U.S. Foreign Disaster Assistance," https://perma .cc/E9FD-PYRL.

53. USAID Bureau for Humanitarian Assistance website, 2021, https://perma .cc/72DX-5KKE.

54. USAID Office of U.S. Foreign Disaster Assistance Fact Sheet, 2019, https://perma .cc/LRR2-VKB8.

55. US State Department, Foreign Assistance Manual 2 FAM 060: "International Disaster and Humanitarian Assistance," last revised February 21, 2019, https://perma.cc /Z8ZX-9V7D.

56. US State Department, Foreign Assistance Manual 2 FAM 060.

57. US Department of Defense Directive 5100.46: "Foreign Disaster Relief (FDR)," in Joint Publication 3-29, "Foreign Humanitarian Assistance," May 14, 2019, https://perma .cc/PUJ2-HH6Q.

58. *An Act to Amend the Act of January 5, 1905, to Incorporate the American National Red Cross* (1947).

59. ARC, "International Disaster Response, Calendar Year 2018," https://perma.cc /HP5W-2MCN; ARC, International Disasters and Crises" (2022), https://perma.cc /PK2F-LPFL.

60. Charter of the ACVFA, May 14, 1946, as revised on April 12, 2019, https://perma .cc/J8KC-ZQ5S.

61. USAID, "Streamlining the Registration Process for Private Voluntary Organizations," July 31, 2019, 84 *Fed. Reg.* 147 (July 31, 2019): 37079–81.

62. National Research Council, *US. Government Foreign Disaster Assistance Program*, v.

63. Samuel Krakow to James Collins, November 15, 1966, bx/V-65, acc/200-85-33, rg /ANRC, NARA-II.

64. Alston, "Climate Change and Poverty."

Bibliography

ARCHIVAL SOURCES

Italy

Food and Agricultural Organization of the United Nations Archives, Rome
    RG 16: Records of the World Food Programme

Switzerland

Archives of the International Committee of the Red Cross, Geneva
    B-AG: Archives générales, 1951–1965
    P-UIS: Union internationale de secours [International Relief Union], 1920–1990
Archives of the International Federation of Red Cross and Red Crescent Societies, Geneva
    Assorted processed and unprocessed records
    Bulletins and Information Circulars of the League of Red Cross Societies, 1919–1939

United Kingdom

The National Archives, Kew
    ADM: Records of the Admiralty, Naval Forces, Royal Marines
        ADM-116/2843: Cases, Record Office
    FO: Records created or inherited by the Foreign Office
        FO-371/81396: General Correspondence, 1906–66, Political Departments
        FO-809/2: General Correspondence of the
                British Consulate and Legation, Managua, Nicaragua
        FO-908/6: Miscellaneous Papers, Yokohama (Japan) Consulate
    HO: Records created or inherited by the Home Office,
                Ministry of Home Security, and related bodies
        HO-286/12: Civilian Gallantry Awards, Home Office Files

United States

Dwight D. Eisenhower Presidential Library, Abilene, KS
    Dennis A. Fitzgerald Papers, 1945–1969
    Dwight D. Eisenhower, Records as President
        White House Central Files, 1953–61
            Official Files
                113: Disaster
        White House Office Files
            Office of the Staff Secretary, 1952–1961
            International Series
    John Foster Dulles Papers, 1951–1959
        Subject Series
Franklin D. Roosevelt Presidential Library, Hyde Park, NY
    Franklin D. Roosevelt, Papers as President
        Official Files
            20: Department of State
            48h: England/India
            124: American Red Cross
            146: Mexico, Government of
            159: Cuba, Government of
Harry S. Truman Presidential Library, Independence, MO
    Harry S. Truman Presidential Papers
        White House Central Files
            Confidential Files
                State Department Correspondence
            Official Files
                124: American Red Cross
                426: Foreign Relief
Herbert Hoover Presidential Library, West Branch, IA
    Herbert Hoover Presidential Papers
        Foreign Affairs Files, 1923–1933
            997-4: China Flood (August 1931)
            998-9: Dominican Republic Hurricane (1930)
            1008-8: Nicaragua Earthquake (1931)
        Frances White Letters Collection
John F. Kennedy Presidential Library, Boston, MA
    Papers of John F. Kennedy, Presidential Papers
        National Security Files
            Ralph A. Dungan Files
            Subjects: Foreign Aid
        White House Central Files
            DI: Disasters
Library of Congress, Washington, DC
    MSS-24208: Lloyd C. Griscom Papers

MSS-27912: Nelson T. Johnson Papers
MSS-31989: Frank Ross McCoy Papers
MSS-40923: Charles Stillman Sperry Papers
MSS-42218: Charles P. Taft Papers
Lyndon B. Johnson Presidential Library, Austin, TX
Papers of Lyndon B. Johnson, President, 1963–1969
National Security Files
Country Files
White House Central Files
CO: Countries
DI: Disasters
Marine Corps History Division, Quantico, VA
Archives Branch
Collection 3615: Major Thomas E. Watson Papers
Historical Reference Branch
Nicaragua: Earthquake
China: Legation, 1928–1936 (Oversized)
National Archives and Records Administration I, Washington, DC
RG 24: Records of the Bureau of Naval Personnel
Entry 88: General Correspondence, 1903–1915
Entry 118G: Logs of US Naval Ships and Stations, 1801–1946
RG 26: Records of the US Coast Guard
Entry 82A: General Correspondence, 1910–1935
RG 38: Records of the Office of the Chief of Naval Operations
Entry 98: Naval Attaché Reports, 1886–1939; Intelligence Division
RG 45: Records Collection of the Office of Naval Records and Library
Entry 286: Communications concerning the January 14,
1907, Earthquake at Kingston, Jamaica
Entry 500: Area File of the Naval Records Collection, 1775–
1910, Microfilm M-625; Area 4, 1908–1909
Entry 520: Subject File of the Naval Records Collection,
1911–1927; Ship Name Subject File
RG 80: General Records of the Department of the Navy, 1798–1947
Entry 19: Office of the Secretary of the Navy; General Correspondence:
1897–1915, 1916–1926, and 1926–1940
RG 94: Records of the Adjutant General's Office, 1780s–1917
Entry 25: Document Files, 1890–1917
RG 127: Records of the US Marine Corps
Entry 43A: Marine Corps Units in Nicaragua, 1927–1933
RG 313: Records of Naval Operating Forces
Entry UD-8B: Atlantic Fleet, 1905–1909, USS *Connecticut*,
Record of Wireless Relays, 1908–1909
National Archives and Records Administration II, College Park, MD
Freedom of Information Act Electronic Ready Room, General CIA Records
RG 24: Records of the Bureau of Naval Personnel

Deck Logs of US Navy Ships and Stations
    Entry 118B: 1951–1955
    Entry 118C: 1956–1959
    Entry 118D: 1961–1965
RG 59: Records of the Department of State
    Central Decimal Files of the Department of State
        1910–1929
        1930–1939
        1940–1944
        1945–1949
        1950–1954
        1955–1959
        1960–1963
    Despatches from US Consuls
        Entry A1–85: Despatches from the US Consuls in St. Pierre,
            Martinique, 1790–1906, Microfilm T-431
    Numerical and Minor Files of the Department of State, 1906–1910, Microfilm M-862
        File 241: Chile
        File 3892: Jamaica
        File 4001: Jamaica
        File 17191: Italy
        File 17192: Italy
    Records of the Bureau of Inter-American Affairs
        Entry A1-3137: Office of West Coast Affairs; Records
            Relating to Chile, 1958–1963
        Entry P-163: Office of Inter-American Political Affairs
    Records of the Bureau of International Organization Affairs
        Entry UD-07D-81: Office of International Economic and Social Affairs,
            1953–1970; World Health Organization Subject Files, 1944–1963
    Records of the Office of the Assistant Secretary for Congressional Relations
        Entry A1-1537B: Studies on Foreign Aid, 1945–1959
    Records of the Office of European Regional Affairs; Bureau of European Affairs
        Entry A1-1175D: Subject Files, 1946–53
    Records of the Office of Under Secretary of State
        Entry P-188: Office of the Deputy Coordinator for Mutual Security, 1959–1961
    Subject Numeric Files of the Department of State
        SOC-10, 1963
        SOC-10, 1964–1966
RG 84: Records of Foreign Service Posts of the Department of State
    Records of Consular Posts, 1789–1912
        Entry UD-240: Records of the Catania (Italy) Consulate
        Entry UD-914: Records of the Valparaiso (Chile) Consulate
    Records of Consular Posts, 1912–1935
        Entry UD-401: Records of the Guatemala City (Guatemala) Consulate
        Entry UD-472: Records of the Kobe (Japan) Consulate
        Entry UD-603: Records of the Nanking (China) Consulate

Entry UD-855: Records of the Tampico (Mexico) Consulate

Entry UD-897: Records of the Tientsin (China) Consulate

Entry UD-948: Records of the Yokohama (Japan) Consulate

Records of Diplomatic Posts, 1789–1912

Entry UD-9: Despatches to Department of State, 1905–1907

Entry UD-35: Records of the US Embassy, Italy

Records of Diplomatic Posts, 1912–1935

Entry UD-10: Records of the US Legation, China

Entry UD-13: Records of the US Embassy, Cuba

Entry UD-16: Records of the US Legation, Dominican Republic

Entry UD-26: Records of the US Embassy, Greece

Entry UD-27: Records of the US Legation, Guatemala

Entry UD-36: Records of the US Embassy, Japan

Entry UD-41: Records of the US Embassy, Mexico

Entry UD-45: Records of the US Legation, Nicaragua

Records of Embassy Posts, 1936–1963

Entry UD-2378A: Classified General Records of the
US Embassy in Czechoslovakia

Entry UD-2554: Records of the US Embassy in Chile

Entry UD-2710A: Classified General Records of the US Embassy, India

Entry UD-2828: General Records of the US Embassy, Japan

Entry UD-2828A: Classified General Records of the US Embassy, Japan

Entry UD-3350: Classified General Records of the US Embassy in Yugoslavia

RG 165: Records of the War Department General and Special Staffs

Entry NM-84/310: Records of the Historical Section relating to
the History of the War Department, 1900–1940

RG 286: Records of the US Agency for International Development

Entry P-211: Country and Subject Files relating to Natural Disasters, 1956–1961

Entry UD-WW-862: Foreign Disaster Relief Case Files, 1960–1964

Bureau for Management

Entry P-115: Records relating to Organizations and Functions, 1952–1984

USAID Bureau for Far East

Entry P-368: Office of Indonesian Affairs; Closed Project Files, 1955–1964

USAID Mission to Iran

Entry P-491: Community Development Division; Subject Files, 1961–1962

Entry P-512: Public Health Division, Subject Files, 1959–1965

USAID Mission to Korea

Entry P-590: Rural Development Division; Public
Works Branch; Subject Files 1955–1962

USAID Mission to Peru

Entry P-891: Executive Office; Records Relating to Disaster Relief, 1970–1972

RG 395: Records of US Army Overseas Operations and Commands, 1870–1942

Entry NM-94/5964: US Army Troops in China, 1912–
1938; Special Orders, 1914–1929

RG 407: Records of the Adjutant General's Office, 1917–1981

Central Decimal Files

Entry PI-17/37-A: 1917–1925
Entry PI-17/37-C: Bulky Files, 1917–1925
Entry PI-17/37-H: 1926–1939
Entry NM-3/363-C: 1949–1950
Entry NM-3/363-E: 1953–1954
RG 469: Records of US Foreign Assistance Agencies, 1948–1961
Records of the Foreign Operations Administration
Mission to Greece
Entry UD-1220: Food and Agriculture Division;
Classified Subject Files, 1947–1954
Office of the Deputy Director for Operations
Entry UD-359: Office of European Operations, Greece
Division; Subject Files, 1948–1954
Records of the International Cooperation Administration
Office of the Director
Entry UD-180: Subject Files of the Director, 1948–1966
Point IV Mission to Haiti, Washington Correspondence
Entry UD-1229: Food and Agriculture Division; Subject Files, 1947–1957
US Operations Mission to Haiti
Entry P-260: Executive Office; Classified Central Subject Files, 1950–1962
Entry P-262: Executive Office; Unclassified Central Subject Files, 1950–1962
US Operations Mission to Korea
Entry P-319: Community Development Division, Office of
Government Services; Unclassified Subject Files, 1954–1961
Entry UD-1276: Executive Office; Subject Files (Central Files)
RG ANRC (formerly RG 200): Records of the American National Red Cross
Entry P-78, League of Red Cross Societies Records, 1919–1999
Entry P-105: Historical Personnel Files, 1856–2008
Series I: 1881–1916
Series II: 1917–1934
Series III: 1935–1946
Series IV: 1947–1964
Series V: 1965–1979
Naval History and Heritage Command Library, Washington, DC
ZO (Operations) Files
Richard M. Nixon Presidential Library, Yorba Linda, CA
Papers of Richard M. Nixon, Presidential Materials
White House Central Files
Subject Files
Rutgers University Library Special Collections and University Archives, New Brunswick,
NJ
MC 655: American Council of Voluntary Agencies for Foreign Service Records
Board of Directors Files
Material Resources Committee/Disaster Files
United States Government Files

PUBLISHED PRIMARY SOURCES

Books

American National Red Cross. *The Report of the American Red Cross Commission to China.* Washington, DC: American National Red Cross, 1929.

Belknap, Reginald. *American House Building in Messina and Reggio.* New York: G. P. Putnam's Sons, 1910.

Bicknell, Ernest. *Pioneering with the Red Cross: Recollections of an Old Red Crosser.* New York: Macmillan, 1935.

Davies, Joseph E., Charles P. Taft, and Charles Warren. *Voluntary War Relief during World War II.* Washington, DC: Government Printing Office, 1946.

Kennan, George. *The Tragedy of Pelée: A Narrative of Personal Experience and Observation in Martinique.* New York: Outlook, 1902.

Manning, William R., ed. *Diplomatic Correspondence of the United States concerning the Independence of the Latin American States.* Vol. 2. New York: Oxford University Press, 1925.

Miller, James Martin, and John Stevens Durham. *The Martinique Horror and St. Vincent Calamity: Containing a Full and Complete Account of the Most Appalling Disaster of Modern Times.* Philadelphia: National Publishing Company, 1902.

National Research Council. *The U.S. Government Foreign Disaster Assistance Program.* Washington, DC: National Academy of Sciences, 1978.

The Office of Foreign Relief and Rehabilitation Operations, Department of State. Washington, DC: Division of Public Information, OFFRO, 1943.

Todd, Frank Morton. *The Story of the Exposition: Being the Official History of the International Celebration Held at San Francisco in 1915 to Commemorate the Discovery of the Pacific Ocean and the Construction of the Panama Canal.* Vol. 4. New York: G. P. Putnam, 1921.

Government Documents

Annals of Congress
Congressional Record
Papers Relating to the Foreign Relations of the United States

Alston, Philip. *Climate Change and Poverty: Report of the Special Rapporteur on Extreme Poverty and Human Rights.* Report to the United Nations Human Rights Council, Forty-First Session, A/HRC/31/39. July 17, 2019.

Consular Reports: Commerce, Manufactures, Etc. Vol. 60. Washington, DC: Government Printing Office, 1899.

Cumulative Supplement to the Code of Federal Regulations of the United States of America. Washington, DC: Government Printing Office, 1943.

Papers Relating to Foreign Affairs. Vol. 2. Washington, DC: Government Printing Office, 1869.

Public Papers of the Presidents of the United States: Richard Nixon, 1971. Washington, DC: Government Printing Office, 1999.

Register of the Department of State. Washington, DC: Government Printing Office, 1911.

United Nations General Assembly Resolution A/Res/69/283. "Sendai Framework for Disaster Risk Reduction 2015–2030." June 3, 2015.

US Library of Congress. Congressional Research Service. *The Congressional Charter of the American National Red Cross: Overview, History, and Analysis.* By Kevin R. Kosar. RL33314. 2006.

Legislation

Act for International Development of 1961. PL-87-195. US Statutes at Large 75 (1961): 424–65.

An Act for the Acquisition of Buildings and Grounds in Foreign Countries. 19th Cong., 2nd sess., chap. 643. US Statutes at Large 60 (1946): 663.

An Act for the Relief of Citizens of Italy. 16th Cong., 2nd sess., chap. 7. US Statutes at Large (1909): 584.

An Act for the Relief of Citizens of the French West Indies. 57th Cong., 1st sess., chap. 787. US Statutes at Large (1902): 198.

An Act for the Relief of Sufferers from Earthquake in Japan. 68th Cong., 2nd sess., chap. 297. US Statutes at Large (1925): 963–64.

An Act for the Relief of the Citizens of the Island of Jamaica. 59th Cong., 2nd sess., chap. 154. US Statutes at Large (1907): 850.

An Act for the Relief of the Citizens of Venezuela. 12th Cong., 1st sess., chap. 79. US Statutes at Large (1812): 730.

An Act to Amend the Act of January 5, 1905, to Incorporate the American National Red Cross. 80th Cong., 1st sess., chap. 50. US Statutes at Large 61 (1947): 80–83.

An Act to Incorporate the American Red Cross. 56th Cong., 1st sess., chap. 784. US Statutes at Large (1900): 277–80.

An Act to Incorporate the American Red Cross. 58th Cong., 3rd sess., chap. 23. US Statutes at Large (1905): 599–602.

An Act to Provide for Assistance in the Development of Latin America and in the Reconstruction of Chile. PL-86-735. US Statutes at Large 74 (1960): 869–79.

Agricultural Act of 1949. PL-439. US Statutes at Large 63 (1949): 1058.

Agricultural Trade and Development Assistance Act of 1954. PL-480. US Statutes at Large 68 (1954): 454–59.

Economic Cooperation Act of 1948. PL-472. US Statutes at Large 62 (1948):153–54.

Federal Disaster Relief Act. PL-81-875. US Statutes at Large 64 (1950): 1109–11.

Foreign Assistance Act of 1962. PL-87-565. US Statutes at Large 76 (1962): 261.

Foreign Assistance Act of 1965. PL-89-171. US Statutes at Large 79 (1965): 659–60.

Foreign Economic Assistance Act of 1950. PL-535. US Statutes at Large 64 (1950): 202.

Inter-American Social and Economic Cooperation Program Appropriation. PL-87-41. US Statutes at Large 75 (1961): 86–87.

International Development and Food Assistance Act of 1975. PL-94-161. US Statutes at Large 89 (1975): 849–50.

Mutual Security Act of 1951. PL-165. US Statutes at Large 65 (1951): 373–87.

Mutual Security Act of 1952. PL-400. US Statutes at Large 66 (1952): 147.

Mutual Security Act of 1954. PL-665. US Statutes at Large 68 (1954): 843.

Mutual Security Act of 1956. PL-726. US Statutes at Large 70 (1956): 557–60.

Mutual Security Act of 1958. PL-85-477. US Statutes at Large 72 (1958): 268.
Reorganization Plan No. 7 of 1953. US Statutes at Large 67 (1953): 639–42.

Newspapers and Periodicals

American Red Cross Bulletin
American Red Cross Magazine
Appeal (St. Paul, MN)
Baptist Missionary Magazine
Charities
Christian Observer
Sunday Denver Post
Devex
Frank Leslie's Popular Monthly
Front Lines
Guthrie (OK) Daily Leader
McClure's Magazine

New York Times
Peking Times
Reader's Digest
Red Cross Courier
Scribner's Magazine
Time
Washington (DC) Daily News
Washington Herald
Washington Post
Weekly Register (Baltimore)
World Affairs

SECONDARY SOURCES

Books

Albergh, Kristin. *Transplanting the Great Society: Lyndon Johnson and Food for Peace.* Columbia: University of Missouri Press, 2009.

Alvandi, Roham, ed. *The Age of Aryamehr: Late Pahlavi Iran and Its Global Entanglements.* London: Gingko Library, 2018.

Auslin, Michael R. *Pacific Cosmopolitans: A Cultural History of U.S.-Japan Relations.* Cambridge, MA: Harvard University Press, 2011.

Avaria, Bárbara Silva, and Alfredo Riquelme Segovia. *Una identidad terremoteada: Comunidad y territorio en el Chile de 1960.* Santiago: Ediciones Universidad Alberto Hurtado, 2018.

Baer, George W. *One Hundred Years of Sea Power: The U.S. Navy, 1890–1990.* Palo Alto, CA: Stanford University Press, 1996.

Balogh, Brian. *The Associational State: American Governance in the Twentieth Century.* Philadelphia: University of Pennsylvania Press, 2015.

Bandopadhyay, Saptarishi. *All Is Well: Catastrophe and the Making of the Normal State.* New York: Oxford University Press, 2022.

Bankoff, Greg. *Cultures of Disaster: Society and Natural Hazards in the Philippines.* London: Routledge, 2003.

Barnett, Michael. *Empire of Humanity: A History of Humanitarianism.* Ithaca, NY: Cornell University Press, 2011.

———, ed. *Humanitarianism and Human Rights: A World of Difference?* New York: Cambridge University Press, 2020.

Barry, John M. *Rising Tide: The Great Mississippi Flood of 1927 and How It Changed America.* New York: Simon and Schuster, 1998.

Bass, Gary. *Freedom's Battle: The Origins of Humanitarian Intervention.* New York: Knopf, 2008.

Baughan, Emily. *Saving the Children: Humanitarianism, Internationalism, and Empire.* Berkeley: University of California Press, 2021.

Bhattacharyya, Debjani. *Empire and Ecology in the Bengal Delta: The Making of Calcutta.* Cambridge: Cambridge University Press, 2018.

Button, Gregory V., and Mark Schuller, eds. *Contextualizing Disaster.* New York: Berghahn Books, 2016.

Cabanes, Bruno. *The Great War and the Origins of Humanitarianism, 1918–1924.* Cambridge: Cambridge University Press, 2014.

Chang, Gordon. *Fateful Ties: A History of America's Preoccupation with China.* Cambridge, MA: Harvard University Press, 2015.

Choate, Mark. *Emigrant Nation: The Making of Italy Abroad.* Cambridge, MA: Harvard University Press, 2008.

Citino, Nathan. *Envisioning the Arab Future: Modernization in US-Arab Relations, 1945–1967.* New York: Cambridge University Press, 2017.

Clancey, Gregory. *Earthquake Nation: The Cultural Politics of Japanese Seismicity, 1868–1930.* Berkeley: University of California Press, 2006.

Coen, Deborah R. *The Earthquake Observers: Disaster Science from Lisbon to Richter.* Chicago: University of Chicago Press, 2013.

Cohen, Gerard Daniel. *In War's Wake: Europe's Displaced Persons in the Postwar Order.* New York: Oxford University Press, 2013.

Colby, Jason. *The Business of Empire: United Fruit, Race, and U.S. Expansion in Central America.* Ithaca, NY: Cornell University Press, 2013.

Connelly, Matthew. *Fatal Misconception: The Struggle to Control World Population.* Cambridge, MA: Harvard University Press, 2008.

Cornebise, Alfred Emile. *The United States 15th Infantry Regiment in China, 1912–1938.* Jefferson, NC: McFarland, 2004.

Courtney, Chris. *The Nature of Disaster in China: The 1931 Yangzi River Flood.* New York: Cambridge University Press, 2018.

Cullather, Nick. *The Hungry World: America's Cold War Battle against Poverty in Asia.* Cambridge, MA: Harvard University Press, 2010.

Curti, Merle. *American Philanthropy Abroad.* New Brunswick, NJ: Rutgers University Press, 1963.

Curtis, Heather D. *Holy Humanitarians: American Evangelicals and Global Aid.* Cambridge, MA: Harvard University Press, 2018.

Dauber, Michele Landis. *The Sympathetic State: Disaster Relief and the Origins of the American Welfare State.* Chicago: University of Chicago Press, 2012.

Davey, Eleanor. *Idealism beyond Borders: The French Revolutionary Left and the Rise of Humanitarianism, 1954–1988.* Cambridge: Cambridge University Press, 2015.

Davidann, Jon Thares. *Cultural Diplomacy in U.S.-Japanese Relations, 1919–1941.* New York: Palgrave Macmillan, 2007.

Davies, Andrea Rees. *Saving San Francisco: Relief and Recovery after the 1906 Disaster.* Philadelphia: Temple University Press, 2011.

Davis, Mike. *Late Victorian Holocausts: El Niño Famines and the Making of the Third World.* London: Verso, 2001.

Demmer, Amanda. *After Saigon's Fall: Refugees and U.S.-Vietnamese Relations, 1975–1995.* New York: Cambridge University Press, 2021.

Dickie, John. *Una catastrofe patriottica, 1908: Il terremoto di Messina.* Rome: Laterza, 2008.

Dosal, Paul. *Doing Business with the Dictators: A Political History of United Fruit in Guatemala, 1899–1944.* Lanham, MD: Rowman and Littlefield, 1993.

Dulles, Foster Rhea. *The American Red Cross: A History.* New York: Harper and Brothers, 1950.

Dyl, Joanna. *Seismic City: An Environmental History of San Francisco's 1906 Earthquake.* Seattle: University of Washington Press, 2017.

Ekbladh, David. *The Great American Mission: Modernization and the Construction of an American World Order.* Princeton, NJ: Princeton University Press, 2011.

Engerman, David. *The Price of Aid: The Economic Cold War in India.* Cambridge, MA: Harvard University Press, 2018.

Farré, Sébastien. *Colis de guerre: Secours alimentaire et organisations humanitaires, 1914–1947.* Rennes: Presses Universitaires de Rennes, 2014.

Fassin, Didier. *Humanitarian Reason: A Moral History of the Present.* Berkeley: University of California Press, 2012.

Fassin, Didier, and Mariella Pandolfini. *Contemporary States of Emergency: The Politics of Military and Humanitarian Interventions.* New York: Zone Books, 2013.

Field, Thomas C. *From Development to Dictatorship: Bolivia and the Alliance for Progress in the Kennedy Era.* Ithaca, NY: Cornell University Press, 2014.

Fiori, Juliono, Fernando Epsada, Andrea Rigon, Bertrand Taithe, and Rafia Zakaria, eds. *Amidst the Debris: Humanitarianism and the End of Liberal Order.* London: Hurst, 2021.

Fitz, Caitlin. *Our Sister Republics: The United States in an Age of American Revolutions.* New York: W. W. Norton, 2016.

Foster, Gaines M. *The Demands of Humanity: Army Medical Disaster Relief.* Washington, DC: Center for Military History, US Army, 1983.

Francis, Michael J. *The Limits of Hegemony: United States Relations with Argentina and Chile during World War II.* Notre Dame, IN: University of Notre Dame Press, 1977.

Frey, Marc, Sönke Kunkel, and Corinna R. Unger, eds. *International Organizations and Development, 1945–1990.* London: Palgrave Macmillan, 2014.

Fuller, Pierre. *Famine Relief in Warlord China.* Cambridge, MA: Harvard University Press, 2019.

Garcia, Maria Cristina. *State of Disaster: The Failure of U.S. Migration Policy in an Age of Climate Change.* Chapel Hill: University of North Carolina Press, 2022.

Gatrell, Peter. *The Making of the Modern Refugee.* Oxford: Oxford University Press, 2013.

Gobat, Michel. *Confronting the American Dream: Nicaragua under U.S. Imperial Rule.* Durham, NC: Duke University Press, 2005.

Granick, Jaclyn. *International Jewish Humanitarianism in the Age of the Great War.* Cambridge: Cambridge University Press, 2021.

Grego, Caroline. *Hurricane Jim Crow: How the Great Sea Island Storm of 1893 Shaped the Lowcountry South.* Chapel Hill: University of North Carolina Press, 2022.

Healey, Mark. *The Ruins of the New Argentina: Peronism and the Remaking of San Juan after the 1944 Earthquake*. Durham, NC: Duke University Press, 2011.

Heerten, Lasse. *The Biafran War and Postcolonial Humanitarianism: Spectacles of Suffering*. Cambridge: Cambridge University Press, 2017.

Helleiner, Eric. *Forgotten Foundations of Bretton Woods: International Development and the Making of the Postwar Order*. Ithaca, NY: Cornell University Press, 2014.

Hewitt, Kenneth. Hewitt, *Regions of Risk: A Geographical Introduction to Disasters*. London: Routledge, 1997.

Hogan, Michael J. *A Cross of Iron: Harry S. Truman and the Origins of the National Security State, 1945–1954*. New York: Cambridge University Press, 1998.

———. *The Marshall Plan: America, Britain and the Reconstruction of Western Europe, 1947–1952*. New York: Cambridge University Press, 1987.

Hong, Young-sun. *Cold War Germany, the Third World, and the Global Humanitarian Regime*. New York: Cambridge University Press, 2017.

Horowitz, Andy. *Katrina: A History, 1915–2015*. Cambridge, MA: Harvard University Press, 2020.

Humbert, Laure. *Reinventing French Aid: The Politics of Humanitarian Relief in French-Occupied Germany, 1945–1952*. Cambridge, UK: Cambridge University Press, 2021.

Hutchinson, John F. *Champions of Charity: War and the Rise of the Red Cross*. Boulder, CO: Westview, 1996.

Immerwahr, Daniel. *How to Hide an Empire: A History of the Greater United States*. New York: Farrar, Straus and Giroux, 2019.

———. *Thinking Small: The United States and the Lure of Community Development*. Cambridge, MA: Harvard University Press, 2015.

Iorizzo, Luciano J., and Ernest Rossi, eds. *Italian Americans: Bridges to Italy, Bonds to America*. New York: Teneo, 2010.

Irwin, Julia F. *Making the World Safe: The American Red Cross and a Nation's Humanitarian Awakening*. New York: Oxford University Press, 2013.

Johnson, Chalmers. *The Sorrows of Empire: Militarism, Secrecy, and the End of the Republic*. New York: Metropolitan Books, 2005.

Johnson, Jennifer. *The Battle for Algeria: Sovereignty, Health Care, and Humanitarianism*. Philadelphia: University of Pennsylvania Press, 2015.

Johnson, Richard A. *The Administration of United States Foreign Policy*. Austin: University of Texas Press, 1971.

Jones, Marian Moser. *The American Red Cross from Clara Barton to the New Deal*. Baltimore: Johns Hopkins University Press, 2012.

Keller, Richard C. *Fatal Isolation: The Devastating Paris Heat Wave of 2003*. Chicago: University of Chicago Press, 2015.

Kelman, Ilan. *Disaster by Choice: How Our Actions Turn Natural Hazards into Catastrophe*. Oxford: Oxford University Press, 2020.

Kierner, Cynthia. *Inventing Disaster: The Culture of Calamity from the Jamestown Colony to the Johnstown Flood*. Chapel Hill: University of North Carolina Press, 2019.

Klinenberg, Eric. *Heat Wave: A Social Autopsy of Disaster in Chicago*. Chicago: University of Chicago Press, 2002.

Klose, Fabian, ed. *The Emergence of Humanitarian Intervention: Ideas and Practices from the Nineteenth Century to the Present*. Cambridge: Cambridge University Press, 2015.

———. *In the Cause of Humanity: A History of Humanitarian Intervention in the Long Nineteenth Century*. Cambridge: Cambridge University Press, 2022.

Knowles, Scott. *The Disaster Experts: Mastering Risk in Modern America*. Philadelphia: University of Pennsylvania Press, 2013.

Laderman, Charlie. *Sharing the Burden: The Armenian Question, Humanitarian Intervention, and Anglo-American Visions of Global Order*. New York: Oxford University Press, 2019.

LaGumina, Salvatore. *The Great Earthquake: America Comes to Messina's Rescue*. Youngstown, NY: Teneo, 2008.

Lal, Priya. *African Socialism in Postcolonial Tanzania: Between the Village and the World*. New York: Cambridge University Press, 2015.

Lancaster, Carol. *Foreign Aid: Diplomacy, Development, Domestic Politics*. Chicago: University of Chicago Press, 2007.

Latham, Michael. *The Right Kind of Revolution: Modernization, Development, and U.S. Foreign Policy from the Cold War to the Present*. Ithaca, NY: Cornell University Press, 2011.

Laycock, Joanne, and Francesca Piana, eds. *Aid to Armenia: Humanitarianism and Intervention from the 1890s to the Present*. Manchester: University of Manchester Press, 2020.

Leffler, Melvin P. *A Preponderance of Power: National Security, the Truman Administration, and the Cold War*. Palo Alto, CA: Stanford University Press, 1992.

LeoGrande, William, and Peter Kornbluh. *Back Channel to Cuba: The Hidden History of Negotiations between Washington and Havana*. Chapel Hill: University of North Carolina Press, 2014.

Lipman, Jana. *In Camps: Vietnamese Refugees, Asylum Seekers, and Repatriates*. Berkeley: University of California Press, 2020.

Lorenzini, Sara. *Global Development: A Cold War History*. Princeton, NJ: Princeton University Press, 2019.

Macekura, Stephen. *Of Limits and Growth: The Rise of Global Sustainable Development in the Twentieth Century*. New York: Cambridge University Press, 2016.

Macekura, Stephen, and Erez Manela, eds. *The Development Century: A Global History*. New York: Cambridge University Press, 2018.

Mauch, Christof, and Christian Pfister, eds. *Natural Disasters, Cultural Responses: Case Studies toward a Global Environmental History*. Lanham, MD: Lexington Books, 2009.

McCleary, Rachel. *Global Compassion: Private Voluntary Organizations and U.S. Foreign Policy since 1939*. New York: Oxford University Press, 2009.

McDonald, Bryan L. *Food Power: The Rise and Fall of the Postwar American Food System*. New York: Oxford University Press, 2017.

McPherson, Alan. *The Invaded: How Latin Americans and Their Allies Fought and Ended U.S. Occupations*. New York: Oxford University Press, 2014.

McVety, Amanda Kay. *The Rinderpest Campaigns: A Virus, Its Vaccine, and Global Development in the Twentieth Century*. New York: Cambridge University Press, 2018.

Meyerowitz, Joanne. *A War on Global Poverty: The Lost Promise of Redistribution and the Rise of Microcredit*. Princeton, NJ: Princeton University Press, 2021.

Millett, Allan R., Peter Maslowski, and William Feis. *For the Common Defense: A Military History of the United States from 1607 to 2012*. 3rd ed. New York: Free Press, 2012.

Moyn, Samuel. *Humane: How the United States Abandoned Peace and Reinvented War*. New York: Farrar, Straus and Giroux, 2021.

Mukherjee, Janam. *Hungry Bengal: War, Famine, and the End of Empire*. New York: Oxford University Press, 2015.

Nunan, Timothy. *Humanitarian Invasion: Global Development in Cold War Afghanistan*. New York: Cambridge University Press, 2016.

Offiler, Ben. *US Foreign Policy and the Modernization of Iran: Kennedy, Johnson, Nixon and the Shah*. London: Palgrave, 2015.

Offner, Amy C. *Sorting Out the Mixed Economy: The Rise and Fall of Welfare and Developmental States in the Americas*. Princeton, NJ: Princeton University Press, 2019.

O'Sullivan, Kevin. *The NGO Moment: The Globalisation of Compassion from Biafra to Live Aid*. Cambridge: Cambridge University Press, 2021.

Parrinello, Giacomo. *Fault Lines: Earthquakes and Urbanism in Modern Italy*. Oxford: Berghahn Books, 2015.

Paulmann, Johannes, ed. *Dilemmas of Humanitarian Aid in the Twentieth Century*. Oxford: Oxford University Press, 2016.

Pavilack, Jody. *Mining for the Nation: The Politics of Chile's Coal Communities from the Popular Front to the Cold War*. University Park: Pennsylvania State University Press, 2011.

Pérez Jr., Louis A. *Cuba under the Platt Amendment, 1902–1934*. Pittsburgh, PA: University of Pittsburgh Press, 1991.

Pietz, David. *Engineering the State: The Huai River and Reconstruction in Nationalist China, 1927–1937*. New York: Routledge, 2002.

——— . *The Yellow River: The Problem of Water in Modern China*. Cambridge, MA: Harvard University Press, 2015.

Porter, Stephen. *Benevolent Empire: U.S. Power, Humanitarianism, and the World's Dispossessed*. Philadelphia: University of Pennsylvania Press, 2016.

Rabe, Stephen G. *Eisenhower and Latin America: The Foreign Policy of Anticommunism*. Chapel Hill: University of North Carolina Press, 1988.

——— . *The Most Dangerous Area in the World: John F. Kennedy Confronts Communist Revolution in Latin America*. Chapel Hill: University of North Carolina Press, 2014.

Rakove, Robert B. *Kennedy, Johnson, and the Non-Aligned World*. New York: Cambridge University Press, 2012.

Reckner, James R. *Teddy Roosevelt's Great White Fleet*. Washington, DC: Naval Institute Press, 2001.

Remes, Jacob A. C. *Disaster Citizenship: Survivors, Solidarity, and Power in the Progressive Era*. Urbana: University of Illinois Press, 2015.

Remes, Jacob A. C., and Andy Horowitz, eds. *Critical Disaster Studies*. Philadelphia: University of Pennsylvania Press, 2021.

Robert, Patrick S. *Disasters and the American State: How Politicians, Bureaucrats, and the Public Prepare for the Unexpected*. New York: Cambridge University Press, 2013.

Rodogno, Davide. *Against Massacre: Humanitarian Interventions in the Ottoman Empire, 1815–1914*. Princeton, NJ: Princeton University Press, 2011.

——— . *Night on Earth: A History of International Humanitarianism in the Near East, 1918–1930*. Cambridge: Cambridge University Press, 2022.

Rohland, Eleonora. *Changes in the Air: Hurricanes in New Orleans from 1718 to the Present.* New York: Berghahn Books, 2018.

Roorda, Eric. *The Dictator Next Door: The Good Neighbor Policy and the Trujillo Regime in the Dominican Republic, 1930–1945.* Durham, NC: Duke University Press, 1998.

Rosenberg, Emily. *Spreading the American Dream: American Economic and Cultural Expansion, 1890–1945.* New York: Hill and Wang, 1982.

Rozario, Kevin. *The Culture of Calamity: Disaster and the Making of Modern America.* Chicago: University of Chicago Press, 2007.

Salvatici, Silvia. *A History of Humanitarianism, 1755–1989: In the Name of Others.* Manchester: University of Manchester Press, 2019.

Sanders, Christopher. *America's Overseas Garrisons: The Leasehold Empire.* New York: Oxford University Press, 2000.

Scarth, Alwyn. *La Catastrophe: The Eruption of Mount Pelée, the Worst Volcanic Disaster of the 20th Century.* New York: Oxford University Press, 2002.

Schencking, J. Charles. *The Great Kantō Earthquake and the Chimera of National Reconstruction in Japan.* New York: Columbia University Press, 2013.

Schmitz, David F. *Thank God They're on Our Side: The United States and Right-Wing Dictatorships, 1921–1965.* Chapel Hill: University of North Carolina Press, 1999.

Schwartz, Stuart. *Sea of Storms: A History of Hurricanes in the Greater Caribbean from Columbus to Katrina.* Princeton, NJ: Princeton University Press, 2015.

Segalla, Spencer. *Empire and Catastrophe: Decolonization and Environmental Disaster in North Africa and Mediterranean France since 1954.* Lincoln: University of Nebraska Press, 2020.

Sherry, Michael. *In the Shadow of War: The United States since the 1930s.* New Haven, CT: Yale University Press, 1997.

Simms, Brenda, and D. J. B. Trim, eds. *Humanitarian Intervention: A History.* Cambridge: Cambridge University Press, 2015.

Simpson, Bradley. *Economists with Guns: Authoritarian Development and U.S.-Indonesian Relations, 1960–1968.* Stanford, CA: Stanford University Press, 2008.

Skilton, Liz. *Tempest: Hurricane Naming and American Culture.* Baton Rouge: Louisiana State University Press, 2019.

Sobocinska, Agnieszka. *Saving the World? Western Volunteers and the Rise of the Humanitarian-Development Complex.* New York: Cambridge University Press, 2021.

Sparrow, James T. *Warfare State: World War II Americans and the Age of Big Government.* New York: Oxford University Press, 2011.

Squires, Gregory and Chester Hartman, eds. *There Is No Such Thing as a Natural Disaster: Race, Class, and Hurricane Katrina.* New York: Routledge, 2006.

Staples, Amy. *The Birth of Development: How the World Bank, Food and Agricultural Organization, and World Health Organization Have Changed the World, 1945–1965.* Kent, OH: Kent State University Press, 2006.

Steinacher, Gerald. *Humanitarians at War: The Red Cross in the Shadow of the Holocaust.* New York: Oxford University Press, 2017.

Steinberg, Ted. *Acts of God: The Unnatural History of Natural Disaster in America.* New York: Oxford University Press, 2000.

Taffet, Jeffrey F. *Foreign Aid as Foreign Policy: The Alliance for Progress in Latin America.* New York: Routledge, 2007.

Tanielian, Melanie. *The Charity of War: Famine, Humanitarian Aid, and World War I in the Middle East*. Palo Alto, CA: Stanford University Press, 2017.

Thompson, John A. *A Sense of Power: The Roots of America's Global Role*. Ithaca, NY: Cornell University Press, 2015.

Thornton, Christy. *Revolution in Development: Mexico and the Governance of the Global Economy*. Berkeley: University of California Press, 2021.

Tillman, Ellen D. *Dollar Diplomacy by Force: Nation-Building and Resistance in the Dominican Republic*. Chapel Hill: University of North Carolina Press, 2016.

Tsui, Brian. *China's Conservative Revolution: The Quest for a New Order, 1927–1949*. New York: Cambridge University Press, 2018.

Tyrrell, Ian. *Reforming the World: The Creation of America's Moral Empire*. Princeton, NJ: Princeton University Press, 2010.

Valencius, Conevery Bolton. *The Lost History of the New Madrid Earthquakes*. Chicago: University of Chicago Press, 2013.

Van Dijk, Boyd. *Preparing for War: The Making of the 1949 Geneva Conventions*. Oxford: Oxford University Press, 2022.

Van Vleck, Jenifer. *Empire of the Air: Aviation and the American Ascendancy*. Cambridge, MA: Harvard University Press, 2013.

Veeser, Cyrus. *A World Safe for Capitalism: Dollar Diplomacy and America's Rise to Global Power*. New York: Columbia University Press, 2002.

Vine, David. *Base Nation: How U.S. Military Bases Abroad Harm America and the World*. New York: Metropolitan Books, 2015.

Watenpaugh, Keith David. *Bread from Stones: The Middle East and the Making of Modern Humanitarianism*. Berkeley: University of California Press, 2015.

Werking, Richard Hume. *The Master Architects: Building the United States Foreign Service, 1890–1913*. Lexington: University Press of Kentucky, 1977.

Wertheim, Stephen. *Tomorrow, the World: The Birth of U.S. Global Supremacy*. Cambridge MA: Harvard University Press, 2020.

Wieters, Heike. *The NGO Care and Food Aid from America, 1945–80: "Showered with Kindness"?* Manchester: Manchester University Press, 2017.

Wylie, Neville, Melanie Oppenheimer, and James Crossland, eds. *The Red Cross Movement: Myths, Practices and Turning points*. Manchester: University of Manchester Press, 2020.

Young, Alden. *Transforming Sudan: Decolonization, Economic Development, and State Formation*. New York: Cambridge University Press, 2017.

Zebrowski, Ernest. *The Last Days of St. Pierre: The Volcanic Disaster That Claimed 30,000 Lives*. New Brunswick, NJ: Rutgers University Press, 2002.

Zunz, Olivier. *Philanthropy in America: A History*. Princeton, NJ: Princeton University Press, 2012.

Dissertations

Mather, Joshua Hideo. "Citizens of Compassion: Relief, Development, and State-Private Cooperation in U.S. Foreign Relations, 1939–1973." PhD diss., St. Louis University, 2015.

Močnik, Josip. "United States–Yugoslav Relations, 1961–80: The Twilight of Tito's Era and the Role of Ambassadorial Diplomacy in the Making of America's Yugoslav Policy." PhD diss., Bowling Green State University, 2008.

Poster, Alexander. "A Hierarchy of Survival: The United States and the Negotiation of International Disaster Relief, 1981–1989." PhD diss., Ohio State University, 2010.

Romero, E. Kyle. "Moving People: Refugee Politics, Foreign Aid, and the Emergence of American Humanitarianism in the Twentieth Century." Ph.D. diss., Vanderbilt University, 2020.

Schemper, Lukas. "Humanity Unprepared: International Organization and the Management of Natural Disaster, 1921–1991." PhD diss., Graduate Institute of International and Development Studies, 2016.

Journal Articles

Ansari, Ali M. "The Myth of the White Revolution: Mohammad Reza Shah, 'Modernization' and the Consolidation of Power." *Middle Eastern Studies* 37, no. 3 (2001): 1–24.

Arthus, Wien Weibert. "The Challenge of Democratizing the Caribbean during the Cold War: Kennedy Facing the Duvalier Dilemma." *Diplomatic History* 39, no. 3 (2015): 504–31.

Bankoff, Greg. "Time Is of the Essence: Disasters, Vulnerability and History." *International Journal of Mass Emergencies and Disasters* 22, no. 3 (2004): 23–42.

Borland, Janet, and J. Charles Schencking. "Objects of Concern, Ambassadors of Gratitude: Children, Humanitarianism, and Transpacific Diplomacy following Japan's 1923 Great Kantō Earthquake." *Journal of the History of Childhood and Youth* 13, no. 2 (2020): 195–225.

Covert, Lisa Pinley. "Planeamiento urbano post-desastre: Patrimonio cultural y el templo de Santiago en Cusco." *Turismo y Patrimonio* 13 (2019): 85–98.

Davies, Gareth. "Dealing with Disaster: The Politics of Catastrophe in the United States, 1789–1861." *American Nineteenth Century History* 14, no. 1 (2013): 53–72.

———. "The Emergence of a National Politics of Disaster, 1865–1900." *Journal of Policy History* 26, no. 3 (2014): 305–26.

Gil, Magdalena. "Disasters as Critical Junctures: State Building and Industrialization in Chile after the Chillán Earthquake of 1939." *Latin American Research Review* 57, no. 1 (2022): 1–19.

Hutchinson, John. "Disasters and the International Order: Earthquakes, Humanitarians, and the Ciraolo Project." *International History Review* 22, no. 1 (2000): 1–36.

———. "Disasters and the International Order II: The International Relief Union." *International History Review* 23, no. 3 (2001): 253–98.

Irwin, Julia F. "Raging Rivers and Propaganda Weevils: Transnational Disaster Relief, Cold War Politics, and the 1954 Danube and Elbe Floods." *Diplomatic History* 40, no. 5 (2016): 893–921.

Oppenheimer, Melanie, et al. "Resilient Humanitarianism? Using Assemblage to Re-evaluate the History of the League of Red Cross Societies." *International History Review* 43, no. 3 (2020): 1–20.

Reinisch, Jessica. "Internationalism in Relief: The Birth (and Death) of UNRRA." *Past and Present* 210, no. 6 (2011): 258–89.

Schencking, J. Charles. "Giving Most and Giving Differently: Humanitarianism as Diplomacy Following Japan's 1923 Earthquake." *Diplomatic History* 43, no. 4 (2019): 729–57.

Sobocinska, Agnieszka. "New Histories of Foreign Aid." *History Australia* 17, no. 4 (2020): 1–16.

Staiti, Claudio. "Il terremoto di Messina del 1908 nei giornali italiani di New York." *Altreitalie* 58 (2019): 5–40.

Tilchin, William. "Theodore Roosevelt, Anglo-American Relations, and the Jamaica Incident of 1907." *Diplomatic History* 19, no. 3 (1995): 385–406.

Williamson, Fiona, and Chris Courtney. "Disasters Fast and Slow: The Temporality of Hazards in Environmental History." *International Review of Environmental History* 4, no. 2 (2018): 5–11.

Williford, Daniel. "Seismic Politics: Risk and Reconstruction after the 1960 Earthquake in Agadir, Morocco." *Technology and Culture* 58, no. 4 (2017): 982–1016.

Wolfe, Mikael. "'A Revolution Is a Force More Powerful Than Nature': Extreme Weather and the Cuban Revolution, 1959–64." *Environmental History* 25, no. 3 (2020): 469–91.

Index

American Red Cross (ARC) (*continued*)
and World War I, 89, 90–91, 111; and
World War II, 175, 180–83
American Red Cross Central Division, 102
American Red Cross Department of
Insular and Foreign Operations, 111,
132, 141
American Red Cross Flood Relief Camp,
94–95
American Red Cross Orphanage, 62, 67
American Red Cross Relief Committee: in
China, 93, 96, 97; in Guatemala, 99–102,
104, 107–8; in Japan, 118, 119, 120
American Red Cross (ARC) relief efforts:
to Chile, 43–45, 169–71, 228–29, 239–40;
to China, 92–98, 153, 156–58, 167–68;
to Cuba, 164–66; Cuban rejection
of, 254–55; to Dominican Republic,
140–51; to Ecuador, 193; to Guatemala,
99–109, 104, 193; to Haiti, 207; to
Honduras, 167; to Iran, 246; to Italy,
53–57, 60–67, 63, 69; to Jamaica, 45–46,
48; to Japan, 115–30, 199, 222; to Mexico,
166–67, 220; to Netherlands, 200–201;
to Nicaragua, 140–51; to Peru, 193;
to San Francisco, 40–41; to Somalia,
243–44; to Taiwan, 221; to Yugoslavia,
249, 251
American Relief Administration, 111, 131, 159
American Relief Committee: in Italy,
56–59, 67; in Japan, 120
Anderson, Earl, 114, 118, 120–21
anticolonial nationalism, 184–85, 190, 191,
218, 236
anti-immigration prejudice. *See* prejudices
Apfel, Herbert, 103–5, 107
ARC. *See* American Red Cross (ARC);
American Red Cross (ARC) relief
efforts
Arévalo, Juan José, 195
armed forces, US: aircraft in Nicaragua
and Dominican Republic, 140–41; Cold
War remobilization of, 176; criticism
of, 274; Haitian hurricane response,
207; Moroccan earthquake response,

225; as pillar in disaster aid, 3, 36, 135,
272–73; postwar expansion and spread
after World War II, 174, 175–76; World
War I growth, 89. *See also* military
installations; *individual branches*
armed forces, US, and ARC: ARC
International Relief Board, 54;
domestic aid in nineteenth century, 27;
in Dominican Republic and Nicaragua,
138–51; partnership establishment,
39–40; partnership under Taft and
Wilson, 70–71
Armour, Norman, 169–70
Army, US, 38, 170: Corps of Engineers, 140,
142–43; Eleventh Infantry, 46; Fifteenth
US Infantry Regiment, 93, 94–95, 97,
98; First Cavalry, 199; growth and
expansion during World War II, 89, 176;
Mexican hurricane relief, 220; Seventh
Engineer Brigade, 257
Army, US, and ARC: Chilean earthquake
relief, 169, 170; in Dominican Republic
and Nicaragua, 140, 142–43; Japanese
earthquake relief, 115, 116, 122–23, 128,
199; San Francisco earthquake relief,
40–41; Tientsin floods response,
92–98
Asiatic Fleet, 45, 75, 114, 115, 117, 120–22,
156, 158
Assam, India, earthquake, 196–97
associational state, 6–7, 25, 36, 180
Atlantic Fleet (Great White Fleet), 52, 60
authoritarianism: of Batista in Cuba, 166;
of Duvalier in Haiti, 252; of Estrada
Cabrera in Guatemala, 101–2; of Trujillo
in Santo Domingo, 151
Aymé, Louis, 30, 31, 34

Baker, Henry, 134
Baker, James Earl, 158
Ball, George, 250, 251, 253
Barall, Milton, 208
Barton, Clara, 27, 38
Batista, Fulgencio, 165, 166
Bayern (ship), 57

Commodity Credit Corporation, 201
communications technologies, 26, 27–28
communism: in China, 191, 198, 218, 223; in
    Greece, 203; in Latin America, 190, 191,
    194–95, 218, 227–28; in South Asia, 196;
    and Yugoslavian aid restrictions, 251
Communist Party of China, 155, 161, 163
Concepción and Chillán, Chile,
    earthquake, 168–71
Congress, US, legislation: Agricultural
    Marketing Act, 159; ARC charter, 38;
    ARC charter revisions, 181; Disaster
    Relief Acts, 268; Federal Disaster Relief
    Act, 191–92; Foreign Assistance Act,
    240–42, 251, 257, 267–69, 271; Foreign
    Economic Assistance Act, 198; Foreign
    Operations Administration, 202;
    Johnson-Reed Immigration Act, 129;
    Lend-Lease Act, 178; National Flood
    Insurance Act, 268; National Security
    Act, 177; Public Law 86-735, 231, 232;
    Public Law 94-161, 269; Public Law 480,
    202, 204
Congress, US, relief appropriations:
    Chilean earthquakes, 42, 171,
    231–33, 238; to China, 92, 198; Italian
    earthquake, 53, 61, 63; Jamaican
    earthquake, 46, 53; Japanese
    earthquake, 124; Martinique volcano,
    31–33, 53; nineteenth-century, 22–24,
    26–27, 29; postwar relief and recovery,
    179; San Francisco earthquake, 41
Connecticut (ship), 52, 60
consuls. *See* Foreign Service, US; State
    Department, US: personnel; *and names
    of individuals*
Contingency Fund (ICA), 215, 228, 230, 231,
    238, 241, 244, 263
Coolidge, Calvin, 115–17, 132, 133
Corbin, H. C., 33
Corwine, William R., 30, 33, 34
Costa Rica earthquake, 75, 76,
    80–81, 82
Crabbs, J. T., 33
criminality, disasters and, 78, 147

Cuba: hurricane relief, 134, 135–36, 164–66,
    254–56; Revolution, 218; War for
    Independence, 26, 28
Cuban National Reconstruction
    Commission, 134
Cuban Red Cross, 165, 255
Cuban Relief Committee, 134
Culgoa (ship), 52
cultural prejudice. *See* prejudices
Curtis, Charles, 139, 140, 142
Cusco, Peru, earthquake, 193–95
Cutting, Bayard, 56–60

Daniels, Josephus, 166
Dardanelles Strait of Turkey earthquake,
    75, 81, 82
Dauber, Michele Landis, 152
Davies, Gareth, 25
Davis, Charles H., 46–49
Davis, George, 71
DDT, 194
decolonization: in Asia, 195; effects of,
    1, 11, 16, 190, 200, 212, 218–19, 236; in
    Morocco, 224, 227
Defense Department, US: establishment
    of, 177; "Foreign Disaster Relief
    Operations," 262; frustrations of, 216;
    guidelines, 217, 262; Haitian hurricane
    relief, 253; Mexican hurricane relief,
    220; and NCWC, 197; postwar, 178, 192,
    215; reforms affecting, 263, 272–73
developing countries, 12–13
development assistance: in Chile, 229–33,
    238–40; in China, 94–98, 159–60, 162;
    in Dominican Republic and Nicaragua,
    143–51; emergency vs long-term aid,
    7–9, 51, 267, 272, 274; in Guatemala, 99,
    102–9; influence in China, 99, 104, 108,
    109, 111, 112, 146; in Iran, 246–48; in Italy,
    51, 60–67; in Japan, 111–30; Kennedy's
    emphasis on, 237, 240–42; postwar
    growth under Eisenhower, 214; postwar
    relief and recovery, 24, 88, 178–85;
    Yugoslavia, 250–52, 256–58
Dickover, Erle, 114, 127

federal government, US, and ARC: ARC International Relief Board, 54; associational state, 6–7, 36; Depression-era policies, 163–64; partnership consolidation, 38–40; partnership under Taft and Wilson, 69–72; World War II changes, 180–83

floods: Algeria, 225; China (1911, Yangzi), 75, 76, 77–78; China (1917, Tientsin), 91–98; China (1924), 136; China (1931), 154–63; China (1935), 167–68; China (1950, Huai River), 198; China (1959), 223; Eastern Europe, 204–7; Guatemala, 192–95; India, 196–97; Mexico, 76–77, 80, 136; Netherlands, 200–201; Pakistan, 196–97; Somali Republic, 242–43; Taiwan, 221; US (Great Mississippi Flood of 1927), 131, 144, 146, 148, 154, 158

FOA (Foreign Operations Administration), 202, 206, 207–9, 212

Food Administration, US, 159

food distribution: in Dominican Republic and Nicaragua, 145–46; in Guatemala, 103; in Yugoslavia, 251. *See also* commodities, surplus; Food for Peace Program; Public Law 480

Food for Peace Program, 237, 238, 246, 263

Ford, Gerald, 264, 269

Foreign Assistance Act, 240–42, 251, 257, 267–69, 271

foreign disaster aid: aircraft introduced into, 141–42 (*see also* military airlifts); allocation of, 10–15; centralization of, 16, 190, 212, 216, 236, 243; and Cold War politics (*see* Cold War); communism as threat to (*see* communism); coordination of, 16, 190, 212, 216, 236, 243, 260; criticism of, 162–63, 267–68, 274–75; as diplomacy (*see* foreign policy, US); emergency funding, 262; expansion and frequency, 178–80, 192, 236, 270–71; and Foreign Assistance Act, 240–42, 251, 257, 267–69, 271; future trends, 270, 275–76; geographical limitations and developments,

24–25, 26, 27, 28, 69; guidelines and standards, 217–18, 261–62; historical lessons, 275; ideological forces, 25, 28; legislative reforms, 259, 264, 266, 268–69; and long-term development assistance, 7–9, 51, 267, 272, 274 (*see also* development assistance); moral aspects of, 2, 28, 42, 70, 76, 149, 158, 170, 238; origins of diplomacy and, 1–3, 6, 9, 23; postwar revolution in, 189–92; scope of coverage, 3–5, 8, 14–15; and sovereignty, 11, 46–49; state-private partnerships in, 6–7, 182, 273; structural reforms, 202–3, 211–17, 236–38, 240–42, 260–68; and surplus commodities (*see* commodities, surplus); systemic problems, 185, 236, 241–44, 258, 260–61, 264, 266–68; technological limitations and developments, 24–28; terminology, 11–12; three-pillared system, 3, 14, 184, 271–72; and US intervention, 143–44, 149–51; and US reputation, 28, 69, 76–78, 149–51, 194, 205–6, 229, 249–50; war relief and recovery, 178–85. *See also specific geographic locations*

foreign disaster aid chronology: 1812–1902, from Venezuela to Martinique, 21–34; 1902–8, casting pillars, 35–49; 1908–9, Italy, 50–67, 63; 1909–16, cementing pillars, 68–86; 1917–18, World War I period, 87–109; 1919–24, Japan, 110–30, 116, 119; 1924–31, Caribbean Basin, 131–51; 1931–39, Great Depression, 152–72; 1939–47, World War II period, 173–88; 1948–54, new mechanisms, 189–210; 1955–60, standardizing, 211–34; 1961–63, development decade, 235–58; 1964–76, federal reforms, 259–76

"Foreign Disaster Emergency Relief" (USAID manual), 261

Foreign Disaster Emergency Relief Account (FDRC), 217, 238

Foreign Disaster Relief Fund, 268

"Foreign Disaster Relief Operations" (US government manual), 217, 262

Foreign Economic Assistance Act of 1950, 198

Foreign Operations Administration (FOA), 202, 206, 207–9, 212

foreign policy, US: with Africa, 26, 218–19, 224, 227, 237; ARC as part of, 70–71; with Caribbean Basin, 28, 34, 37, 132, 138–51; with Chile, 41–42, 169, 229; with China, 92, 155–63, 198–99, 218; and Cold War politics (*see* Cold War); with Cuba, 166, 218, 227, 254–56; and decolonization (*see* decolonization); and development assistance, 237; and diplomats' motives for aid, 76–78; disaster aid as central to, 259–60, 275; disaster aid vs foreign aid, 5; dollar diplomacy, 69; with Dominican Republic, 138–51; factors considered for disaster aid, 10; with France, 224–25; and global expansion, 25–28, 69, 174; Good Neighbor Policy, 154, 164, 166–71; with Great Britain, 23, 26, 45–49; with Guatemala, 98–99, 101; gunboat diplomacy, 69, 141; with Haiti, 208–9, 252–54; intervention and policing in, 143–44, 149–51; with Iran, 245–47; with Italy, 51, 54–55; with Jamaica, 45–49; with Japan, 112, 117, 118–23, 125–30; with Latin America, 190, 191, 194–95, 218, 227–29; with Mexico, 220–21; with Middle East, 218; with Nicaragua, 139–51; under Nixon and Ford, 265; origins of diplomacy and disaster aid, 1–3, 6, 9, 23; Roosevelt administration's initiatives, 36–49; with Soviet Union, 218, 219; with Taiwan, 221–22; and WWI developments, 88–90, 108–9; World War II's effect on, 173–85; with Yugoslavia, 235, 248–52, 256–58

Foreign Service, US: ARC's reliance on, 135; and Chinese flood response (1931), 155–63; emergency funds for, 262; postwar expansion of after World War II, 176–78; postwar role of, 192, 215; reforms to, 272; Rogers Act and, 90

France: and Martinique relief, 30–31, 34; nineteenth-century aid to, 24; US relations with, 224–25

Frank Leslie's Popular Monthly (magazine), 30

Freeman, Orville, 251, 257

Frei, Eduardo, 229

General Electric, 113

General Welfare Clause of Constitution, 152

Geneva Convention, 27

Giolitti, Giovanni, 65

Global South, 12–13, 270

Good Neighbor Policy, 154, 164, 166–71, 194

Grand Hotel Regina Elena, 64, 65

Grau San Martín, Ramón, 165, 166

Great Alaskan earthquake (1964), 268

Great Depression, 132, 137, 152–53, 163–64, 168

Great Kantō earthquake, 110–30, *116*, *119*; development assistance, 122–30; diplomatic approach toward, 112, 117, 118–23, 125–30; disaster aid, 111–17; enormity of response to, 110–13, 124; and foreign refugees, 113, 114, 127; Japanese government's response, 113, 114, 120–30

Great Mississippi Flood (1927), 131, 144, 146, 148, 154, 158

Great White Fleet (Atlantic Fleet), 52, 60

Greek earthquake, 202–3

Greene, Roger, 93, 94, 96, 98

Griscom, Elizabeth, 56, 57, 58, 63, 67

Griscom, Lloyd: on International Relief Board, 71; and Italian earthquake relief, 55–60; and Italian recovery and rebuilding, 60–67

Gruenther, Alfred, 228

Guam, annexation of, 26

Guantánamo Bay: Jamaican response from naval base in, 46, 48; US control of, 37

Guardia Nacional, 138, 139, 142, 143, 151

Guatemala: floods, 193–95; Guatemala City
earthquake, 98–109, *104*
Guatemalan Red Cross, 102, 103
gunboat diplomacy, 69, 141

Hague Convention of 1907, 37
Hai River floods, 91–98
Haitian Army, 208
Haitian hurricanes, 135, 136, 207–9, 252–54
Haitian Red Cross, 208, 254
Hale, Eugene, 49
Hanihara Masanao, 117, 125
Hanna, Matthew, 139, 140, 142, 143, 145–50
Hanna, Philip, 76–77, 80
Hannah, John, 265
Harding, Warren, 115
Hawaiʻi, 11, 26
Hay, John, 31
health and sanitation projects, 273; in
China, 77–78, 94–95; in Cuba, 135;
in Dominican Republic, 144–45; in
Ecuador, 193–94; in Guatemala, 102–7,
193; in Iran, 246; in Italy, 63; in Japan,
113–15, 122–24, 128; in Nicaragua, 141,
144–45, 148, 149, 150; in Peru, 193–94; in
San Francisco, 41
Hicks, John, 42, 44
highway construction, 96–98
Hollister, John B., 233
Holmes, Julius, 245, 246, 247, 248
Honduran hurricane, 167
Hong Kong typhoon, 45
Hoover, Herbert: Dominican Republic
and Nicaraguan relief under, 139, 140,
141, 148–49; humanitarian reputation,
131–32, 137, 194; Japanese earthquake
response, 115; novel form of aid by, 153
Hoover, Herbert, Jr., 206
hospital construction: in Guatemala, 105,
107; in Italy, 63, 67; in Japan, 122–24; in
Nicaragua, 148
hotel construction, Messina, Italy, 63–64,
65, 66
housing construction: Iran, 247–48;
Italian rebuilding projects, 61–63,

63; in Dominican Republic, 148; in
Yugoslavia, 256–57. *See also* refugee
camps
Howe, Walter, 227, 228, 229, 230, 232
Hsiung Hsi-ling (Xiong Xiling), 94, 97
Huai River Commission, 159
Hughes, Charles Evan, 117, 125
Hull, Cordell, 165, 167
humanitarian aid: aircraft used in, 140–41,
169; ARC recognized as tool for, 27;
diplomatic advantages of, 76–77;
Italian recovery as model of, 62, 67;
partnerships with Americans abroad,
81–82; postwar expansion of, 176;
selling surplus crops, 158–60, 162–63;
terminology, 11–12; World War II's effect
on, 174, 178–85. *See also* development
assistance; foreign disaster aid
humanitarian crises, 3–5, 24, 77, 174. *See
also* disasters, natural
Hungarian Revolution of 1956, 218
hurricanes: in Cuba, 134, 135–36, 164–66,
252, 254–56; in Dominican Republic,
132, 138–51; in Haiti, 135, 136, 207–9,
252–54; in Honduras, 167, 265; in
Mexico, 166–67, 219–21; in Puerto Rico,
29; in United States, 268
Hussey, Henry, 94

ICA. *See* International Cooperation
Administration
Illinois (ship), 52
IMF (International Monetary Fund), 250
immigration: Chinese, 92; Dutch, 201;
Italian, 55, 58, 67; Japanese,
112, 129
imperialism, 28, 30, 37
India: earthquake and floods, 196–97;
famine relief, 29
Indo-Pakistan War (1971), 265
inequality, 12–13, 270
instability, disasters and, 77–78
Institute of Inter-American Affairs
(IIAA), 194
Insular and Foreign Division, ARC, 111

Marine Corps, US: criticism of, 149–51; in Dominican Republic and Nicaragua, 138–39, 140, 142, 143–44, *144*, 149–51; in Greece, 203; modernization and expansion of, 37; in Nicaragua, *144*; policing by, 143–44, 149–51; Second Marine Brigade, 140, 142; World War I growth, 89

Marshall Plan, 179, 189, 191, 206

Martinique volcano eruption, 21–22, 30–36

MATS (Military Air Transport Service), 197, 215, 216, 225

MATS C-54 Skymaster, 197

McClure's (magazine), 30

McCoy, Frank, 115, 118, 120, 121, 122, 123, 126

McDonnell, R. T., 97

McNamara, Robert, 249

medical care. *See* health and sanitation projects

Meigs (ship), 115, 120

Merritt (ship), 115, 120

Messina, Italy. *See* Italy, Messina and Reggio earthquake

Mexican Red Cross, 166

Mexico: floods, 76–77, 80, 136; hurricanes, 166–67, 219–21

Middle East, 26, 181, 201, 218, 237, 245

military, US. *See* armed forces, US; military airlifts; military installations; *and individual branches*

military airlifts: to Chile, 169, 228; to Dominican Republic, 140–41; to Ecuador, Peru, and Guatemala, 193; to Haiti, 207, 253; to India and Pakistan, 197; to Iran, 246; to Morocco, 225; to Nicaragua, 140–41; postwar increases in, 184, 185, 215; reforms affecting, 263; to Somalia, 242; to Yugoslavia, 249

Military Air Transport Service (MATS), 197, 215, 216, 225

military installations: in Caribbean, 169; in Germany, 246, 249; in Guam, 266; in Italy, 266; in Japan, 199; in Morocco, 225; role of, 83; in Singapore, 266; World War II growth of, 176, 178, 184

Ministry for Foreign Affairs (China), 159

Ministry for Public Works (Italy), 64, 65

missionaries, American, 26; in China, 29, 81–82, 92; in Japan, 29

Mississippi River (Great Mississippi Flood), 131, 144, 146, 148, 154, 158

Moncada, José María, 139, 141–43, 144, 146, 148–49, 150

Monterrey, Mexico, floods, 76–77, 80

Moody, William, 31

Moroccan earthquake, 224–27, *226*

Morrow, Charles, 94–95

Morrow, Dwight, 134

Mount Pelée volcano eruption, 21–22, 30–34

Müller, Walter, 230

Mutual Security Act, 215, 216, 238, 241. *See also* Contingency Funds

Mutual Security Agency, 191, 213

Nagoya, Japan, typhoon, 222

Naples, 57, 62

National Catholic Welfare Council (NCWC), 197

National Flood Insurance Act, 268

National Flood Relief Commission (China), 154, 158, 162

National Geographic Society, 33

National Guard, US, 38

nationalism, anticolonial, 184–85, 190, 191, 218

Nationalist Party (Kuomintang), 154–55, 160–61, 162, 198

National Research Council's Committee on International Disaster Assistance, 270

National Security Act of 1947, 177

National Security Council: and Cuban aid, 255; establishment of, 177; and Haitian relief, 253

national sovereignty, 11–12

NATO (North Atlantic Treaty Organization), 201, 255

Navy, US: Asiatic Fleet, 45, 75, 114, 115, 117, 120–22, 156, 158; Atlantic Fleet, 52, 60; Ceylon flood relief, *214*; Defense

Skopje, Yugoslavia, earthquake, 235, 248–52, 250, 256–58
Somali Republic floods, 242–43
Somme (ship), 115
Somoza, Anastasio García, 143, 149
Soong, T. V., 154–55, 156, 158, 160, 162
Southeast Pacific, US expansion in, 28
South Korean typhoon, 221–23
Soviet Union. *See* USSR (Union of Soviet Socialist Republics)
Spain: Spanish-American War, 26, 29; Spanish Civil War, 168
Sperry, Charles, 52, 57, 59, 60
Stalin, Josef, 218
Standard Oil, 113
State Department, US: and ACVFA, 182–83; Chilean earthquake response, 228–32, 239; Chinese flood response, 136, 155–63; Committee on War Relief Agencies, 181; coordination of efforts studies by, 244; Cuban hurricane response, 254–55; Division of Eastern Affairs, 159; Dulles's funding and instructions for, 217, 241; Haitian hurricane response, 253–54; and ICA, 212, 213, 217, 218; Iranian earthquake response, 245–48; Japanese typhoon relief, 222; Mexican hurricane relief, 220–21; Moroccan earthquake relief, 224, 225; personnel, 70, 71, 75–83, 90; as pillar in disaster aid, 3, 36, 134; pivotal role under Taft and Wilson, 69, 70, 71, 74; postwar role, 192; reforms affecting, 263, 272; Somalian aid, 242–44; Soviet Bloc aid, 206–7; Taiwan flood relief, 221; and USAID, 241, 244; World War II expansion, 174, 175, 176–78, 181–82; World War I restructuring, 90; Yugoslav earthquake response, 249–52, 257. *See also* Foreign Service, US; *and names of specific individuals*
State Department, US, and ARC: Chilean earthquake response, 44–45, 169–71; Chinese flood response, 92–94, 167–68; Cuban hurricane response, 164–66; global reach of, 133; and Good

Neighbor Policy, 166–68; Guatemalan earthquake response, 99–101, 107; Indian relief, 197; Italian earthquake response, 54–56, 66–67; Japanese earthquake response, 111–30; postwar changes, 181; promotional efforts, 71–72
State Department Policy Planning Council, 262
Steinberg, Ted, 41
Stewart (ship), 117
stockpiles of emergency supplies, 262, 266, 268, 273
Strait of Messina region. *See* Italy, Messina and Reggio earthquake
Struse, Alvin, 105, 107
Stuart, Edward, 102, 106
Supreme Court, US, 129
Swettenham, Alexander, 46–48
Swift, Ernest J., 141–45, 147, 149, 170

Taft, William Howard: ARC's growth under, 69–72; as ARC's honorary president, 72; Italian relief response and namesake, 54, 67; Roosevelt's support for, 51; sympathetic communications by, 74
Taiwan floods, 221
Tampico, Mexico, hurricanes, 166–67, 219–21
technology: as early block to aid response, 24–25; late nineteenth-century advances in, 26, 27
Third Pan-American Conference, 42
Third World: disparities between US and, 12–13; Kennedy's focus on, 237–38; and Point Four Program, 191
Thurston, Raymond, 253
Thurston, Walter, 99–101, 102
Tiburon Peninsula, Haiti, 207
Tientsin (Tianjin), China, floods, 91–98
Time magazine, 263–64
Tito, Josip Broz, 249, 251, 256, 257
Tokugawa Iesato, 120
transportation, nineteenth-century advances in, 24–26, 27
Treaty of Peace, 199

Tripp, Stephen R., 261–64, 265

Trujillo, Rafael, 139, 141, 142, 144, 145, 148–49, 151

Truman, Harry; disaster aid under, 192, 193, 194, 195, 196, 200; foreign aid initiatives, 176, 177, 179, 182, 191

tsunamis: in Italy (*see* Italy, Messina and Reggio earthquake); in Japan, 29, 199; in Peru, 24

Turkish earthquake, 75, 81, 82

Turkish Red Crescent Society, 82

typhoons: in Hong Kong, 45; in Japan, 222; in South Korea, 221–23; Typhoon Gloring (Rita, 1972), 265; Typhoon Sarah, 221–23; Typhoon Vera, 222

underdeveloped nations. *See* Third World

Underwood, Oscar, 32

Union of Soviet Socialist Republics. *See* USSR (Union of Soviet Socialist Republics)

United Fruit Company, 99, 102

United Nations Human Rights Council, 276

United Nations Relief and Rehabilitation Administration (UNRRA), 179

Upham, Frank, 81

urbanization, 270

US Agency for International Development (USAID): centralization of aid under, 236, 241; contingency funds, 263; creation and scope of, 240–41; Disaster Coordination and Information Exchange Center, 266; "Foreign Disaster Emergency Relief" manual, 261; Haitian hurricane relief, 253; Iranian earthquake relief, 245–46, 247; Office of Foreign Disaster Assistance, 269, 271; reforms of, 261–66, 272; relief cooperation and coordination studied by, 244; Somalian flood relief, 242–43; Task Force on Earthquake Rehabilitation, 246; Task Force on Enhancement of the A.I.D. Disaster

Relief Function, 267; Yugoslav earthquake relief, 251, 257

US Caribbean Command, 193, 207, 228

US European Command, 246

US Shipping Board, 116, 118, 120

USSR (Union of Soviet Socialist Republics): Soviet Bloc floods, 204–7; US relations with, 218, 219; Yugoslav ties, 249. *See also* Cold War

Valdivia, Chile, earthquake, 227–33, 238–40

Valparaíso, Chile, earthquake, 41–45

Venezuela, Caracas, earthquake, 22–23

Villagio Regina Elena, 63

Vittorio Emanuele III (king), 52

volcanic eruptions, Martinique, 21–22, 30–36

voluntary aid organizations: and ACVFA, 182–83; in Chilean earthquake relief, 228–29; conferences for, 266; criticism of, 274–75; East Asian aid, 221–22; in Haitian hurricane relief, 253, 254; increase and expansion of, 174, 175, 180–83, 215–16; Martinique relief response, 33; nineteenth-century reliance on, 25, 27; as pillar in disaster aid, 36, 134; reforms involving, 262–63, 266, 273; Tientsin flood response, 93; and USAID, 241; and WRCB, 182; Yugoslav relief, 249, 251. *See also* American Red Cross (ARC); *and names of organizations*

Wadsworth, Eliot, 148

Wake Islands, annexation of, 26

war, humanitarian crises caused by, 24, 90, 111, 131, 174, 178, 183–84

War Department, US: Defense Department created, 177; disaster aid under Taft and Wilson, 74–75; Japanese relief expenditures, 124; as pillars in disaster aid, 36, 70

War Department, US, and ARC: ARC International Relief Board, 54; Chilean earthquake relief, 170; domestic aid by, 26–27; financial arrangement, 39; in Guatemala, 102; Martinique aid by, 33; as pillars in disaster aid, 36, 70; under Taft and Wilson, 71. *See also* armed forces, US, and ARC

War on Terror, 271

War Relief Control Board (WRCB), 182

Washington Naval Conference, 112

Watson, Thomas E., 142, 145–48

Welles, Sumner, 164–66

West Prospect (ship), 114

White House Conference on International Cooperation, 262

White Revolution (Iran), 245, 247–48

Wilson, Huntington, 71–72

Wilson, Woodrow, 69; as ARC's honorary president, 72; sympathetic communications by, 74; World War I policies and actions, 90, 99

Winthrop, Beekman, 71

Wood, Leonard, 114, 115

Woods, Cyrus, 113–15, 117–21, 123–26, 128

World Bank, 250

World War I period, 87–109; ARC's reorganization and growth, 89–91, 111; armed forces growth during, 89; foreign policy during, 88–90, 108–9; State Department's restructuring, 90

World War II period, 173–85; ARC during, 175, 180–83; armed forces growth during, 89, 174–76, 184; federal government expansion during, 174, 175, 178–80; foreign policy and, 173–85; natural disasters during, 174; State Department's expansion, 174–78, 181–82

Yamamoto Gonnohyōe, 114, 121

Yangzi and Huai Yellow River Basin floods, 154–63

Yangzi and Yellow River floods, 167–68

Yangzi River floods, 75, 76, 77–78

Yankton (ship), 52

Yost, Charlie, 224, 225

Yugoslavian earthquake, 235, 248–52, 250, 256–58

Yugoslavian Red Cross, 235